Paracelsian Moments

Habent sua fata libelli

SIXTEENTH CENTURY ESSAYS & STUDIES SERIES

GENERAL EDITOR
RAYMOND A. MENTZER
University of Iowa

EDITORIAL BOARD OF SIXTEENTH CENTURY ESSAYS & STUDIES

ELAINE BEILIN Framingham State College	ROGER MANNING Cleveland State University, Emeritus
MIRIAM U. CHRISMAN University of Massachusetts, Emerita	MARY B. MCKINLEY University of Virginia
BARBARA B. DIEFENDORF Boston University	HELEN NADER University of Arizona
PAULA FINDLEN Stanford University	CHARLES G. NAUERT University of Missouri, Emeritus
SCOTT H. HENDRIX Princeton Theological Seminary	THEODORE K. RABB Princeton University
JANE CAMPBELL HUTCHISON University of Wisconsin–Madison	MAX REINHART University of Georgia
CHRISTIANE JOOST-GAUGIER University of New Mexico, Emerita	JOHN D. ROTH Goshen College
RALPH KEEN University of Iowa	ROBERT V. SCHNUCKER Truman State University, Emeritus
ROBERT M. KINGDON University of Wisconsin, Emeritus	NICHOLAS TERPSTRA University of Toronto

MERRY WIESNER-HANKS
University of Wisconsin–Milwaukee

Paracelsian Moments

Science, Medicine, & Astrology in Early Modern Europe

EDITED BY Gerhild Scholz Williams & Charles D. Gunnoe, Jr.

SIXTEENTH CENTURY
ESSAYS & STUDIES
VOLUME 64

Copyright © 2002 by Truman State University Press
100 East Normal Street, Kirksville, Missouri 63501-4221 USA
http://tsup.truman.edu
All rights reserved

Library of Congress Cataloging-in-Publication Data

Paracelsian moments : science, medicine, and astrology in early modern Europe /
   Edited by Gerhild Scholz Williams and Charles D. Gunnoe, Jr.
       p. cm. — (Sixteenth century essays & studies ; v. 64)
   Includes bibliographical references and index.
   ISBN 1-931112-11-8 (pbk. : alk. paper) — ISBN 1-931112-12-6 (casebound :
alk. paper)
       1. Paracelsus, 1493–1541. 2. Science—Europe—History—16th
century. I. Scholz Williams, Gerhild. II. Gunnoe, Charles D., 1963–
III. Sixteenth Century Studies Conference (1999 : St. Louis, Mo.) IV. Series.

R147.P2 P385 2002
509.031—dc21   2002018742

Cover art: Paracelsus Surgery Macrocosmos, from Paracelsus, *Der grossen Wundarzney* and
Prognostications: Rest, from Paracelsus, *Prophecien und Weissagungen*, permission of Becker
Medical Library, Washington University, St. Louis, Missouri.

Cover design by Lechelle Calderwood
Printed by Thomson-Shore, Dexter, Michigan
Set in Adobe Minion with FCaslon display type

No part of this work may be reproduced or transmitted in any format by any means, electronic or mechanical, including photocopying and recording or by an information storage or retrieval system, without permission in writing from the publisher.

∞The paper in this publication meets or exceeds the minimum requirements of the American National Standard—Permanence of Paper for Printed Library Materials, ANSI Z39.48 (1984).

In memory of Jeffrey Brent Weber and Michael W. Johnson

Contents

Acknowledgments................................ix

Introduction xi

Paracelsiana

Paracelsus's Biography among His Detractors 3
   Charles D. Gunnoe, Jr.

Paracelsus and the Boundaries of Medicine
in Early Modern Augsburg......................... 19
   Mitchell Hammond

To Be or Not to Be a Paracelsian 35
Something Spagyric in the State of Denmark
   Jole Shackelford

"A Spedie Reformation" 71
*Barber-Surgeons, Anatomization, and the Reformation
of Medicine in Tudor London*
   Lynda Payne

Seeing "Microcosma" 93
Paracelsus's Gendered Epistemology
   Hildegard Elisabeth Keller

Paracelsus on Baptism and the Acquiring
of the Eternal Body 117
   Dane Thor Daniel

Paracelsus and van Helmont on Imagination........ 135
Magnetism and Medicine before Mesmer
   Heinz Schott

Natural Magic and Natural Wonders

Unholy Astrology........................... 151
Did Pico Always View It That Way?
   Sheila Rabin

Wine and Obscenities 163
Astrology's Degradation in the Five Books of Rabelais
   Dené Scoggins

**Robert Boyle, "The Sceptical Chymist,"
and Hebrew** 187
   Michael T. Walton

Johannes Praetorius......................... 207
Early Modern Topography and the Giant Rübezahl
   Gerhild Scholz Williams

Demons, Natural Magic, and the Virtually Real 223
Visual Paradox in Early Modern Europe
   Stuart Clark

**Selected Bibliography of Paracelsiana
and Early Modern Science** 247

Index 257

Acknowledgments

This volume presents a selection of essays chosen from the History of Science presentations at the annual Sixteenth Century Studies Conference in St. Louis (October 1999). The panels centered around Paracelsus and were held in conjunction with activities highlighting the fine collection of Paracelsiana at Washington University's Medical School Library. We wish to express our appreciation to Lilla Vekerdy, medical librarian, for her help in selecting the cover illustration, and for arranging a stimulating exhibit of primary and secondary materials associated with Paracelsus.

We are grateful to the anonymous readers for their criticisms and helpful suggestions; the volume owes much to their diligence and expertise. We wish to thank Julian Paulus for his contributions to the selected bibliography devoted to Paracelsus and for the history of science, which was added to provide a reference tool for the reader. Our most heartfelt appreciation goes to Michaela Giesenkirchen and John Morris for their patient queries to contributors and editors and their expert editing. Furthermore, the editors wish to thank Washington University and Aquinas College for the financial support that helped bring the essays to publication.

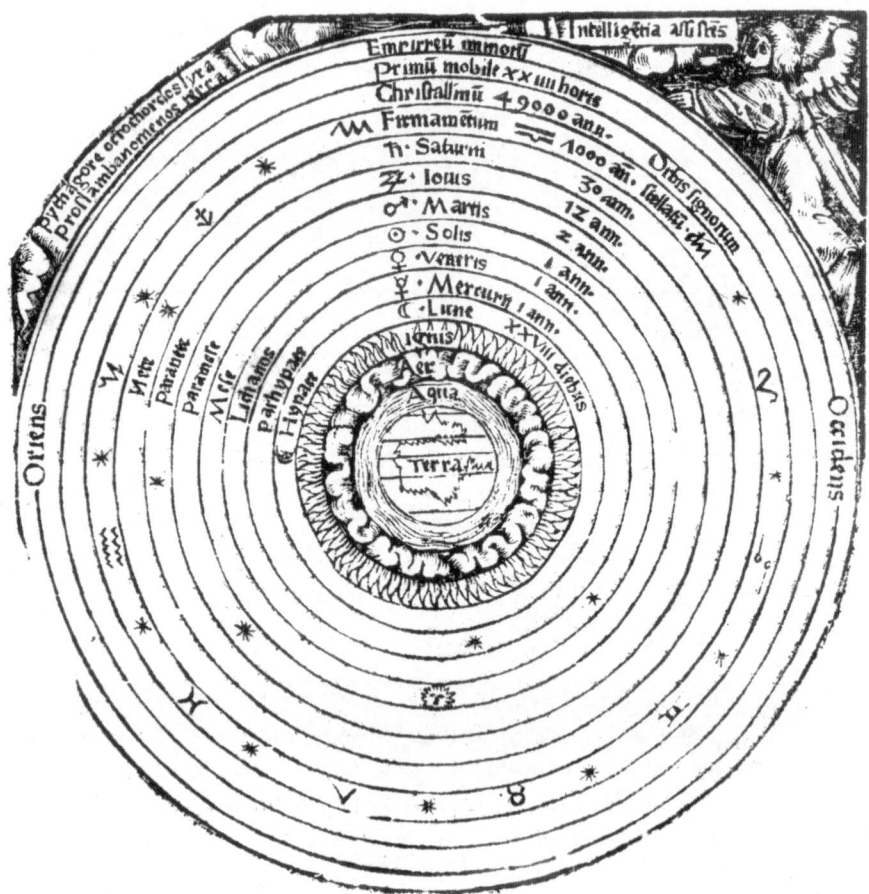

Paracelsus Surgery Macrocosmos, from Paracelsus, *Der grossen Wundarzney: das erst Buch* (Augsburg: Steyner, 1536), xxii verso, permission of Becker Medical Library, Washington University, St. Louis, Missouri.

Introduction

In 1527, Theophrastus Bombastus von Hohenheim, better known today as Paracelsus, a peripatetic physician with somewhat doubtful university credentials, made his way to Basel, Switzerland. Basel was at the center of the Northern Renaissance, housing a vital publishing trade and serving as the sometime residence of Desiderius Erasmus, the prince of the humanists. The Reformation, too, was beginning to make inroads in the town, though the old faith still had support among the university faculty and wealthy burghers. In matters of religious controversy, the Baselers had not fully shown their hand, and would not make any definitive move for some time. Despite sympathies for the radical reformers, Paracelsus never left the Catholic faith. He did, however, reject what he called the *Mauerkirche*, the walled church, in favor of a more direct, evangelical relationship with God.[1] Almost upon his arrival in Basel, Paracelsus scored a major success by effecting a cure on the wounded leg of the famed printer Johannes Froben. From this triumph, our physician became the toast of the town. Heedless of the resistance of the hidebound faculty members but spurred on by influential humanists, the town council made a place for him at the city's university.

There are moments in history when person, place, and time fall into sublime synchronicity. Paracelsus in Basel during 1527 appeared to be such a juncture. Paracelsus, however, did not carefully plot his career strategies as to how best to make an impact on the Basel academic scene. He continued to construct his philosophy of medicine based on experience and curiosity (*experientia* and *curiositas*)

---

1. Gerhild Scholz Williams, "Paracelsus," in *The Oxford Encyclopedia of the Reformation* (New York: Oxford University Press, 1996), 3:212.

and to explore chemical cures in lectures at Basel University, delivered in his native German. As if the latter fact alone were not enough to proclaim his scientific and medical iconoclasm, he allegedly destroyed the *Canon* of Avicenna at a public book burning on 24 June. This attack on the medical authorities of antiquity signaled to the medical establishment in Basel and beyond that a revolution in medicine was afoot, a revolution that was to raise violent clashes among the university-trained physicians. Still, Paracelsus did enjoy support among some of the brightest minds of the European Renaissance.

If this was Paracelsus's moment, he squandered it recklessly. The academic opposition gathered its strength; events took a bad turn. Froben died. The offended university faculty and their allies struck back, ridiculing Paracelsus and his new science in print as "Cacophrastus." Furthermore, Paracelsus got into a financial dispute with a wealthy burgher over the amount owed for treatment. As he had done so many times before, Paracelsus resorted to flight to extract himself from these increasing difficulties. His wanderings through a good part of central Europe would continue until his death in 1541, at which point only a fraction of his vast output of writings had been printed.[2]

Paracelsus's notoriety preceded his fame by several decades. His moment was not to come during his lifetime, but, to the extent that there was a genuine Paracelsian moment, it began during the second half of the sixteenth century. It expanded to a movement that continued to flourish into the early decades of the seventeenth century. Scholars such as Michael Toxites, Adam von Bodenstein, and Gerhard Dorn scoured central Europe in search of Paracelsus's manuscripts. By 1570 a wide-ranging Paracelsian revival was under way. Learned academics such as Theodor Zwinger and Gunther von Andernach began to take Paracelsus seriously. At the turn of the century, Paracelsus's ideas were central to debates about pharmacology and medicine across northern and central Europe; his theories on the nature of healing, elemental spirits (*Elementargeister*), and the relationship between mankind and the cosmos were ceaselessly rehearsed, with much acclamation and occasional derision, in the burgeoning early modern literature on witches and wonders.

Beginning with the tremendous success of the *Grosse Wundarznei* (1536), Paracelsus's work enjoyed an increasingly vigorous reception that continued to grow substantially for many years, especially in the decades after his death in 1541. His influence reached, considerably altered, into the twentieth century. Paracelsus became the favorite subject of the emerging discipline of the history of

---

2. Robert-Henri Blaser, *Paracelsus in Basel* (Muttenz: St. Argobast Verlag, 1979).

Introduction

medicine in Germany, owing, for the most part, to the pathbreaking editorial work of Karl Sudhoff, who devoted his professional life to assessing the manuscript and print history of Paracelsus's works and publishing the scholarly editions that remain authoritative to this day.[3] Despite this auspicious start, the assessment of Paracelsus's life and work has, in many ways, been more problematic in the twentieth century than during the early modern period. This development was due in part to Sudhoff's and his collaborators' conviction that the Paracelsian corpus would reach the modern reader and scholar to the best advantage if it were neatly divided into two halves, broadly speaking: a scientific and a religious one. The important and influential religious aspects of his worldview were relegated to the specialists in religion and intellectual and social history, leaving the empirical tracts to the researchers in medical history.[4] Furthermore, the ideologically driven trilogy of Erwin Guido Kolbenheyer based on Paracelsus's life greatly shaped the popular image of Paracelsus in Germany.[5] Presented as a hero to right-wing nationalists, Paracelsus was coopted by the Third Reich for its own propaganda purposes, a phenomenon that reached its ahistorical culmination in the 1943 Ufa film *Paracelsus*, directed by G. W. Pabst.[6]

The reception of Paracelsus across the Channel and across the Atlantic never reached the level of enthusiasm that it did in German-speaking territories. He had an impact on pharmacology, and some of his alchemical writings influenced the development of chemistry, but relatively few English speakers were familiar with more than the occasional concept from the vast body of his writings.[7] Movements such as the radical spiritual vision associated with Paracelsus, the "Theophrastia

---

3. Karl Sudhoff, *Versuch einer Kritik der Echtheit der Paracelsischen Schriften, I. Theil: Bibliographia Paracelsica: Besprechung der unter Hohenheims Namen 1527–1893 erscheinenen Druckschriften* (Berlin, 1894; repr., Graz, 1958).

4. See Peter Dilg, "Paracelsus-Forschung gestern und heute: Grundlegende Ergebnisse, gescheiterte Versuche, neue Ansätze," in *Resultate und Desiderate der Paracelsus-Forschung*, ed. Peter Dilg and Hartmut Rudolph, Sudhoffs Archiv Beihefte, 31 (Stuttgart: Franz Steiner, 1993), 2–24.

5. Erwin Guido Kolbenheyer, *Die Kindheit des Paracelsus* (1917), *Das Gestirn des Paracelsus* (1921), *Das Dritte Reich des Paracelsus* (1926).

6. See Udo Benzenhöfer, "*Ecce Ingenium Teutonicum*: Bemerkungen zur Paracelsus-Romantrilogie Erwin Guido Kolbenheyers," in *Paracelsus: Das Werk—die Rezeption*, ed. Volker Zimmermann (Stuttgart: Franz Steiner, 1995), 161–70; and idem, "Zum Paracelsus-Film von Georg W. Pabst," in *Parega Paracelsica: Paracelsus in Vergangenheit und Gegenwart*, ed. Joachim Telle (Stuttgart: Franz Steiner, 1991), 359–377.

7. This point has recently been energetically made by Stephen Pumfrey, "The Spagyric Art; or the Impossible Work of Separating Pure from Impure Paracelsianism: A Historiographical Analysis," in *Paracelsus*, ed. Ole Peter Grell (Leiden: Brill, 1998), 21–51. See Allen G. Debus, *The English Paracelsians* (New York: Franklin Watts, 1966).

Sancta," took hold among some of the radical religious groups that appeared during the English Civil War (1642–49). At the time when Paracelsus was enjoying a scholarly reappraisal on the Continent, the study of Paracelsus in the Anglophone world remained largely in the hands of occultists and spiritualists such as Arthur Edward Waite.[8] While interest in Paracelsus increased within the medical-historical community in Germany, Switzerland, and Austria, more cautious Anglo-American historians such as Lynn Thorndike remained dismissive of his scientific contribution. After 1945, Third Reich refugees like Henry Sigerist and Owsei Temkin inaugurated more serious research on Paracelsus in the United States. But it was not until Walter Pagel and Allen Debus published their seminal work that Paracelsus became one of the icons in the history of early modern science among English-language scholars. In recent years, the integration of Paracelsus into the larger narrative of the scientific revolution has been further advanced by Charles Webster. Scholars such as Jole Shackelford and Ole Peter Grell have followed Debus's lead in pursuing the influence of Paracelsianism beyond Germany.[9]

At this point in the development of the history of science and medicine, Paracelsus is attracting the attention of scholars the world over. In addition, a number of regional societies devoted to the study of his work are furthering the publication of new and exciting Paracelsus research.[10] The most salutary development of the post war era has been the appearance of a scholarly edition of Paracelsus's religious corpus, a task undertaken by Kurt Goldammer and now carried forward by Urs Leo Gantenbein. Scholars such as Robert Jütte, Bruce Moran, and Joachim Telle have investigated and continue to offer new insights into the socio-historical aspects of Paracelsianism. Bea Lundt, Hildegard Elisabeth Keller, Gunhild Pörksen, and Katharina Biegger, among others, have advanced our understanding of issues of gender, sexuality, and marriage in Paracelsus's writings. Scholars such as Udo Benzenhöfer, Ute Gause, Carlos Gilly, and Hartmut Rudolph have also turned their attention, to great effect, to the religious dimension of Paracelsus's thought.[11]

---

8. See Andrew Cunningham, "Paracelsus Fat and Thin: Thoughts on Reputations and Realities," in *Paracelsus*, ed. Grell, 53–77.

9. Walter Pagel, *Paracelsus: An Introduction to the Philosophical Medicine in the Era of the Renaissance*, 2d ed. (Basel: Karger, 1982); Debus, *The English Paracelsians*; idem, *The French Paracelsians* (Cambridge: Cambridge University Press, 1991); and Charles Webster, *From Paracelsus to Newton: Magic and the Making of Modern Science* (Cambridge: Cambridge University Press, 1982).

10. These included the Austrian-based Salzburger Beiträge zur Paracelsusforschung, the Swiss series Nova Acta Paracelsica, and the more recent Heidelberger Studien zur Naturkunde der frühen Neuzeit.

11. See especially the respective studies in *Paracelsus*, ed. Grell; and Heinz Schott and Ilana Zinguer, eds., *Paracelsus und seine internationale Rezeption in der frühen Neuzeit* (Leiden: Brill, 1998). Paracelsus,

Furthermore, the growing interest in Paracelsus's oeuvre bore fruit in a series of conferences and publications in 1993 and 1994 celebrating the five-hundredth anniversary of Paracelsus's birth.[12] These publications, welcome as they are, also point to the abiding desideratum of Paracelsus studies, namely the challenge of publishing Paracelsus's complete works and the related task of producing adequate scholarly translations and study and research aids. The translations of his writings frequently focus on the most esoteric works or those of questionable authenticity, and they must be used with great caution.[13] Finally, there has been no lack of popular interest in Paracelsus, as can be seen in a number of biographies and mythologizing entries in recent reference works.

The collection of essays assembled here takes its place among those studies that seek further to broaden access to Paracelsus's works and aid the understanding of his influence. This book is not about Paracelsus only, but about the wide range of issues explored by Paracelsus himself and taken up by many who were directly or indirectly affected by the same mental universe that sustained his thought and his writings. This volume's title, *Paracelsian Moments*, is meant to articulate a subject considerably larger than Paracelsus himself. Rather than suggest that Paracelsus's ideas were at the heart of whatever was progressive or interesting in early modern natural philosophy, magic, mysticism, and healing, these essays probe the early modern discourse where it concerns itself with the world as it presented itself from 1490 to 1680. Men like Giovanni Pico della Mirandola, François Rabelais, Heinrich Cornelius Agrippa, Athanasius Kircher, and Johannes Praetorius can, upon closer inspection, be put forward as exemplars of this movement that touched Paracelsus or was touched by him. Because of the intense interest his work has generated, both in the early modern world and among contemporary scholars, Paracelsus is the most appropriate figure to be selected as the standard-bearer of this volume. His obscure medical dicta, his belief in the reality of elemental spirits such as nymphs and mountain spirits, his pronouncements on procreation and sexuality, and his quirky Marian piety make him very much a pre-Enlightenment figure. Furthermore, his interdependent vision of gender, his

---

Sämtliche Werke. II. Abteilung: Theologische und religionsphilosophische Schriften, ed. Kurt Goldammer (Wiesbaden: Franz Steiner, 1955–).

12. E.g., Peter Dilg and Hartmut Rudolph, eds., *Resultate und Desiderate der Paracelsus-Forschung* (Stuttgart: Franz Steiner, 1993); Heinz Dopsch et al., *Paracelsus (1493–1541): "Keines andern Knecht..."* (Salzburg: Verlag Anton Pustet, 1993); and Joachim Telle, ed., *Analecta Paracelsica* (Stuttgart: Franz Steiner, 1994). See bibliography for a more complete listing.

13. Among the most pressing needs of English-language Paracelsus scholarship is a comprehensive assessment of all English-language translations of Paracelsus's works.

espousal of alternative medicine, and his deeply spiritual, nonconfessional piety mark him as a thinker and experimenter worthy of continued scholarly attention.

This volume also reflects the fertile international dialogue over the last decades among scholars of early modern science, history, gender, and literature. Recent scholarly trends have created a nonpositivistic, interdisciplinary climate where challenging connections among disciplines can be made. Thus, our *Paracelsian Moments* is less a celebration of Paracelsus à la Sudhoff than an attempt to present the conversation that is now occurring among scholars of literature, social history, and the history of science. It aims to illuminate a singular phenomenon, Paracelsus and his mental universe, through a multiplicity of disciplinary strategies.

"Paracelsian moments" is also meant to invoke a way of thinking and writing about the natural world that had roots in Italian Neoplatonism, crossed the Alps into the Northern Renaissance, and found its most memorable and inspiring, if somewhat idiosyncratic, expression in the writings of Paracelsus. The decades after 1570 witnessed a full-fledged Paracelsian reception that played a critical role in the scientific debates that excited early modern Europe before Paracelsianism lost its paradigm-shaping power to the mechanical philosophy of nature in the late seventeenth century. Long after his hypotheses were losing their "scientific" prestige, the type of religious-alchemical concerns elucidated by Paracelsus enjoyed a significant afterlife in popular writers like Johann Praetorius as well as among the Faust adaptors Georg Widman and Johannes Pfitzer. To this must be added the European encounter with the New World, which forced early moderns to ask fundamental questions about the nature of humanity itself.[14] Equally challenging to the early modern mental universe were the multifaceted reform movements of the sixteenth century, which constructed new visions of biblical and spiritualist authority.[15] All of these factors worked together to provoke the most profound speculations on fundamental questions of early modern intellectual life.

Paracelsus having been placed at the center of the story, Charles Gunnoe's essay on the anti-Paracelsian attempt to discredit Paracelsus examines the controversial nature of Paracelsus in his era. Gunnoe offers a study of Thomas Erastus's efforts to damage Paracelsus's reputation by forming a composite image of Paracelsus in

---

14. See Ingo Schütze, *Zur Ficino Rezeption bei Paracelsus*, in *Parega Paracelsica*, ed. Joachim Telle (Stuttgart: Franz Steiner, 1991), 39–44.

15. See Andrew Weeks, *Paracelsus: Speculative Theory and the Crisis of the Early Reformation* (Albany: State University of New York Press, 1997).

INTRODUCTION

his four-volume *Disputations on the New Medicine of P. Paracelsus* (1571–73). Erastus was a medical professor at the University of Heidelberg and one of the more important Protestant natural philosophers of the late sixteenth century. He represents the extreme of the common tendency of humanistically trained academic physicians to repudiate Paracelsus's innovations. With excellent connections at the imperial court and in Paracelsus's Swiss homeland, Erastus was able to gather numerous biographical morsels concerning Paracelsus. Through careful management of his sources, Erastus constructed a biography that emphasized Paracelsus's heretical-to-demonic beliefs, his undependable and dissolute personal habits, and his ignorantly prescribed and frequently lethal medical therapies. Given the clearly partisan slant of Erastus's work, it must be used with care as a historical source. Nevertheless, it gives a clear indication of how threatened many academic physicians felt by new medicine as the Paracelsian revival reached its apogee.

The late-sixteenth-century controversy over Paracelsian medicine is also on display in Mitchell Hammond's investigation of the Augsburg medical community. Hammond provides something unusual in Paracelsian historiography: a social history of the reception of Paracelsian practitioners in one German town. Although Augsburg had been the home of an early and vocal supporter of Paracelsus, in the person of the city physician, Wolfgang Thalhauser, Augsburg's learned physicians initially had the upper hand in limiting the activity of Paracelsian practitioners in their city. In 1582, a College of Medicine had been sanctioned by the town council to have wide jurisdiction in regulating medical practitioners. If Hammond's account were limited to the legal history, we would leave it thinking the university-trained supporters of traditional Greek medicine à la Hippocrates were well armed to keep Paracelsianism at bay. Hammond has been able to find a number of cases, however, where Paracelsian practitioners gained the sanction of the city council to treat a large pool of patients even when they were opposed by the College of Medicine. The alliances made by these alternative practitioners with burghers from both the upper and middle levels of the city enabled them to outmaneuver the College of Medicine in conflicts with the city council. From the cases Hammond brings to light, it appears that the acute need of the townspeople for medical treatment usually trumped the authority of learned physicians. This pathbreaking archival study sheds much light on the sociohistorical paths that Paracelsian knowledge traveled.

Jole Shackelford's study is informed by the recent discussion led by Stephen Pumfrey and others on whether the term *Paracelsianism* has been so broadly defined in recent scholarship as to be largely meaningless.[16] Shackelford continues this line of inquiry by asking the basic question of whether the prominent natural Danish philosophers Caspar Bartholin and Ole Worm can fairly be considered Paracelsians. There is evidence that both figures, following in the wake of their famous countryman Petrus Severinus, had more than a passing interest in aspects of Paracelsian theory and practice. However, the apparent surface Paracelsianism of Bartholin and Worm disappears under closer scrutiny. All the progressive trends of their scholarship cannot be reckoned upon the ledger of Paracelsian influence. The tendency toward basing knowledge on experience was nourished by neo-Aristotelian, Galenic, and natural-historical schools, and in fact these scholars frequently repudiated Paracelsus's own most distinctive ideas. Shackelford offers a nuanced vision of the interaction between the schools, and his work goes a long way to explaining why the Aristotelian model remained persuasive and continued to make substantial contributions to the knowledge of the natural world in the early modern era.

Moving across the water to England, Lynda Payne investigates the many ways in which the Company of Barber-Surgeons solidified its professional, social, and economic status in sixteenth-century England. As in all social ascents, the trip was not accomplished without difficulty, and Payne devotes special attention to the manner in which the barber-surgeons' expertise in anatomy could have become a social and legal liability. Turning a potential weakness into a strength, the barber-surgeons asserted their monopoly over autopsy as a hallmark of their medical superiority over both unlearned quacks and university-trained physicians. Here a special group of learned surgeons waged a multifront campaign as they made their case in print, sought and obtained the enhanced professional status from the state, and celebrated their victorious rise in skillful artistic self-promotion. Rather than advocate the radical vision of medical renewal associated with Paracelsus that was so widespread on the Continent, the learned surgeons drew from the same sanctioned Greek sources of knowledge that inspired their university-trained rivals. As the century drew to a close, a few celebrated episodes of the unveiling of notorious quacks further solidified the public confidence in and respectability of the Company of Barber-Surgeons.

From the professionalization of the medical profession in early modern England, we turn to Hildegard Keller's investigation of the gendered aspects of Paracelsian medicine. Paracelsus's "gendered epistemology" details his thinking about the world, humankind, and the relationship of males and females to both. Scholars have been fascinated and not a little confused by Paracelsus's unconventional approach to gendered medicine, specifically in the context of male and

---

16. Pumfrey, "Spagyric Art," 21–51.

Introduction

female pathologies and their association with perception, the human "seeing" of the macrocosm. A worthy physician does not only see the microcosm (man and woman) reflected in the macrocosm and vice versa; according to Paracelsus, he also knows that traditional humoral medicine is insufficient to recognize the difference between the "smallest world" (woman) and the "small world" (man), between their illnesses, and how best to treat them. Accordingly, the physician does not "see" the woman's illness just by looking at her outward appearance. The illness lies hidden within woman, in her matrix, which itself is hidden within her body. Far from being threatening or dangerous, this mystery can and will be recognized and treated by the physician who understands the relationship between the diagnostic gaze ("the physician's eyes") and the macrocosm and the microcosm. Before his gaze, the secrets (*Heimlichkeiten*) of human nature, specifically of woman, will be revealed. Once he "sees" with his own eyes the secrets of the matrix and its relationship to the cosmos, he can treat and heal, making use of experience and knowledge.

Turning from Paracelsus's medical preoccupations and following in the footsteps of continental scholars such as Kurt Goldammer and Hartmut Rudolph, Dane Thor Daniel has begun the journey of coming to terms with Paracelsus's unique religious vision and its relationship to natural philosophy. Daniel contributes to the knowledge of Paracelsus's sacramental theology through a study of Paracelsus's conception of baptism. While Paracelsus composed a number of treatises on the sacraments that still await publication, his sacramental thought cannot be seen in isolation from his more widely known works of natural philosophy such as the *Astronomia magna*. In both genres Paracelsus endeavored to develop his unique vision of the "immortal philosophy." The sacraments are critical components of this vision, as it is through them that the spiritual flesh that will inherit the kingdom of God is vouchsafed to the Christian believer. While maintaining a high vision of baptismal efficacy, Paracelsus develops a broadly evangelical focus on what makes a legitimate baptism. Daniel proves yet again how difficult it is to force a confessional label on Paracelsus. Daniel's study also adds weight to the interpretation of scholars such as Charles Webster and Erik Midelfort that Paracelsus's worldview was not driven by Hermetic, Neoplatonic, or even Gnostic concerns but rather was largely determined by his own literal, if idiosyncratic, reading of the Bible.[17]

Heinz Schott turns to arcana of a different kind. He explores one of the most important and influential ideas of natural philosophy, the idea that all bodies,

---

17. See H.C. Erik Midelfort, *A History of Madness in Sixteenth-Century Germany* (Stanford: Stanford University Press, 1999); and Charles Webster, *From Paracelsus to Newton: Magic and the Making of Modern Science* (Cambridge: Cambridge University Press, 1982).

including the human organism, are connected to each other by a network of magnetic influences, called magnetism. The focus here is on the relationship of magnetism to the power of imagination that joins humankind to the whole of nature, the microcosm to the macrocosm. An essential concept in Paracelsian epistemology, imagination was also central to the discussion of white and black magic, from Ficino to van Helmont to Mesmer. The close relationship between the Paracelsian imagination and magnetism is evident in the writings of many early modern natural philosophers, including William Gilbert, Johannes Kepler, and Athanasius Kircher. Their studies imported some of the most original and influential ideas of Paracelsus into the scientific and philosophical discourses of the sixteenth and seventeenth centuries. However, while both imagination and magnetism were fully accepted as objects of scientific exploration, it was not until the late eighteenth century that Franz Mesmer combined the two into his psychosomatic theories, thereby ensuring the survival of some essentially Paracelsian ideas into the modern period.

Astrology, a way of "seeing" the heavens, understanding the universe, and knowing God's plan for mankind, was a consuming passion of Renaissance people, learned and lay alike. It loomed large in Paracelsus's medical and philosophical writings, revealing the influences of Italian natural philosophers on scholars and writers north of the Alps. Sheila Rabin explores the seemingly contradictory pronouncements on astrology's role advanced by Giovanni Pico, first in his *900 Theses*, and, toward the end of his life, in the *Disputations against Judicial Astrology*, where he vigorously inveighed against the practice on religious, philosophical, technical, and historical grounds. Pico rejected the idea that stars can be manipulated to serve mankind in its undertakings; he was critical of the use of astral magic. Pico noted that stars have physical effects: they provide light, heat, and motion. Upon closer inspection of his writings, it appears that Pico moved in the direction of a materialistic view of the universe, a universe that is controlled by God, but not necessarily by angelic or satanic intelligences. He suggests that the alleged powers of the stars over people and events need to be proven, not just asserted, and that the true order of the planets cannot be established by astrologers.

Some forty years later, Rabelais ridiculed the divinatory arts in his *Histories of Gargantua and Pantagruel*, even though, as Dené Scoggins demonstrates, he was widely known and respected as a writer of almanacs and prognostications, being in fact much better known as a physician and astrologer than as a poet. His expertise and success in the field and his negative assessment of the astrological arts puzzles the modern reader. Scoggins's careful review of Rabelais's pronouncements on astrology in the five books of *Gargantua* reveals that he did not reject astrology but rather that he criticized those who placed greater trust in stellar signs than in

Introduction

the more direct path provided by the divinely inspired knowledge of the human body itself. He devotes much humorous discussion to the very difficult question that lies at the heart of all astrological writings, whether people can penetrate the unknowable and gain access to the future by practicing astrology. At the core of Rabelais's thinking about astrology lies a quintessential Paracelsian paradigm: the body as microcosm reflects the universe as macrocosm. Thus, any self-respecting physician would treat the human body according to the concordances that are organized in layers of cosmic correspondences. The body, with its physical and spiritual needs, is the universal hub on which cosmic influences focus the energies that are then turned back toward the universe. Paracelsus's plague metaphor comes to mind. The divine cannot be known through signs, although knowledge of the body, of human physiology, can lead to knowledge of the greater world, allowing for a new kind of human exploration, in human anatomy. God's secrets lie as much in the human body as in the cosmos, and the philosopher must strive for the understanding of both.

With his study of Robert Boyle's use of Hebrew, Michael Walton examines the intellectual context of both the emerging discipline of chemistry and Christian Hebraism in seventeenth-century England. One of the intellectual giants of the seventeenth century, Boyle is remembered for achievements such as removing the magical-occult associations—a direct inheritance from Paracelsus—from chemistry, his espousal of a corpuscular view of matter, his contribution to the Royal Society, and, not least, Boyle's Law. Boyle's experimentalism was coupled with a firm belief in the relevance of biblical revelation both for knowledge of salvation and for insight into the natural world. Given the continued relevance of the creation accounts in Genesis for scholarly inquiry, Boyle developed a passionate desire to master Hebrew in order to understand the biblical texts better. That he became quite proficient is shown by his use of Hebrew literature and by the high praise bestowed on him by noted Orientalists of his era. Boyle employed his exegetical insights in support of a reading of Genesis that supported his corpuscular understanding of matter. Given Boyle's seminal role in seventeenth-century science, Walton argues that understanding Boyle's Christian Hebraism supplies a crucial context for understanding the cultural milieu of the scientific revolution.

One of the most enduring of Paracelsian theories, his ideas about elemental spirits, looms large in Johannes Praetorius's (1630–1680) writings on strange and wondrous people. Gerhild Williams looks at Praetorius's three-volume study on the Silesian giant Rübezahl, the *Daemonologia Rubinzalii*. Like Paracelsus, Praetorius had a passion for observing, collecting, categorizing, and explaining the wondrous, the preternatural, in God's creation. Giants in general, and Rübezahl in particular, are specimens of the early modern marvelous that border, on occasion, on the demonic. Inhabiting the Silesian mountains of the Riesengebirge, and thus at the intersection between wilderness and early modern civilization, Rübezahl is the purveyor of great wealth, a keeper of secrets, a shape shifter, a temperamental weather maker, a magician, and a spirit privy to human frailty, pride, and foolishness. Much like Paracelsus, Praetorius explores the Book of Nature and writes about wonders to satisfy the curious minds of his readers and gain insights into the secrets of creation. Sublunar or monstrous spirits such as Rübezahl dwell in their own "chaos," as do other elemental spirits. The *Daemonologia* provides an example of the popular and semiscientific reception of Paracelsian theories about natural philosophy and cosmology.

Rounding off this collection of essays with a contribution to the sixteenth-century debates about the reliability of visual perception, Stuart Clark explores the question whether human vision provides trustworthy access to the real world. The idea that reality is constructed by human consciousness and that this construction can go far in deceiving the senses affected early modern views on witchcraft, the power of the Devil, optical observations, astronomy, and experimental science. It influenced the reception of skeptical thought introducing radical doubt about the reliability of senses, specifically the eyes, into the debates about the efficacy of Catholic and Protestant religiosity. To Keller's "medical seeing," Clark adds the "sacramental seeing" of sixteenth-century worship. The confrontation of visual experience with reality in the case of witchcraft, for example, can deceive the senses to such a degree that any concept of the real becomes subject to disconcertingly "radical visual indecision." Demonology, the notion that the Devil as the inventor of virtual (occult) reality wields power granted to him by God, confirms that human perception can be successfully deceived. Aside from recommending that witches who fell prey to the Devil's dissimulations be burned, early modern intellectuals, natural scientists, and/or demonologists wrote vast and learned tomes on the construction of satanic reality, on optics, and on all sciences. This learned preoccupation with seeing and making visible oftentimes brought them perilously close to magic and satanic subterfuge themselves.

The essays assembled in this volume confirm once again that the depth and breadth of the influence of Paracelsus on many aspects of early modern culture continue to offer exciting challenges to archivists, scholars, teachers, and students. This volume is intended as a part of the continuing intellectual conversation about the man, his work, and the amazing longevity of his ideas.

Paracelsiana

Paracelsus's Biography among His Detractors

Charles D. Gunnoe, Jr.

> *I was able to perceive nothing of piety or erudition in him beyond a marvelous quickness in preparing medicine.... During the two years that I lived with him his days and nights were given over to drunkenness and dissipation, so that it was hardly possible to discover him sober from one hour to the next.*
> —Johannes Oporinus on Paracelsus

With these words the effort to discredit the reputation of sixteenth-century Europe's greatest medical innovator had begun. Although Paracelsus's former assistant Johannes Oporinus may have begun the process unwittingly, the campaign would pick up steam as the Paracelsian revival advanced in the 1560s and 1570s and would enroll the likes of Konrad Gesner, Johannes Crato von Krafftheim, and Heinrich Bullinger, men conventionally considered among the most irenic figures of sixteenth-century Europe. In the early 1570s, Thomas Erastus, a medical professor at the University of Heidelberg, would collect the various biographical tidbits that were current among the anti-Paracelsian party and splice them into his magisterial four-volume *Disputations on the New Medicine of Philipp Paracelsus*.[1] This study will use the stories selected by Erastus to draw a composite picture of Paracelsus as constructed by the anti-Paracelsian community.

---

1. Thomas Erastus, *Disputationum de medicina nova Philippi Paracelsi Pars Prima: In qua, quae de remediis superstitiosis & Magicis curationibus ille prodidit, praecipue examinantur* (Basel: P. Perna, [1571]); *Disputationum de nova Philippi Paracelsi medicina Pars Altera: In qua Philosophiae Paracelsicae Principia & Elementa explorantur* ([Basel], P. Perna, 1572); *Disputationum de nova Philippi Paracelsi Medicina Pars Tertia* ([Basel]: P. Perna, 1572); and *Disputationum de nova medicina Philippi Paracelsi Pars Quarta et Ultima* (Basel: P. Perna, 1573) (hereafter collectively cited as *De medicina nova*, with the volume designated by a roman numeral). See also the translation of *De medicina nova*, I (without the appendix), Franz Josef Schmidt, ed., *Disputationen über die neue Medizin des Philippus Paracelsus* (Hamm: self-published, 1978).

Charles D. Gunnoe, Jr.

Thomas Erastus and the *Disputations* on the New Medicine of P. Paracelsus

Erastus finds an entry in most dictionaries and encyclopedias for lending his name to "Erastianism," the concept that the state should exercise control over the church. While Erastus was an important player in sixteenth-century church politics, he was also significant as a wide-ranging natural philosopher who tackled topics from soil science to witchcraft.[2] After backing the losing side in a dispute over the proper form of church discipline in the Palatinate in the late 1560s, Erastus refocused his scholarly efforts on topics more proper to his position as medical professor at the University of Heidelberg. The first major project of the later phase of his academic career would be his *Disputations on the New Medicine of P. Paracelsus*.[3] These anti-Paracelsian disputations offered an exhaustive refutation of Paracelsus's theology, medical philosophy, and therapeutic techniques at the crest of the late-sixteenth-century Paracelsian revival.[4]

One of the most enduring and often most entertaining aspects of Erastus's *Disputations* was its depiction of Paracelsus's life and customs for future generations. The fact that Erastus printed so many discrediting stories about Paracelsus caused him to be viewed as something of a pariah by early modern Paracelsians and by Germanic founders of the history of medicine. Setting aside the hero worship of Paracelsus, one can appreciate that the unflattering stories that Erastus no doubt assembled for a polemical purpose hold in most cases genuine biographical kernels that should be seriously considered by scholars seeking to reconstruct Paracelsus's biography. Erastus assaulted Paracelsus not through misrepresentation and fabrication but through selection. Erastus may have been inclined to believe and report the worst of Paracelsus, but there is no reason to think that he was dishonest. Where his accounts can be verified with other sources, it is clear

---

2. Regarding Erastus's biography, see Ruth Wesel-Roth, *Thomas Erastus: Ein Beitrag zur Geschichte der reformierten Kirche und zur Lehre von der Staatssouveränität* (Lahr/Baden: M. Schauenburg, 1954); and Charles D. Gunnoe, Jr., "Thomas Erastus in Heidelberg: A Renaissance Physician during the Second Reformation, 1558–1580" (Ph.D. diss., University of Virginia, 1998).

3. For further details on the background and content of these disputations, see Charles D. Gunnoe, Jr., "Erastus and His Circle of Anti-Paracelsians," in *Analecta Paracelsica: Studien zum Nachleben Theophrast von Hohenheims im deutschen Kultergebiet der frühen Neuzeit*, Heidelberger Studien zur Naturkunde der frühen Neuzeit, 4, ed. Joachim Telle (Stuttgart: Franz Steiner, 1994), 127–48; and idem, "Erastus and Paracelsianism: Theological Motifs in Thomas Erastus' Assault on Paracelsianism," in *Reading the Book of Nature: The Other Side of the Scientific Revolution*, Sixteenth Century Essays and Studies, 41, ed. Allen G. Debus and Michael Walton (Kirksville, Mo.: Sixteenth Century Journal Publishers, 1998), 45–65.

4. For a general overview of the Paracelsian revival, see Lynn Thorndike, *A History of Magic and Experimental Science* (New York: Columbia University Press, 1923–1958), 5:617–51.

that he recounted the sordid tales he collected with great accuracy.[5] Erastus was the first person to print extended excerpts from Oporinus's description of Paracelsus's style of life, a text widely regarded as the most important biographical document from Paracelsus's famous Basel period.[6] Erastus also printed excerpts of Gesner's correspondence that offered a shrill warning concerning the deviant nature of both Paracelsus's medical theory and his theology.[7] Because these texts were preserved in other manuscript or later printed versions, they are especially useful to this study as yardsticks for Erastus's methods and accuracy as an editor. At the sentence level, Erastus reproduced these texts verbatim, but he was not averse to paring down and rearranging a text to achieve his stylistic and rhetorical goals. As becomes most clear in Erastus's treatment of Oporinus's letter, Erastus was a selective editor but not prone to alter the actual content of his source material or invent material out of whole cloth.[8]

---

5. I have previously assessed the flow of information from Gesner and Oporinus into Erastus's *Disputations*, so I will limit my discussion of their role in this study; see Gunnoe, "Erastus and His Circle," 132–37.

6. Erastus's excerpts from Oporinus's letter with English translation are printed as an appendix in Gunnoe, "Erastus and His Circle," 146–48. See also Sepp Domandl, "Paracelsus, Weyrer, Oporin [sic]. Die Hintergründe des Pamphlets von 1555," in *Paracelsus, Werk und Wirkung: Festgabe für Kurt Goldhammer zum 60. Geburtstag*, Salzburger Beiträge zur Paracelsusforschung, 13, ed. Sepp Domandl (Vienna: Verband der wissenschaftlichen Gesellschaften Österreichs, 1975), 53–70, 391–92; and Udo Benzenhöfer, "Zum Brief des Johannes Oporinus über Paracelsus: Die Bislang älteste bekannte Briefüberlieferung in einer 'Oratio' von Gervasius Marstaller," *Sudhoffs Archiv* 73 (1989): 55–63.

7. Erastus, *De medicina nova*, III:14–15: "His scriptis reperi propositiones a te missas nebulonis illius N.N. [Alexander von Suchten] de cuius in re medica, causis & curationibus morborum imperitia, vt nihil dicam (neque enim opus est apud te). Vide quem nobis conspicuum filium Dei faciat: non alium scilicet, quam spiritum mundi & naturae, eundemque corporis nostri (mirum quod non etiam asini ac bouis addit) qui artificio Discipulorum Theophrasti a materia seu corporibus elementorum separari potest. Si quis ipsum vrgeat; dicet se Philosophorum sententiam retulisse, non suam. At qui ita recitat, vt laudet. Et scio alios quoque Theophrasteos talia suis scriptis aspergere, vnde Christi diuinitatem eos negere facile appareat. Ipsum quidem Theophrastum Arianum fuisse omnino mihi constat. Hoc agunt, vt Christum omnino nudum hominem fuisse persuadeant, nec alium in eo spiritum quam in nobis esse. Scio in Polonia & dudum Arianos multos fuisse, & nuper ab Italis quibusdam eo profectis ipsorum numerum augeri. Itaque & hanc ob causam, & vt artis medicae nostrae fundamenta ac methodum secundum Hippocratem & Galenum verissime traditam tuearis, oro te & hortor, vir doctissime, vt omnibus modis te illis calumniatoribus, magis, & Arianis opponas. Qua in re si quid ego etiam potero, non deero." See Charles Webster, "Conrad Gessner and the Infidelity of Paracelsus," in *New Perspectives on Renaissance Thought: Essays in the History of Science, Education and Philosophy*, ed. J. Henry and S. Hutton (London: Duckworth, 1990), 13–23.

8. Erastus's quotations and translations from the works of Paracelsus appear to be quite accurate. Jole Shackelford has also been able to find very precise excerpts from Petrus Severinus's *Idea Medicinæ* in Erastus's text. Jole Shackelford, "Early Reception of Paracelsian Theory: Severinus and Erastus," *Sixteenth Century Journal* 26 (1995): 127–28.

Erastus went to great lengths to collect information on Paracelsus. Around 1570, he asked his contacts in Switzerland to gather information regarding Paracelsus's family but was not able to connect Paracelsus to any living Swiss relative. This fact was some consolation to Erastus, who wanted to distance Paracelsus from his beloved homeland; as he writes in *Disputations*, "I can hardly believe he was Swiss."[9] Erastus incorrectly assumed that Paracelsus's given name had been "Philipp Bombast" and that he was born in Hohenheim in Southwestern Germany. He noted that Paracelsus had lived in the province of Carinthia in Austria for some years as a child. An unnamed source informed Erastus that Paracelsus, while tending geese as a young boy, was castrated by a soldier.[10] Erastus suggested that this story corroborated Oporinus's claim that Paracelsus had shown little interest in women, but it ultimately seems too much calculated as a partisan insult to be credible. Later Paracelsus allegedly traveled to Spain, where he first learned magic, then chemistry. This account may represent the vestige of a pro-Paracelsian tradition that sought to connect Paracelsus with the tradition of the great Spanish alchemical theorists Ramon Lull and Arnald of Villanova.[11]

Johannes Crato von Krafftheim: Paracelsus in Moravia and Vienna

The most significant biographical gem that Erastus was able to uncover was a report by the Moravian nobleman Berthold von Leipa (d. 1575), relayed to Erastus by Johannes Crato von Krafftheim (1519–1585).[12] Crato was an imperial physician

---

9. Erastus, *De medicina nova*, I:237: "Helvetium fuisse vix credo." All translations are my own.

10. Erastus, *De medicina nova*, I:238: "Hoc in loco, narratum mihi est exectos ei testes fuisse à milite, dum anseres pasceret... Eunuchum fuisse cum alia multa, tum facies indicant: & quod, Oporino teste, feminas prorsus despexit." The examination of Paracelsus's skeleton has revealed the existence of certain female characteristics on his pelvis and cranium. This fact has led some scholars to suggest that Paracelsus may have suffered from a congenital hormonal imbalance related to the abnormal processing of androgens. A firm diagnosis of this condition would certainly shed light on Paracelsus's psychological state and his sexual history. See Herbert Kritscher, G. Hauser, C. Reier, Johann Szilvassy, and W. Vycudelik, "Forensisch-anthropologische Untersuchungen der Skelettreste des Paracelsus," in *Paracelsus (1493–1541): Keines andern Knecht...*, ed. Heinz Dopsch, Kurt Goldammer, and Peter F. Kramm4 (Salzburg: Verlag Anton Pustet, 1993), 53–61.

11. Erastus, *De medicina nova*, I:238.

12. Erastus, *De medicina nova*, IV:159–60. The Leipa were a prominent Moravian noble house, with strong connections to the Bohemian Brethren, who held the hereditary title of marshal of the kingdom of Bohemia. The name "Leipa" stems from the Bohemian town Ceska Lípa and is often falsely given as "Leipnik" after a Moravian town of that name. Vladimir Zapletal, "Paracelsus-Tradition in der Tschechoslowakei," in *Gestalten und Ideen um Paracelsus* (Vienna: Verlag Notring, 1972),

who served the Habsburg emperors Ferdinand I, Maximilian II, and Rudolf II. He possessed impeccable credentials both as an orthodox Protestant (he had studied at Wittenberg with Martin Luther and Philipp Melanchthon) and as a humanist physician (he was the protégé of Paduan humanist Giovanni Battista da Monte). One day in 1570, while Crato was talking with two of Emperor Maximilian's other physicians, Giulio Alessandrini and Nicolaus Biese in the imperial chambers in Prague,[13] the nobleman Berthold von Leipa approached them and told them of Paracelsus's misadventures in Moravsky Krumlov.[14] Berthold recounted that his father, Johann von Leipa, had summoned Paracelsus to his bed because of Paracelsus's reputation for treating arthritis. Though Paracelsus attended Johann for two years, he was not able to improve his condition. Other noble patients were not to be so fortunate. Paracelsus's treatment caused Berthold himself to lose sight in one eye, and the medication he gave to the wife of Baron Johann von Zerotín caused her to break into seizures and promptly expire.[15] This disaster was enough to trigger one of Paracelsus's characteristic flights, this time into Hungary. Berthold also informed Crato that Paracelsus had written quite a lot in Moravsky Krumlov—one of the volumes of the Astronomia Magna is in fact dated from that time—though Berthold added that he accomplished this in a drunken state.[16] Evidently Paracelsus was able to recover the manuscripts that he deposited in Moravsky Krumlov, perhaps through the intervention of the emperor.

158–60; Augustin Tschinkel, "Paracelsus bei König Ferdinand I," in *Paracelsus, Werk und Wirkung*, 340; and J.H. Zedler and C.G. Ludovici, *Grosses vollständiges Universal-Lexicon aller Wissenschaften und Künste* (Halle, 1732–50), 16:1644–45.

13. Julius Alexandrinus von Neustein (Italian, Guilio Alessandrini, 1506–90), personal physician to Emperors Ferdinand I and Maximilian II from Trent, and Nicolaus Biese (1516–72), a Galenic physician from Ghent, who likewise served as personal physician to Maximilian II.

14. A Moravian city (German, Mährisch-Kromau) located southwest of Brno. Zapletal judged that Paracelsus's stay in Moravsky Krumlov probably lasted five to six months. Zapletal, "Paracelsus-Tradition in der Tschechoslawakei," 161; and Karl Pisa, *Paracelsus in Österreich* (St. Pölten/Vienna: Niederösterreichisches Pressehaus, 1991), 116–20.

15. Perhaps the wife of Johann III von Zerotín. According to Zedler, Johann III was married to Johanna von Leipa, the daughter of Johann III von Leipa and Mariana von Sternberg. If this identification is correct, Paracelsus must have journeyed to the Zerotín residence in Straznice, a Moravian city located near the current Czech-Slovak border. Zedler, *Universal-Lexicon*, vol. 62, cols. 1555–56; and Zapletal, "Paracelsus-Tradition in der Tschechoslowakei," 161. Like the Leipa with whom they were intermarried, the Zerotín were a prominent Moravian noble house and supporters of the *Unitas Fratrum*. See R.J.W. Evans, *Rudolf II and His World* (Oxford: Clarendon Press, 1973), 41–42.

16. The conclusion of the *Astronomia Magna* is dated 22 July 1537. Paracelsus, *Sämtliche Werke. I. Abteilung: Medizinische, naturwissenschaftliche und philosophische Schriften*, ed. Karl Sudhoff, 14 vols. (Munich and Berlin, 1922–33), 12:vi, 12:273.

How does one assess such a tale? On the one hand, it was told some thirty years after the events that it purports to describe, and it was passed from Berthold to Crato and then to Erastus. Given the care with which Erastus handled other texts by his humanist colleagues, we can confidently assume that Erastus printed Crato's account accurately. On the other hand, it is the story of an eyewitness (or rather, a one-eyed witness) of Paracelsus's activities in Moravia. It correlates well with Paracelsus's known movements of the time, and the individuals described were historical personages. Though the story of a dissatisfied patient as recorded by a disparager of Paracelsus, the account possesses a fundamental historicity and has been used by Paracelsus scholars, with less than direct acknowledgment, as a source for Paracelsus's activities in Moravia.[17]

In addition to Berthold's story of events which transpired in Moravia, Crato was also able to describe the circumstances of Paracelsus's supposed meeting with King Ferdinand in Vienna. This is one of two versions of the alleged meeting that we possess. The other is an utterly fanciful Czech account, *Colloquium Ferdinandi regis cum D. Theophrasto Paracelsus Svevo*.[18] In that version of events, despite the murderous opposition of the "learned physicians," Paracelsus treated the ailing king with potable gold and immediately improved his condition. After the successful cure, Paracelsus and the king sat down to dinner, during which Paracelsus discoursed on the requisite moral qualities of a successful alchemist. Paracelsus was fêted by the king and given a golden chain on his departure. About the only similarity between this Czech account and the one that appeared in Erastus's pages is the hostility between Paracelsus and the court physicians. In the Crato-Erastus text, King Ferdinand could smell a quack and denounced Paracelsus as a "lying and impudent impostor." Not content to conclude with Paracelsus's humiliation at the imperial court, the Crato-Erastus text contained the additional insult: "Paracelsus always had traffic with Jews and the most vile men."[19]

---

17. The basic work here is Zapletal's article, which contains no citations. A close reading of this work reveals a heavy dependence on the Crato text. This reliance is rather ironic, as Zapletal had complained of Berthold's reminiscences, "die jedoch nun mehr verschwommen waren" (162). Pisa (likewise without citations) brought over much of Zapletal's content into his book. Zapletal, "Paracelsus-Tradition in der Tschechoslowakei"; and Pisa, *Paracelsus in Österreich*.

18. A German translation of this Czech text is printed in Georg Sticker, "Ein Gespräch des Königs Ferdinand mit Paracelsus," *Nova Acta Leopoldina* 10 (1941): 267–79. Tschinkel critiques this translation in "Paracelsus bei König Ferdinand I." This story is also hinted at more faintly in the early-seventeenth-century "Rhapsodia vitæ Theophrasti Paracelsi." See Sven Limbeck, "Paracelsus in einer frühneuzeitlichen Historiensammlung: Die 'Rhapsodia vitæ Theophrasti Paracelsi' von Peter Payngk," in *Analecta Paracelsica*, 1–58.

19. *De medicina nova*, IV:160: "Semper illi negotium fuisse cum Iudæis & vilissimis hominibus." Alternatively, the "Colloquium Ferdinandi regis cum D. Theophrasto Paracelsus Svevo" places an

Crato's version of this story was recorded earlier and possesses more historically verifiable details than the later Czech account. It is difficult to escape the conclusion that the Crato-Erastus text must be given priority over the *Colloquium* as an historical source. Both texts tell more about the polarization of the Paracelsus image that had occurred by the late sixteenth century, however, than about any putative meeting between Ferdinand and Paracelsus. In the one account, Paracelsus was the heroic alchemist, whose success at chrysopoeia clearly led his list of credits. In the other account, Paracelsus was a plain charlatan whose greatest offense was his refusal to enter a meaningful dialogue with Ferdinand's humanist physicians. Armed with the firsthand information from Berthold von Leipa, Crato was able to construct a distinct image of a shiftless, drunken, and deliberately obscure quack that harmonized perfectly with the other negative testimony that Erastus assembled and published.

HEINRICH BULLINGER: PARACELSUS IN ZURICH

Another significant account of Paracelsus's life was relayed to Erastus by Heinrich Bullinger (1504–1575). Bullinger's chief claim to fame consists in being the successor of Huldrych Zwingli as the head of the Reformed Church in Zurich, Switzerland. Bullinger had likely met Paracelsus in 1527, when Bullinger had taken leave of his teaching position in Kappel to spend a few months hearing lectures under Zwingli in Zurich.[20] He reports that he had the opportunity to have a number of conversations with Paracelsus, generally about theological topics. He proclaimed that he could not detect any conventional religiosity in Paracelsus but that he did discern a great predilection for magic. Bullinger included the standard criticism of Paracelsus's dress and hygiene and, like Oporinus, noted Paracelsus's propensity to carouse with the *hoi polloi*. Sometimes Paracelsus drank to such

---

anti-Semitic comment in the mouth of Paracelsus. See G. Sticker, "Ein Gespräch des Königs Ferdinand mit Paracelsus," 275. For a recent assessment of Paracelsus's anti-Semitism, see Udo Benzenhöfer and Karin Finsterbusch, "Antijudaismus in den medizinisch-naturwissenschaftlichen und philosophischen Schriften des Paracelsus," in *Paracelsus und seine internationale Rezeption in der frühen Neuzeit. Beiträge zur Geschichte des Paracelsismus,* ed. Heinz Schott and Ilana Zinguer (Leiden: Brill, 1998), 96–109.

20. Bullinger was in Zurich from the middle of June until the middle of November. Fritz Blanke, *Der Junge Bullinger*, in *Heinrich Bullinger: Vater der reformierten Kirche,* ed. Fritz Blanke and Immanuel Leuschner (Zurich: Theologischer Verlag, 1990), 67; *Heinrich Bullingers Diarium (Annales vitae) der Jahre 1504–1574*, ed. Emil Egli (Basel: Basler Buch- und Antiquariatshandlung Vormals Adolf Geering, 1904), 11.

excess that he could not even make it back to the inn where he was staying and would sleep off his hangover on the nearest bench. Bullinger summed up his portrait with the frank censure: "In short, he was a dirty and vulgar man in every respect. He seldom if ever attended church services, and seemed to care little for God and holy things."[21]

Bullinger's account must be taken seriously since it is one of the few eyewitness accounts of Paracelsus available. However, it must be conceded that Bullinger's engagement with Paracelsus must have been brief. Likewise, the long time span between when he likely met Paracelsus (ca. 1527) and when he probably composed this reminiscence (late 1570) undermines this document's value as source. Still, given the general reliability of Bullinger and the account's congruence with other stories of Paracelsus, it must be given at least a modicum of credence. Bullinger's report contains a clear insinuation of magic and a direct slight of Paracelsus's piety. These accusations, paired with Gesner's attacks, revealed unanimity in the Zurichers' assessment: Paracelsus had practiced forbidden black magic. This was the inescapable conclusion of the combined testimonies of Gesner, Oporinus, and Bullinger, as Erastus expressed in his reply to Bullinger after receiving his report: "It is certain that he was a magician and a confederate with demons."[22] Bullinger's account offered yet another testimony, from a person who was an unrivaled authority in Reformed Protestant circles, to the heretical nature of Paracelsianism.

Georg Vetter: A Companion of Paracelsus

If Erastus is to be believed, not only Paracelsus's enemies but also his friends alleged that Paracelsus toyed with demonic magic. Although not as provocative as

---

21. *De medicina nova*, I:239–40: "Contuli cum eo, inquit, semel & iterum de rebus varijs etiam Theologicis vel Religionis. Sed ex omnibus eius sermonibus pietatis nihil intelligere licuit, Magiae vero, quam ille nescio quam fingebat, plurimum. Si eum vidisses, non Medicum dixisses, sed Aurigam: & sodalitio Aurigarum mirifice delectabatur. Ergo dum viueret hic in diversorio Ciconiae, observabat aduentantes in hoc hospitium Aurigas: & cum his homo spurcus vorabat & perpotabat: ita nonnunquam vino sopitus, ut se in proximum scamnum collocaret, crapulam que foedam edormiret. Deinde interiectus quibusdam de habitu & vestitu eius, qualia Oporinus etiam habet, sic concludit. Breviter sordidus erat per omnia & homo spurcus. Rarò aut nunquam ingredibiebatur coetus sacros, ac visus est Deum & res Divinas leviter curare." This passage is probably quoted from a lost letter from Bullinger to Erastus which was written between 29 October 1570 and 3 January 1571.

22. Erastus to Bullinger, 3 January 1571. Zentralbibliothek Zürich, Ms. F 62 fol. 204v: "Ago ingentes gratias, quod mihi Paracelsum depinxisti. Idem f[ecit] Oporinus, quo usus fuit amamense per biennium. Magum fuisse et Cacodaemonis confoederatum cer[tum] est."

the stories from Crato and Bullinger, of equal historical interest is the testimonial by Paracelsus's purported travel companion, Georg Vetter. Vetter did not leave a large mark upon the historical record. Karl Sudhoff, the dean of modern Paracelsus scholars, reports that Vetter attended Paracelsus's lectures in Basel (1527). Michael Toxites's dedication of the 1574 edition of Paracelsus's testament suggests that Vetter later became a cleric.[23] Erastus recounted that Vetter had traveled with Paracelsus for over two years "through Austria, Transylvania, and other regions." This would suggest that Vetter accompanied Paracelsus sometime during the period from 1537 to 1540. Vetter's testimony is particularly noteworthy because it came from someone who still claimed to be a supporter of Paracelsus. Not surprisingly, the information from Vetter's account largely mirrored the stories of Oporinus, Crato, and Bullinger, including the familiar charge of alcoholism and of impiety. Expressed more strongly in the Vetter account, however, was an accusation of diabolical magic:

> Recently a most pious, learned, and industrious man, Georg Vetter—loving and devoted to Paracelsus—adamantly brought to my attention that [Paracelsus] was extremely devoted to impious magic and that he had befriended an evil spirit. "I feared nothing more," he said, "that when [Paracelsus] was drunk (which was frequently) that he would summon a troop of devils, which again and again he wanted to do by his special art; but at my request, he abandoned it. When I warned Paracelsus (when sober) that he made these utterances, which gravely offended God, and that the Devil would render the final judgment upon his own servants, he responded that he would soon be singing for redemption."[24]

Erastus added the rhetorical comment that he was forced to accept Vetter's account because of Vetter's piety and his genuine devotion to Paracelsus. He also included Vetter's comment that "The total time in which I lived with him, he did not discuss theology nor philosophy nor medicine besides surgery," which not

---

23. Karl Sudhoff, *Bibliographia Paracelsica: Besprechung der unter Hohenheims Namen 1527–1893 erschienenen Druckschriften* (Berlin, 1894), 156, 256. It is noteworthy that Vetter's apparent cooperation with Erastus did not alienate him from Toxites.

24. *De medicina nova,* II:2: "Nuper mihi vir pius, doctus, industrius, D. Georg. Vetterus Paracelsi amans & stusus, asserverantissimè affirmavit, eum Magiæ nefandæ perquam studiosum fuisse, & cacodæmonem haud aliter quàm socium nominare consuevisse. Nihil magis, inquiebat metui, quotiens ebrius erat (erat autem frequenter) quàm ut agmina Diabolorum accerseret: quod sæpenumero facere voluit (speciem artis suæ editurus) sed à me rogatus omisit. Cum sobrium monerem, ut missa isthaec facere, quòd Deum graviter offenderet, quodque ad extremum triste stidpendium solitus esset persolvere famulis suis diabolus: respondebat, se non multò pòst receptui cantaturum esse."

only assaulted Paracelsus's piety, but also lowered him from the ranks of learned physicians to the more humble status of a surgeon.[25]

Too little is known of Vetter or the transmission of this text to Erastus to make a definitive assessment of its merits possible. It is known that Vetter was actually associated with Paracelsus and that Erastus seems to have quoted his sources faithfully in the *Disputations*. It should be noted that the most damning line, "that he was accustomed to call an evil spirit his friend," is not purported to be a quotation from Vetter but is Erastus's own interpolation. Despite these limitations, it is hard to escape the conclusion that Paracelsus considered himself an authentic *magus* and that his practice of these arts was enough to make a more conventionally pious friend very uncomfortable. If one lends additional credence to Vetter, it is possible to speculate that Paracelsus considered himself to be walking, and perhaps occasionally transgressing, the fine line between natural and demonic magic.[26]

Markus Recklau: Paracelsus in Upper Bavaria

Erastus also included the account of the Palatine court physician Markus Recklau, whose testimony focused on Paracelsus's medical reputation. Recklau, a personal physician to the Palatine electors, supplied Erastus with a noteworthy tale of Paracelsus's activities in Upper Bavaria.[27] According to Recklau, Paracelsus was summoned to attend the wife of an Augsburg patrician named von Langenmantel who was then residing in the town of Schongau on the Lech in the foothills of the Alps. Paracelsus was able to restore the appearance of health to the woman, but soon after his departure she expired. From there he traveled down the Lech to Landsberg, Bavaria, where Recklau practiced medicine at that time.

---

25. Ibid.

26. For Paracelsus's own opinion regarding the possibility of Christian magic, see Arlene Miller Guinsberg, "Die Ideenwelt des Paracelsus und seiner Anhänger in Hinsicht auf das Thema des christlichen Magus und dessen Wirken," in *Von Paracelsus zu Goethe und Wilhelm von Humboldt*, Salzburger Beiträge zur Paracelsusforschung (Vienna: Verband der wissenschaftlichen Gesellschaften Österreichs, 1981), 22:27–54.

27. Erastus recounts that Recklau was a personal physician to the electors Ottheinrich (r. 1556–59) and Frederick. He did not specify Frederick II (r. 1544–56) or Frederick III (r. 1559–76) though the time of the composition of Erastus's book in the early 1570s makes Frederick III the obvious candidate. It is possible that Markus Recklau was related to Johann Recklau, who was a member of the church council under Frederick III. Volker Press, *Calvinismus und Territorialstaat: Regierung und Zentralbehorden der Kurpfalz 1559–1619*, Kieler Historische Studien 7 (Stuttgart: Ernst Klett Verlag, 1970), 243.

Recklau currently attended two noble patients of the von Pfeten family—one suffering from edema (dropsy), the other from consumption (phthisis).[28] Upon Paracelsus's advent, a local surgeon, Georg Rausner, informed Recklau that an illustrious physician had arrived who possessed the ability "to cure almost all serious diseases." With Recklau's approval, Paracelsus was taken to see the two noble patients. Recklau recounted that Paracelsus interrogated both him and the noble relative of his patients in a most condescending manner. Paracelsus required Recklau to give a lengthy analysis of the patients' ailments, but after Recklau had finished his discourse, Paracelsus simply remained silent. Later Paracelsus, the surgeon Raunser, and Recklau retired to a local tavern. While having a drink, Paracelsus allegedly spied the town's fine fountain, which spewed water high into the air. According to Recklau, Paracelsus declared the fountain to be a threat to public health. Paracelsus purportedly asserted, "If I were a doctor in this town, I would appeal to the magistrate to have this fountain destroyed,...lest the wind blow water into the faces of the women coming to market."[29] Paracelsus left the next day without saying farewell to Recklau, his noble patron Sebald von Pfeten, or his patients. Recklau added the additional insult that he had been unable to elicit one word of Latin from Paracelsus's tongue.

The Recklau text also included an epilogue recounting Paracelsus's next malpractice, relayed to Recklau by the Paracelsian Johann Vogt. Upon leaving Landsberg, Paracelsus traveled the short distance to Munich, where he treated a certain Monachus, an official in the kitchen of the duke of Bavaria. The story alleges that the ducal physicians Dr. Panthaleon and Dr. Alexander Karthauser were already treating Monachus quite successfully. The patient was taken from their care and entrusted to Paracelsus, who treated the patient with mercury. When Paracelsus saw that his patient, rather than improving, was ebbing towards death, he fled to Austria, leaving his patient to die the next day.[30]

As with Vetter, not enough is presently known of Recklau to assess this account, although his reputed prominence as a court physician for the Palatine Wittelsbachs and perhaps as the city physician of Landsberg am Lech should prompt future research. The events described may have occurred during Paracelsus's travels in 1536, after the publication of *Grosse Wundarznei*, but standard biographical

---

28. The patients were the wife and sister of Sebald von Pfeten.

29. *De medicina nova*, III:213: "[S]i huius oppidi forem medicus, inquit, apud magistratum hoc agerem, ut fons iste tolleretur.... [N]e ventus mulieribus ad mercatum venientibus aquam in faciem impelleret."

30. *De medicina nova*, III:212–14.

accounts of Paracelsus make no mention of a visit to Schongau or Landsberg.[31] It is no great revelation to learn that Paracelsus led a highly itinerate life. The events which occurred in Landsberg, the only ones which Recklau likely observed firsthand, are likewise described from the position of a practitioner understandably jealous of his own status as a medical authority. While much in Recklau's account might smack of professional rivalry, the quixotic figure he describes harmonizes well with the historical Paracelsus. One hopes that future research will shed more light on these obscure episodes.

Brief Accounts

Erastus also included a number of shorter testimonies in the *Disputations on the New Medicine of P. Paracelsus* that likewise undercut Paracelsus's reputation as a medical practitioner and chemical innovator. Crato's assertion that Paracelsus lifted his innovative preparations from medieval manuscripts falls into this category.[32] In a similar spirit is the comment of Balthasar Brauchius of Schwäbisch-Hall that the physicians Wolfgang Thalhauser and Johann Magenbuch had produced similar innovations in medical preparations, though they remained pious men and did not advance the ludicrous claim to be founding a whole new art.[33]

Erastus also claimed to have testimony from the inhabitants of Basel who had witnessed Paracelsus in action. These men did not begrudge credit to Paracelsus in his treatment of ulcers, but they were less laudatory of his treatment of other maladies. Although Paracelsus at first seemed to have cured the printer Johannes Froben, the Baselers (including Erasmus) would quickly learn that his medications had negative side effects. In Erastus's assessment, Paracelsus's remedies were

---

31. E.g., Walter Pagel, *Paracelsus: An Introduction to the Philosophical Medicine in the Era of the Renaissance*, 2d ed. (Basel: Karger, 1982); and Johannes Hemleben, *Paracelsus: Revolutionär, Arzt und Christ*, 2d ed. (Frauenfeld and Stuttgart: Verlag Huber, 1974).

32. *De medicina nova*, IV:300. "Remedia quibus aliquando usus dicitur, non illius esse ex eo certus sum, quòd librum vidi ante 200. ferè annos à monacho quodam Ulmæ scriptum." The text in question has been identified as "Von der Heiligen Dreifaltigkeit," which was composed ca. 1410–19. Authorship of the piece is generally credited to a Minorite monk named Ulmannus. Herwig Buntz, "Deutsche alchimistische Traktate des 15. und 16. Jahrhunderts" (doctoral diss., University of Munich, 1968), 34. See Joachim Telle, in *Lexikon des Mittelalters* (Munich and Zurich: Artemis, 1983), 2:812–13.

33. *De medicina nova*, II:3. See Joachim Telle, "Wolfgang Talhauser: Zu Leben und Werk eines Augsburger Stadtarztes und seinen Beziehungen zu Paracelsus und Schwenckfeld," *Medizinhistorisches Journal* 7 (1972): 1–30; and Joachim Telle and Peter Assion, "Der Nurnberger Stadtarzt Johannes Magenbuch: Zum Leben und Werk eines Mediziners der Reformationszeit," *Sudhoffs Archiv* 56 (1972): 353–421.

very powerful in expelling corrupted humors, but the vestiges of Paracelsus's medicaments that remained in the body proved most toxic. Erastus recounted the rumor, which he alleged had been corroborated to him by many, that everyone who consumed Paracelsus's drugs in Basel perished within a year.[34] In short, Paracelsus's cure was worse than the disease.

Assessment

Although some attempt has been made to appraise the reliability of these stories, main interest here has been to document how Erastus shaped the image of Paracelsus rather than to embark on a quest for the "historical Paracelsus." The representation that Erastus advanced possessed seven basic features: (1) Paracelsus and his followers were dangerous heretics; (2) Paracelsus dabbled in demonic magic; (3) Paracelsus was unlearned and wrote in a most obscure style; (4) as a physician, Paracelsus was at best ineffective and at worst lethal; (5) Paracelsus was a drunkard; (6) Paracelsus was personally unreliable and prone to flight in difficult circumstances; (7) though Paracelsus was a successful chemist, he was not as innovative or perhaps as independent as conventionally thought. Each of these basic points was to be found in more than one story, and most were made clearly enough that Erastus had little need to interpolate more. One exception is the charge of intercourse with demons. Though the theme was certainly present in the stories, Erastus spelled it out more explicitly than did his sources.

While Erastus emphasized these distinct points of Paracelsus's biography, it is doubtful that his readers would have put them together as clearly as they are laid out here, since Erastus scattered these biographical tidbits throughout the four volumes of *Disputations*. It is important to note that almost all of these accounts are given in the voice of a third party. This provided the rhetorical advantage of allowing the reports to appear more authoritative and Erastus to appear more objective. Nevertheless, Erastus was able to manage his sources in such a way as to put forward a coherent image of Paracelsus. By means of judicious selection, he was able to create his own composite narrative from the many eyewitness testimonies he included in the *Disputations*. The narrative Erastus crafted was an exhaustive attack on both Paracelsus's personal integrity and his professional competence.

Whether this narrative was persuasive to Erastus's contemporaries is another question. The anecdotal evidence gathered for this paper allows only a provisional

---

34. *De medicina nova,* IV:253. Erastus had been aware of these stories as early as 1570. Erastus to Johann Jakob Grynaeus, 29 November [1570]. Basel UB, Ms. G II 4, fol. 300-1.

assessment. One the one hand, Erastus's effort was apparently successful in convincing humanist physicians that Paracelsus was a disreputable character. Erastus's work quickly became a source for other anti-Paracelsian writers; for example, Bernhard Dessen von Kronenburg (1510–1574) incorporated biographical details from Erastus's work into his *Medicina veteris et rationalis* in 1573.[35] Later academic physicians such as Andreas Libavius and Daniel Sennert who sought to mediate between Paracelsian and Galenic medicine tended to be most wary of Paracelsus as an individual.[36] Allen Debus likewise has argued that Erastus's critique had a significant impact on the reception of Paracelsianism in England. Debus asserts that "the Paracelsian mystical universe was introduced by the way of a major attack on it, and with very few exceptions this alchemical cosmology became the object of distrust and suspicion during the Elizabethan period."[37]

On the other hand, Erastus's attack on Paracelsus's person does not appear to have had a major impact on the appeal of Paracelsian scientific ideas. For those who were inclined to look favorably on Paracelsus, Erastus's efforts smacked of partisanship and ad hominem overkill. This can be observed in the case of the English Paracelsian Richard Bostocke, who elected to write off his hero's alcoholism as a national defect:

> If Paracelsus some tyme woulde be dronke after his Countrey maner I can not excuse hym no more then I can excuse in some nations glottenie, in others pride, and contempt of all others in comparison of themselves, in others breach of promise and fidelitie, in others dissimulation, triflyng and muche babling, but lett the doctrine bee tried by the worke and successe, not by their faultes in their liues.[38]

What Bostocke was unwilling to concede was that Paracelsus's quirks and peccadilloes somehow invalidated his scientific achievement. To the contrary, the virulence

---

35. Bernhard Dessen von Kronenburg, *Medicina veteris et rationalis* (Cologne: Johannes Gymnicus, 1573), 54, 90. Some of the details of Erastus's accounts also found their way into Christoph Gottlieb von Murr's "Litterargeschichte des Theophrastus Paracelsus," in *Neues Journal zur Litteratur und Kunstgeschichte* 2 (1799): 233–34.

36. See Bruce Moran, "Libavius the Paracelsian? Monstrous Novelties, Institutions, and the Norms of Social Virtue," in *Reading the Book of Nature*, ed. Debus and Walton, 67–79; and Allen G. Debus, "Guitherius, Libavius and Sennert: The Chemical Compromise in Early Modern Medicine," in *Science, Medicine and Society in the Renaissance: Essays to Honor Walter Pagel*, ed. Allen G. Debus (New York: Science History Publications, 1972).

37. Allen G. Debus, *The English Paracelsians* (New York: Franklin Watts, 1966), 49.

38. R.B. [Richard Bostocke], *The difference betwene the auncient Phisicke… and the later Phisicke* (London, 1585), Liiii verso.

of Erastus's attack on Paracelsus's person seemed to betray an a priori unwillingness to entertain Paracelsus's scientific contribution.

Levinus Battus (1545–1591), a medical professor in Rostock, likewise was not impressed by Erastus's assault on Paracelsus's personal integrity. Little is known of Battus, but since he considered Paracelsus the "Luther of medicine," one can confidently label him a Paracelsian.[39] In a letter from 1573, Battus took issue both with the ad hominem nature of Erastus's attack and his source base. Battus dismissed Vetter's account because he was such an obscure figure and the negative stories from Basel because they were likely written by Paracelsus's rivals. He argued that Oporinus's letter hardly proved all that Erastus had attempted to draw from it and that Thalhauser's and Magenbuch's excellence in chemical preparations did not detract from Paracelsus's honor. As for the claim that Paracelsus was not particularly learned or eloquent, Battus cited the adage of Celsus that medicines and not eloquence cured diseases. Finally, he did not take Erastus's charge of diabolism seriously, because he thought that Paracelsus had made proper distinctions between natural and demonic magic in his writings and that Paracelsus's methods—the Kabbalah, chiromancy, physiognomy, the theory of signatures, and the art of the furnace—were not demonic but were derived from the light of nature.[40] In short, the obvious partisan slant made Erastus's work easier to dismiss, and a Paracelsian had little difficulty in perceiving as white what Erastus had seen as black.

Erastus's representations reflected one side of the polarization of the image of Paracelsus in the late sixteenth century. Not unlike Luther, Paracelsus had been increasingly sainted by his admirers and demonized by his enemies.[41] At the extreme end of both positions, Paracelsus was no longer a radical medical innovator with a quirky personality but a man imbued with either magical or demonic power.

---

39. The claim that Paracelsus was the "Luther of medicine" has a long history. See the recent articles by Vivian Nutton and Bernhard Dietrich Haage in *Paracelsus: Das Werk: Die Reception*, ed. Volker Zimmermann (Stuttgart: Franz Steiner, 1995), and Andrew Weeks, *Paracelsus: Speculative Theory and the Crisis of the Early Reformation* (Albany: State University of New York Press, 1997).

40. Levinus Battus, letter to [Henricus Smetius ?], 12 March 1573. Printed in Henricus Smetius, ed., *Miscellanea... Medica* (Frankfurt: Jonas Rhodius, 1661), 691–92. Note that this letter was written before the completion of *De medicina nova*, IV. For more on Battus and Smetius, see Wilhelm Kühlmann and Joachim Telle, "Humanismus und Medizin an der Universität Heidelberg im 16. Jahrhundert," in *Semper Apertus: Sechshundert Jahre Ruprecht-Karls-Universität Heidelberg 1386–1986*, ed. Wilhelm Doerr et al. (Berlin: Springer, 1985), 1:277–81.

41. E.g., see R. W. Scribner, "The Incombustible Luther: The Image of the Reformer in Early Modern Germany," *Past and Present* 110 (1986): 36–88.

Acknowledgments: I would like to express my thanks to Jole Shackelford (University of Minnesota), Joachim Telle (Heidelberg), and the staff of the Historical Division of the National Library of Medicine (Bethesda, Maryland) for their invaluable assistance.

Paracelsus and the Boundaries of Medicine in Early Modern Augsburg

Mitchell Hammond

In 1569, the Braunschweig physician Martin Cop published a treatise warning of the dangers of antimony, the arsenic compound that was the purgative cure-all of his day. His main target was not the concoction itself, which he thought was useful in some circumstances, but its indiscriminate use by "mountebanks and theriac salesmen" who cheated the common people of both money and health. These swindlers were not just a nuisance, they were potentially dangerous. One had even persuaded a friend of Cop's to swallow powdered iron to cure arthritis, causing severe constipation that was relieved only by enemas and suppositories. Fortunately, some of these shady characters were caught and banished by city-appointed physicians and deputies, but Cop vented his frustration that it was so easy for such pretenders to sell themselves: "[T]hey only hold up the books of Theophrastus Paracelsus and allege such skill, and, under this pretense, [they] obtain all the more approval and belief among the unlearned."[1]

Cop's writing is only one of thousands of diatribes against unorthodox medicine penned by early modern physicians intent on shoring up their shaky professional and financial status against interlopers of all kinds.[2] No targets were more frequently singled out than the followers of Paracelsus, whose chemically based

---

1. "[S]ie nu[r] anheben sich *Paracelsi Theophrasti* Buocher vnd kunst zu rhuomen, vnd vnter solchem schein bey den vnwissenden desto mer beyfal vnd glaubens bekomen." Martin Cop, *Das Spissglass Antimonium oderr Stibium genandt* (Braunschweig, 1569), 2r.

2. Conflicts of this kind persisted throughout the early modern period. See the discussion of the late eighteenth century in Mary Lindemann, *Health and Healing in Eighteenth-Century Germany* (Baltimore: Johns Hopkins University Press, 1996), 222–35.

remedies exerted influence far beyond the circles of the learned. But the impact of Paracelsus among everyday people—middle- and lower-class men and women with no stake in alchemy, theology, or medical theory—is an area that we still know very little about.[3] In the last ten years, many scholars have examined Paracelsus's thought, in particular his theology and anthropology, and manuscript studies have mapped an impressive network among physicians and alchemists, ferreting out traces of Paracelsian influence in unlikely places.[4] However, to understand fully what was upsetting Cop and so many other contemporary physicians, we must explore a new range of topics. What were the routes through which Paracelsian ideas or practices passed to people on the streets of early modern cities? Which aspects of his legacy found the greatest reception among common people? How did city governments respond to debates over Paracelsus in their communities? The answers to these questions will show the range of factors that affected the spread of scientific and medical ideas in urban settings, and they will indicate the reach of Paracelsus's intellectual legacy into every segment of society.

As a preliminary foray into this topic, this paper will explore the popular response to Paracelsian healers in the German imperial city of Augsburg in the late sixteenth and early seventeenth centuries. For many reasons, Augsburg is a significant case. With a population approaching forty-five thousand in 1550, Augsburg was the largest city to be significantly influenced by the Protestant reform movement.[5] It was also a center of trade and printing that exerted tremendous cultural

---

3. Relatively few scholars have attempted to integrate the questions and methods of social history into the study of Paracelsus. See Charles Webster, "Paracelsus: Medicine as Popular Protest," in *Medicine and the Reformation*, ed. Ole Peter Grell and Andrew Cunningham (London: Routledge, 1993), 57–77; Robert Jütte, "Valentin Rösswurm: Zur Sozialgeschichte des Paracelsismus im 16. Jahrhundert," in *Resultate und Desiderate der Paracelsus-Forschung*, ed. Hartmut Rudolph and Peter Dilg (Stuttgart: Franz Steiner, 1993), 99–112; and Rudolf Schlögl, "Ansätze zu einer Sozialgeschichte des Paracelsismus im 17. und 18. Jahrhundert," in *Resultate und Desiderate*, 145–62.

4. See, e.g., Ute Gause, *Paracelsus: Genese und Entfaltung seiner frühen Theologie* (Tübingen: Mohr, 1993); Katharine Biegger, "Paracelsica Theologica im Katholischen Gesangbuch Johann Leisentrits von 1567," in *Parerga Paracelsica*, ed. Joachim Telle (Stuttgart: Franz Steiner, 1993), 105–20; and Hartmut Rudolph, "Hohenheim's Anthropology in the Light of His Writings on the Eucharist," in *Paracelsus: The Man and His Reputation, His Ideas and Their Transformation*, ed. Ole Peter Grell (Leiden: Brill, 1998), 187–206.

5. For Augsburg's population figures in the early modern period, consult Barbara Rajkay, "Die Bevölkerungsentwicklung von 1500 bis 1648," in *Geschichte der Stadt Augsburg*, ed. Gunther Gottlieb et al. (Stuttgart: Konrad Theiss, 1985), 252–57. The definitive study of Augsburg for this period is Bernd Roeck, *Eine Stadt in Krieg und Frieden*, 2 vols. (Göttingen: Vandenhoeck & Ruprecht, 1989). Overviews of medicine in the city can be found in Werner Dieminger, "Bader, Barbiere, Wund- und

influence over the German lands. As in many other cities in the 1580s and 1590s, Augsburg's medical elite founded a College of Medicine that tried to control the marketplace for medical goods and services. On a number of occasions thereafter, the College appealed to the city government to stop traveling Paracelsians from treating clients that they had attracted in Augsburg. The surviving records from the College's archive and the city council protocols tell relatively little about the practices at issue and do not help to identify a core of Paracelsian belief. The records do, however, trace the path of Paracelsian medicine from medical theory to problems of social and legal legitimacy which influenced the access to healing in the early modern city.

Even though Augsburg's biggest disputes over Paracelsus did not take place until the later 1580s, Paracelsus himself set the tone for later conflict with his own visit to the city in spring and summer 1536. He came there to publish his surgery handbook, *Die Grosse Wundarznei*, after a disagreement with a publisher in Ulm.[6] Although only a few traces of his visit remain, it is clear that Paracelsus, as usual, aroused strong feelings among Augsburg's physicians. He was received with great enthusiasm by his disciple Wolfgang Thalhauser, who had been in the city since 1529 and had recently been appointed to the position of city physician (*Stadtphysicus*). Together the two gave an open disputation in July which presented some of the major ideas in Paracelsus's forthcoming text. Paracelsus's declaration at the beginning of *Die Grosse Wundarznei* that his work corrected flaws in both traditional theory and current practice cannot have found a receptive audience among the Galenic physicians at work in the city. According to a later account, Paracelsus also administered a healing potion to an adolescent girl without her parents' consent; the cure was successful, but this probably did little to endear him to city officials.[7]

It was, however, Thalhauser's own remarks at the opening of *Die Grosse Wundarznei* that provoked the sharpest response from his medical colleagues. In a prefatory letter, Thalhauser harshly criticized the state of medicine in the Empire, claiming that the art had fallen into such disarray that it was in need of a

---

Zahnärzte in der freien Reichsstadt Augsburg von 1316 bis 1806" (medical diss., Ulm University, 1998); and Gerhard Gensthaler, *Das Medizinalwesen der freien Reichsstadt Augsburg bis zum 16. Jahrhundert* (Augsburg: Hieronymus Mühlberger, 1974), 77–91.

6. The episode is summarized in Joachim Telle, "Wolfgang Thalhauser: Zu Leben und Werk eines Augsburger Stadtarztes und seinen Beziehungen zu Paracelsus und Schwenckfeld," *Medizinhistorisches Journal* 7 (1972), 5–10.

7. Ibid., 6.

doctor itself.[8] Those who used to learn their trade through a long *peregrinatio academica* now contented themselves with recipes learned from books or in the next town.[9] In his view, "the conditions for the sick everywhere are so pitiful and miserable, that I am amazed at how all the authorities can endure or tolerate it."[10] He exhorted the city authorities to oversee medicine and halt the ongoing abuses. Such a reform would be "like a new house that one builds on the street," carried out in spite of the opposition of the medical profession.[11] Thalhauser's seven Augsburg colleagues were infuriated; while at least some of them agreed that city magistrates ought to have a hand in the oversight of medicine, they found it galling to be criticized as if they themselves were unlearned empirics.[12] They eventually confronted Thalhauser and demanded that he print a Latin apology, and his letter was removed from later editions of *Die Grosse Wundarznei*. Thalhauser apparently agreed to write the apology, but his slow progress on it angered the other physicians even more. Finally, on 27 July 1538, Thalhauser and his opponents met in the company of the mayor, who forced Thalhauser to soften his stance and then forbade everyone involved to pursue the matter any further.[13]

While Paracelsus also thought that medicine was badly in need of reform, he believed that correcting abuses was not a task for mortals at all. At the end of the first section of *Die Grosse Wundarznei*, he lamented that medicine had fallen into a state of apostasy and dishonor.[14] But he believed that these seemingly useless practitioners also served a purpose. Just as in the Bible there were false prophets and apostles alongside the true ones, Paracelsus reasoned,

---

8. Paracelsus, *Sämtliche Werke, I. Abteilung: Medizinische, naturwissenschaftliche und philosophische Schriften*, ed. Karl Sudhoff, 14 vols. (Munich and Berlin, 1922–33), 10:12. Hereafter I refer to works using the roman numeral to denote the part of the collected edition of the complete works of Paracelsus followed by the volume and page number: the preceding reference would thus be I, 10:12.

9. Paracelsus, *Sämtliche Werke*, I, 10:13.

10. "[E]s ie der kranken halben so jemerlich uberal und erbermlich zugehet, das mich zum höchsten verwundert, wie es die oberkeit allenthalben nur dulden oder leiden kan." Paracelsus, *Sämtliche Werke*, I, 10:13.

11. "[W]ie ein neues haus, das einer an der straßen bauet...." Paracelsus, *Sämtliche Werke*, I, 10:14.

12. Telle, "Wolfgang Thalhauser," 8–9. Only four years earlier, four doctors (Johann Tiefenbach, Adolf Occo II, Johann Trenklin, and Gereon Sailer) had called for greater oversight of the doctors and apothecaries. Gensthaler, "Medizinalwesen," 78.

13. The text of the city council protocol that summarizes the meeting is transcribed in Telle, "Wolfgang Thalhauser," 26–27.

14. Paracelsus, *Sämtliche Werke*, I, 10:79.

[H]ow can a doctor not patiently permit false doctors to stand next to him? For the false will seek out their doctor, the just [will seek out] a just one; thus God joins together those who belong together, thereby good and evil must live with each other; since of course no wheat has ever grown without rats in it, thus man must let this come to pass.[15]

Paracelsus emphasized that God's plan would transcend any human attempts at intervention and that all people would naturally find their place in that plan. By implication, however, it seems that he disapproved of one doctor's interference with the practice of another, even in potentially dangerous situations. Thus, Paracelsus not only challenged the way that medicine was practiced or the pagan sources from which medical theory was derived. His biblically based belief that only God had authority over medical practice contradicted the objectives of medical and civic officials who sought to protect citizens from chicanery and inappropriate physical risks. This fundamental attack on the legitimacy of medical authority was what lay behind the conflicts over Paracelsian practitioners in cities across the Empire later in the century.

That controversy did not take shape for some decades, however, because Paracelsus's multifaceted writings did not circulate widely until well after his death in 1541. His first great popularizer, Adam von Bodenstein, after he was expelled in 1564 from the medical faculty in Basel, began publishing Paracelsus's works.[16] In 1571, the Danish physician Petrus Severinus published his *Idea medicinæ*, a work that reconciled many of Paracelsus's ideas with the milieu of orthodox, academic medicine. Finally, Johann Huser assembled and collected many of Paracelsus's alchemical and medical works in ten large volumes published in Basel and Cologne between 1589 and 1591. These printed works divested Paracelsus's ideas of much of their religious underpinning; by the turn of the seventeenth century, Paracelsian medicine and alchemy had won supporters in universities and influential courts around Europe. Moreover, Paracelsian ideas inspired a host of medical practitioners, unlearned itinerant healers as well as academically trained

---

15. "[W]ie kan der arzet nicht mit in gedulden falsche arzet neben im zustehen? dan die falschen kranken suchen iren arzet, die gerechten den gerechten; also füget got die zusamen so zusamen gehören. darum so muß guts und böß mit einander leben; dieweil doch nie kein weizen on den ratten gewachsen ist, muß mans also geschehen lassen." Paracelsus, *Sämtliche Werke*, I, 10:79.

16. The following is drawn from Hugh Trevor-Roper, "The Court Physician and Paracelsianism," in *Medicine at the Courts of Europe, 1500–1837*, ed. Vivian Nutton (London: Routledge, 1990), 79–95; and Ole Peter Grell, "The Acceptable Face of Paracelsianism: The Legacy of *Idea medicinæ* and the Introduction of Paracelsianism into Early Modern Denmark," in *Paracelsus: The Man and His Reputation*, 245–68. See also Jole Shackelford, "Paracelsianism in Denmark and Norway in the Sixteenth and Seventeenth Centuries" (Ph.D. diss., University of Michigan, 1989).

physicians, to speak the language of chemical medicine and incorporate chemical remedies into their healing methods.

Increased attention to Paracelsus's writings was not the only factor that encouraged city leaders to take a hard look at the self-styled Paracelsian healers who sojourned in cities across the Empire. In the last quarter of the sixteenth century, many cities expanded their relief programs for the sick poor, drafted new ordinances regulating medicine, and formed professional groups for their physicians.[17] In Augsburg in January 1582, the physicians and the city council incorporated a College of Medicine which advised the city council and oversaw medical activities in the city. Other cities, including Nuremberg, Ulm, Nördlingen, Worms, and Vienna, followed suit and enacted new ordinances of their own. The members of the Augsburg College, seventeen in number in 1582, had a variety of responsibilities.[18] Deputies were appointed to examine midwives and regularly inspect the wares at Augsburg's four apothecary shops; other members of the College attended patients at the Alms House, which treated hundreds of Augsburg's sick poor every year. Delegations of three or more physicians assisted barber-surgeons when autopsies were necessary, and the College as a whole submitted opinions to the city council on matters of public policy. This included regular consultations about petitions from healers and their patients, as well as salespeople, about which goods and practices were allowed and under which circumstances.

Officially, the Augsburg College of Medicine fully supported "the age-old, true, proven Hippocratic Medicine" against all opponents; as its founding charter made clear, its primary purpose was to protect the city from irresponsible healers: the Jews and quacks, old women, and tradespeople "with their harmful, poisonous metallic distillations, and tinctures" who took advantage of the common man.[19] Such false practitioners were like wolves that preyed on unsuspecting sheep, and they made a mockery of the medical profession.[20] Although Paracelsians were not mentioned by name, the reference to metallic remedies was unmistakable, and the city drafted a new apothecary ordinance in 1594 that inveighed against the many "money-thirsty Empirics, Paracelsians, and incompetents" who were active in the

---

17. Alfons Fischer, *Geschichte des deutschen Gesundheitswesens* (Hildesheim: Georg Olms, 1965), 89–92 and 162–88.

18. Dieminger, "Bader, Barbiere, Wund- und Zahnärzte," 32. For biographical information about the physicians in Augsburg in the sixteenth century, consult Elsbeth Martz, "Gesundheitswesen und Ärzte in Augsburg im 16. Jahrhundert" (medical diss., Munich University, 1950).

19. Articles 1 and 17 of *Ordnung zwischen den herren* Doctorn Medicinae *zu Augspurg, mit eines Ersamen Raths daselbsten wissen vnd bewilligung auffgericht* (Augsburg: Schoenigk, 1582), Aii, Bi.

20. Ibid., art. 16.

city.[21] The physicians in the College and the city council deputies who oversaw the apothecary guild used these ordinances to fine many unofficial healers (women, herb salesmen, and others) and often put them out of business.

In short, Paracelsian medicine, with its intrinsic antiauthoritarian streak, reached the peak of its prestige at the same time that cities such as Augsburg created formal institutions to oversee medicine and public health. Upon its founding, the Augsburg College of Medicine began to stringently examine new applicants who wished to join. One of the first casualties of the new procedure was Georg am Wald, a physician with Paracelsian leanings who would later become known for promoting a panacea of his own invention.[22] In his examination, am Wald refused to submit to his questioners; according to one account, after the first question he informed them that he did not consider them worthy to interrogate him since he had healed illnesses that they could not.[23] This bald-faced refusal to acknowledge civic authority gave city leaders only one option: am Wald was told to leave Augsburg and try his luck elsewhere.

Most of the time, however, matters were not so simple. The city council, which was the final arbiter in these matters, had interests other than those of the physicians to consider, including political expedience, legal jurisdiction, and the pragmatic needs of city residents. Frequently the council ruled in favor of sick citizens who believed that unorthodox healers held out the best (or the only) means of a cure. The College of Medicine and the ordinance reforms increased the scrutiny of medical care and the level of debate over medical policy, but they did not enable Galenic doctors to stop Paracelsians from practicing in the city.

The dispute over Sebastian Froben, for which abundant archival records survive, provides an especially clear example of the difficulties that city leaders faced. A native of Passau, Froben first came to the attention of Augsburg's government at the end of July 1588, when he was imprisoned for unpermitted healing. When city councillors questioned him on August 10, he explained that he had come to the city a few days previously at the request of Dr. David Bürgle, a lawyer, who also allowed Froben to stay at his house.[24] Initially, Froben had been granted permission to treat Bürgle and Georg Wagner, a watchmaker, but when other people

---

21. Article 25 of *Eines Ersamen Raths der Statt Augspurg Apotecker Ordnung* (Augsburg: Schoenigk, 1594), Fiv.

22. Wolf-Dieter Müller-Jahncke, "Georg am Wald (1554–1616): Arzt und Unternehmer," in *Analecta Paracelsica*, ed. Joachim Telle (Stuttgart: Franz Steiner, 1994), 213–304. The article contains a complete index of am Wald's extensive publications.

23. Ibid., 219.

24. Stadtarchiv Augsburg (hereafter StAA), Strafamt, "Urgicht Sebastian Froben," 1588b, 10 August (Froben), 1r.

approached him, Froben attempted to help them as well. This was apparently the reason for his imprisonment. Froben claimed that he had received only fifty gulden total for his efforts and that he had helped many people "out of God's will," that is, either for free or for a very modest sum. With this testimony, Froben preempted one of the principal arguments that were used against unofficial healers, namely that he was sapping the livelihood of Augsburg's resident doctors by taking away their customers.

Froben was fortunate that his main patron, and others as well, intervened immediately and forcefully on his behalf. In a petition to the city council, David Bürgle defended Froben and explained his own behavior: "[S]ince his art was familiar to me (for in my youth and adulthood I have read much in the books of Theophrastus, whose pupil he is, and I have found nothing incorrect in there, only great secrets), I let him come and go from my house without hesitation."[25] Bürgle also had sharp words for the physicians of Augsburg, whom he compared to dogs chasing a cornered rabbit, even though they themselves could not help some of the sick. His own daughter was seriously ill, he said, and the other doctors in town would not attend to her. Bürgle's petition was seconded by eight other residents, including Georg Wagner, his maid, and the family of David Miller, who claimed that Froben had helped them with illnesses that the physicians had not wanted to treat. Referring to him as a "medicus und chirurg," they asked that Froben be released and permitted to complete the healing process that he had begun.[26] These arguments persuaded the city council to let Froben go, but, in support of Augsburg's physicians, the council forbade Froben from residing in Augsburg and continuing his healing activities.[27]

But Froben's career in Augsburg was far from over. He first moved to Oberhausen, a small town nearby that fell just outside of Augsburg's legal jurisdiction. From this perch he continued to treat citizens who petitioned the city council to allow his assistance. For instance, a dyesmith named Hans Daigeler claimed that Froben had relieved him of a bad infection in his foot in only three days.[28] Other tradespeople, including a shoemaker, petitioned on behalf of their wives, and the city council felt compelled to grant their requests, although the College of Medicine objected to the infringement on their trade. This state of affairs continued

---

25. "[W]ie sein kunst wol bewust, (dann ich inn mejner Jugend, vnd alter inn deß *Theophrasti* bucher, dessen *discipulus* er ist, vil gelesen, aber nie nichts vnrechts, sondern grosse geheÿnussen darinn gfundt....[)] hab ich ine bej mir, ohne scheüch, aus vnd ein geen lassen." Ibid., 10 August (Bürgle).

26. Ibid., 11 August.

27. StAA, Strafamt, Straffbuch 1588–1596, 18 August 1588.

28. Collegium Medicum (hereafter CM), "Fremde Ärzte," Fasc. 8, 1 October 1588.

for three years, until October 1591, when Froben announced his desire to move to Augsburg permanently. At this point the College intervened with a lengthy brief in which they characterized Froben as "a great enemy and dissenter from Hippocratic and Galenic medicine" and as "a fraud who should once again be stopped."[29] The city council initially followed the Collegium's recommendation and refused Froben's request. Then the healer marshaled his forces among the citizens of Augsburg. On September 19, he requested permission to reside in Augsburg for five or six years. This was an extraordinary request, since the normal visiting time approved by the city was three months, or at most a year. Froben's device for pressuring the council, however, was no less extraordinary: he presented a petition signed by 107 people who attested that he had cured deadly diseases for which they had not found helpful medicine anywhere else.[30] The names and trades of the signers spanned the lower and middle ranks of Augsburg society. They included several women, eight weavers, a schoolmaster, a bookseller, and one of the city's barber-surgeons, Hans Brestle.

Faced with such widespread popular response, the city council reversed its earlier support of the College and granted Froben the right to live in Augsburg for one to two years, although he could only practice healing with the consent of the mayor. This was a humiliating defeat for the College, which had been accustomed to asserting its authority without opposition in the previous nine years. In a letter to the city council nine months later, the physicians fumed that Froben "does not himself know who or what he is, and no one can know how he should be assessed."[31] Many other Galenic physicians around the Empire must have felt the same when they first confronted the rebellious Paracelsians. The Augsburg physicians further claimed that Froben "seems to be more a truth-teller or magus than a doctor," and darkly hinted that he used magic arts.[32] But these protests did no good, and Froben went on to practice in Augsburg not only the next year but for thirty more years.

As with few other cases involving unpermitted medicine, the dispute over Sebastian Froben permits us to reconstruct the avenues by which Paracelsian healing practices coursed through a major city. Froben's entrée into Augsburg

---

29. "[A]in grosse feindt. vnd widersacher der *Hypocratica & Gallenica Medicina*... als ein winckelartzet nochmallen zue exterminirn." StAA, CM, Fremde Ärzte, Fasc. 8, 11 July 1591.

30. Ibid., 19 September 1591.

31. "So er doch selbsten nit waist, wer, od[er] w[as] er ist, auch niemands wissen kan, für wen mann ine aigentlich halten od[er] schätzen solle." StAA, CM, "Pfuscher, hiesigen vnd fremde Ärzte," 11 April 1592.

32. "[Er] mer einen warseger vnd *Magum* alß einen arczet, erzaigt hat." Ibid.

came from a lawyer and a watchmaker, two men who were relatively well off and educated and whose business dealings, in all likelihood, depended on contacts outside the city. Although his first supporters were from these men's households, over three years Froben developed a clientele that was centered among lower-class tradespeople—the weavers, shoemakers, and dyers who made up the majority of Augsburg's population. More than likely, the appeal of his services was a lower price rather than any unorthodox practices, and it is doubtful that many of his clients fully appreciated the nature of his conflict with the College of Medicine. However, the physicians and the city councillors they advised fully understood Froben's opposition to medicine as it was usually practiced. Froben triumphed outright not because he converted a powerful patron to his philosophy, but because he proved his usefulness to society at large in a way that the city council could not gainsay or ignore.

A different range of factors came into play with the curious case of Georg Herdt, an alchemist with a shady past who came to Augsburg in December 1595. Why Herdt first came to the attention of the city council is not clear, but when he was first questioned, he identified himself as "a Theophrastic *doctor* who cures all sort of diseases with his *Theophrastic* art and *medicine*, distillates and confect waters, oil and the like."[33] A few days later, however, a letter arrived from the office of Prince Wilhelm of Bavaria, revealing that Herdt was a wanted man. Earlier that year, Herdt had been among the circle of alchemists at work under Prince Wilhelm of Württemberg, one of the most active supporters of Paracelsian activity at the turn of the seventeenth century. Herdt and some associates had contracted to transmute silver into gold for a princely reward of twelve thousand gulden. After two trials, the experiment still had not worked, and Herdt had disappeared without paying off his expenses. Prince Friedrich had tracked Herdt to Munich and alerted Prince Wilhelm, who had followed Herdt's trail to Augsburg.[34]

Of course, this cast a cloud of suspicion over Herdt's medical claims, but when he was questioned again on January 4, he maintained the kernel of his original story. It was true, he admitted, that he had come to Augsburg five weeks earlier to escape prosecution in Württemberg. However, while staying in a guest house, he had begun healing people; his patients included the wife of a weaver, who suffered from "cancer (*krebs*) in one of her breasts," a young boy with a growth on his eye, and another woman whose leg had been badly injured two years before.[35] The

---

33. "[E]in Theophrastischer *medicus*, curiere allerleÿ khrannckheiten mit seiner *Theophrastischen* khunst vnnd *medicin*, distilier vnnd *Conficiere* wasser, öhl vnd dergleichen." StAA, Strafamt, "Urgicht Georg Herdt," 1596a, 22 December 1595.

34. Ibid. A rough copy of the alleged contract, dated 30 March, is extant in the file.

same day, five weavers submitted a petition that confirmed what Herdt had claimed and asked for his release so that he could continue his treatments. The woman with the injured leg, who was the wife of Jacob Bernbacher, even volunteered that the barber-surgeons at the Alms House had wanted to cut off her leg before Herdt had helped her.

Which claim should the city honor: the charge of a dishonored contract from a distant prince or the desire of half a dozen poor tradespeople for medical care? Despite renewed petitions from Württemberg, the city council decided to let Herdt go and informed Prince Friedrich in a letter dated February 6 that Herdt had answered all questions to their satisfaction. A brief summation of Herdt's proceedings noted that he was not the principal party in the contract with Friedrich and therefore Augsburg would not hold him accountable.[36] But this was certainly a slender justification for ignoring the wishes of Württemberg's prince, and it may be that Augsburg's council was making a show of its sovereign independence with its indifference to his request. In any case, Herdt lingered in Augsburg for at least another six weeks, and when he was questioned again, he claimed that his services were still needed by the woman with cancer. The city council barred him from the city on March 19; however, as noted by an annotation to this ruling, the council reversed the decision when the weavers renewed their petition for Herdt's help four days later.[37]

Although the city council reserved the right to pass judgment on all petitions, it clearly went out of its way to accommodate citizens who put their trust in unorthodox healers. In Herdt's case, the circumstances were especially remarkable because he was a potential criminal with only the slimmest of connections to the city. Once again, Paracelsian medicine found a following in one of the poorest segments of Augsburg society.

In these verdicts and others, the city council generally supported the petitions of Augsburg's sick citizens unless the normal business of medicine was seriously disrupted. In later cases, the College of Medicine expressed disapproval of Paracelsian healers but avoided attacking their privileges directly. For example, in November 1611, Matthias Engelman, who openly advertised "my learned, Theophrastic, extraordinary, secret art and medicine," was instructed to undergo examination by the College after his healing skills were called into question by the mayor.[38] The College's physicians concluded that "he [Engelman] is completely

---

35. Ibid., 4 January 1596.
36. StAA, Strafamt, Straffbuch 1588–1596, 3 February.
37. Ibid., 23 March.
38. StAA, CM, "Fremde Ärzte," Fasc. 8, 11 November 1611.

unlearned in medicine and is also (disregarding that his so-called medicine may have succeeded with some people) inexperienced in the discernment of diseases."[39] Despite their criticism, the physicians agreed that it would be appropriate for Engelman to finish with his current patients, as long as they understood that they took treatment "at their own risk" (*uf ir gefahr*). The parenthetical statement, moreover, was a necessary concession because Engelman already had at least sixteen sponsors—one of whom, interestingly enough, was the Doctor Bürgle who had sponsored Sebastian Froben twenty-three years before. In another case, from April 1613, the College refrained from attacking Hans Fischer, who proudly called himself an empiric and compared his experience to that of an explorer who visited lands instead of reading about them or looking at pictures.[40] Although the affront was obvious, the physicians did not rise to the bait; they merely noted that Fischer's practice violated the apothecary ordinance and that they would not enter into unnecessary disputation with him.[41] The city council permitted Fischer to finish treating the ten patients with whom he had begun and ordered him to consult the mayor before taking on more.

Thus, by the second decade of the seventeenth century, Paracelsian healers had established a procedure (though not exactly a legal one) by which they entrenched themselves in Augsburg. The successful healers usually attracted several clients before they were challenged, and then they defended their skills with testimonies from their patients. Officially, Galenic medicine and approval from the College of Medicine were legal necessities; informally, city councilors often deferred to the wishes of sick citizens or to other circumstances. One additional virtue of this arrangement was that members of the College could publicly support Galenic orthodoxy while following their own intellectual interests. This was an important consideration for at least one Augsburg physician, Karl Widemann, who actively corresponded with many Paracelsians and collected alchemical treatises and recipes for over twenty years.[42] As a member of the Collegium after 1590, Widemann was involved in some or all of the deliberations over Paracelsian medicine in Augsburg. There is no evidence that he ever lent his support to any of the Paracelsian healers that passed through; while he may have done so in conversation, it is more

---

39. "[E]r in der *medicina* gantz vngelernet auch (vnangesehen ime sein berhüembte artzney kunst bey etlich[en] gelungen haben möchte) die kranckheiten zu *discernieren* nicht erfahren ist." Ibid., 17 December 1611.

40. StAA., CM, "Fremde Ärzte," Iome. I., 18 May 1613.

41. Ibid., 23 May 1613.

42. Julian Paulus has superbly reconstructed Widemann's circle of correspondents and linked them to the wider network of Paracelsian practitioners. See Paulus, "Alchemie und Paracelsismus um 1600: Siebzig Porträts," in *Analecta Paracelsica*, 335–406.

likely that he separated his intellectual interest in Paracelsus from the pressing task of defending the College and its business interests against outside threats.

Paracelsians apparently had similar success elsewhere in securing a foothold in the medical marketplace, although the scattered evidence that has come to light usually survives in the protests of their opponents. For example, in 1580 an apothecary in Worms named Despasianus wrote a colleague in Augsburg to complain that his recommendations to the city council were routinely flouted by salesmen from abroad who sold antimony and other dangerous remedies.[43] Even in academic settings, physicians struggled to compete with uncredentialed but persistent Paracelsian healers. Between 1599 and 1613, the medical faculty at Tübingen tried without success to stop Tobias Hess, an avid Paracelsian and collector of prophetic texts, from unpermitted medical practice.[44]

Moreover, it is significant that the usual pattern of Paracelsian healers—an informal gathering of clients, followed by challenges from medical officials and eventual acceptance of city leaders—was followed successfully by other unofficial practitioners and their patients. In particular, women healers in Augsburg used testimonials from their patients as evidence of special expertise in curing diseases of the female body. Such claims were necessary because Augsburg's association of barber surgeons vigorously attacked the treatment of any external bodily condition by women as an infringement on its professional jurisdiction. Despite the barber-surgeons' attempts to intervene, in the first decade of the seventeenth century several women pursued careers as healers with the tolerance of the city council.

Among the most successful of these healers was Ursula Weidner, who was initially assessed a fine of five gulden by the managing board of the barber-surgeons' association in October 1604. In a written defense to the city council, Weidner recounted that she had once herself suffered an affliction in her breast. Rather than have it removed, which was the suggestion made by a barber-surgeon, she undertook to heal it herself with herbs. After her recovery, she began helping people inside and outside the city. In a second submission on November 8, Weidner gave testimony from three of her patients. One woman said she had been unable to nurse her child for two weeks before Weidner came and helped her in ways that no barber-surgeon would agree to; after her treatment, the woman recounted, she had given Weidner a small honorarium, "but [she] had earned ten times more, she Frau Weidner who is a poor woman."[45] Such praise defended Weidner from the

---

43. StAA. CM. Apotheker von 1557 bis 1641, Iome. II, fol. 376r.

44. Universitätsarchiv Tübingen, Unbefügte Medizin. 20/12, nos. 2–5.

45. "[H]atte 10. mal mehr v[er]dient, vnd sye die Widnerin ein armes frawlin." StAA, Handwerkerakten: Bader und Barbierer, 1601–33, 8 November 1604.

familiar allegation that she was swindling women who otherwise would seek out the services of a licensed barber-surgeon. Another of Weidner's patients praised her healing art directly to the city council, and on November 13 the council allowed Weidner to continue her activities with the prior approval of the mayor for each patient.

Eight years later, Weidner was again accused by a master in the barber association, this time for cutting an open wound, a service that was reserved for members of the barbers' guild association. Weidner denied using any of the tools connected with the trade, and noted further that many of her female patients "prefer experienced women rather than men out of modesty and shame."[46] She further submitted the names of thirty-six women whom she had helped, many of whom signed with their own hands. In their rebuttal, the barber-surgeons were reduced to arguing that they used herbal remedies similar to Weidner's themselves and that they were being falsely accused of curing breasts with the knife. But by this point, Weidner was clearly winning the public relations war, and the barber-surgeons were forced to concede Weidner's prerogative as it had been established eight years before. Seven years later, in 1619, Weidner sent a letter to the city council herself to request compensation for her healing activities, and she was awarded a substantial annual salary of twenty gulden from the Office of Public Works.[47]

Augsburg's debates over Paracelsian healers and unofficial medicine suggest two conclusions that should be explored more thoroughly. First of all, these records show the furthest extent of the diffusion of Paracelsian ideas and methods in the urban settings of the Empire. In cities such as Augsburg, the aspects of Paracelsianism with medical applications were most likely to circulate and have a wide impact. This was partly because people of all social groups might need healing and partly because cities in the sixteenth century had developed ordinances and institutions through which Paracelsian claims were challenged and discussed. Georg Herdt proved his usefulness as a healer only after finding himself in hot water as an alchemist. Information about Paracelsian medicine spread by word of mouth, beginning with either an informed layperson or a traveling healer who served as a bridge between a princely court and a weavers' guild, or a learned treatise and the application of a practical recipe.[48] This circulation had a dynamic of

---

46. Ibid., 23 October 1612.

47. Ibid., 12 December 1619.

48. My view has some affinity with Peter Burke's depiction of "carriers" of popular tradition who contributed to the diffusion and transformation of popular culture. See Peter Burke, *Popular Culture in Early Modern Europe* (New York: Maurice Temple Smith, 1978), 91–102.

its own apart from the exchange of recipes and treatises carried on by intellectuals such as Karl Widemann and his correspondents.

Second, with respect to the boundaries of medicine in the city, the group with the most influence with city leaders was, more often than not, the patients rather than academic physicians or unorthodox healers. Although the city council consulted and supported the College of Medicine and other medical practitioners in principle and upheld the ordinances, at the same time it usually allowed sick citizens to try a variety of healing methods. Medical orthodoxy was held in balance with other priorities of civic government, foremost among them the desire to provide a context where the sick, and in particular the sick poor, could find the best way to cure their illnesses. The case of Augsburg thus suggests that the civic response to Paracelsian medicine was part of a general movement toward medical pluralism, in which city officials judged medical treatments as much by their effect on the populace as by their adherence to medical orthodoxy.

The dispute over Paracelsian medicine in Augsburg ranges rather far afield from Paracelsus's impact on a later chemical philosophy, but it suggests how theoretical debates over the art of medicine ultimately influenced city administration and the lives of everyday people. Focusing on the impact that Paracelsian practices had upon weavers and shoemakers, one may take seriously the idea that Paracelsianism was, in part, a popular phenomenon with consequences beyond the realm of protoscientific development.

Acknowledgments: This analysis is a product of research conducted in the Augsburg city archives in the fall of 1998. Citation of archival sources reflects conventions currently in use at the Stadtarchiv Augsburg. All translations are my own.

To Be or Not To Be a Paracelsian

Something Spagyric in the State of Denmark

Jole Shackelford

Historians of science and medicine have been interested in the spread and influence of Paracelsian medicine and its associated philosophical idea-matrix since the pioneering studies by Kurt Sprengel in the eighteenth century. However, in recent decades the religious, political, and social dimensions of Paracelsus's thought have come to be better understood, and the importance of these aspects of Paracelsian theory for the fortunes of Paracelsianism as an intellectual movement and medical practice has increasingly occupied historians. As attention has shifted from the mainly medical and progressive scientific aspects of Paracelsian medicine to a more nuanced and encompassing attempt to understand and document the historical development of Paracelsianism and its place in the history of early modern culture, the problem of defining "Paracelsian" in a meaningful way has become more acute.

Stephen Pumfrey has recently confronted this problem directly, arguing that the manifold meanings that scholars have given to Paracelsianism have sometimes made the term too amorphous for useful analytical work. Historians have considered Paracelsians to be (1) followers of Paracelsus, (2) chemical philosophers (that is, theoreticians), and (3) iatrochemists, that is, those who prepared chemical remedies and used them therapeutically. Members of the first group can be readily identified, if they declare themselves to be Paracelsians or express overt sympathy with or admiration for Paracelsus and his reforms. Identifying Paracelsianism with chemical philosophy, a core of ideas that interpenetrates and competes with other scientific paradigms, is especially useful to those interested in studying Paracelsianism as an intellectual construction, but Pumfrey's admonition that the Paracelsian identity of chemical philosophy becomes increasingly diluted and conflated with other philosophical schools with the passage of time must be heeded.[1] Finally, Pumfrey points out that a close identification of iatrochemistry with Paracelsian medicine becomes untenable in the seventeenth century, when iatrochemists as a group also "included Galenists and others who violently denounced the chemical philosophies which had originally provided explanations" for the actions of chemical drugs.[2] For this period in general, he concludes that "conceptions [of Paracelsianism] based on the prescription of chemical remedies and even on a commitment to chemical principles have the unfortunate consequence of leading away from, rather than through the most significant thickets of controversy. Furthermore, they do not prove to be clear and stable."[3] In other words, to equate iatrochemistry with Paracelsianism renders meaningless the important distinctions between Galenists and Paracelsians, between dogmatic and empirical physicians.

Pumfrey argues cogently that what is of chief interest to seventeenth-century historians are the "thickets of controversy" that characterize intellectual debate in the period, through which Paracelsian studies might serve as guide; that one should not reduce Paracelsianism to mere practice, but "emphasise the philosophical, religious and ideological debates which Paracelsianism occasioned."[4] While one should not wholly ignore the importance of the practical contexts of Paracelsian medicine, one may generally agree with Pumfrey's emphasis on treating Paracelsianism as an *ideology* that underlies multiple modes of practical and intellectual expression. Only this definition is broad enough to help understand the social, political, and intellectual contexts in which Paracelsian ideas are encountered; it helps explain, for instance, why Protestant sectarians might have been drawn to Paracelsian philosophy or why a Paracelsian physician might have found comfort in the heterodox religion of, say, Valentin Weigel or Jacob Boehme. But, as a practical matter, how does an ideology present itself? How is the historian to go about identifying and labeling prospective Paracelsians? One needs a set of criteria to guide one's research, such as those proposed below and field-tested in seventeenth-century Denmark.

---

1. E.g., historians of mid-seventeenth-century English science and medicine prefer the term *Helmontian* to *Paracelsian* because in that context Paracelsian ideas were largely mediated by the Belgian chemist J.B. van Helmont, who interpreted them and modified them to suit his own words. See Piyo Rattansi, "The Helmontian-Galenist Controversy in Restoration England," *Ambix* 12 (1964): 1–23.

2. Stephen Pumfrey, "The Spagyric Art; or the Impossible Work of Separating Pure from Impure Paracelsianism: A Historiographical Analysis," in *Paracelsus: The Man and His Reputation, His Ideas and Their Transformation*, ed. Ole Grell (Leiden: Brill, 1998), 26.

3. Ibid., 27.

4. Ibid., 34.

Paracelsianism in Late Sixteenth and Early Seventeenth Century Denmark

Paracelsianism was brought to Denmark in the 1570s and became fairly well established by 1600. Paracelsianism is defined as a combination of (1) adherence to Paracelsian theoretical concepts, (2) the use of chemical medicines that were characteristically Paracelsian, (3) rejection of the absolute hold of Aristotle and Galen over natural philosophy and medicine, (4) the profession of religious ideas and practices that were associated with Paracelsian theory, and (5) overt expressions of admiration for Paracelsus or for Paracelsians as an identifiable movement or school of thought.[5] These factors serve as operational indicators of an underlying Paracelsian ideology that may find various expressions, depending on the social and intellectual context and the form of the historical source. For example, one would not expect a published academic medical treatise, a formulary or drug inventory, and the record of a court proceeding to speak to the same kinds of intellectual issues. The first might reveal the author's passion for chemical philosophy, the second mutely attest to the manufacture and sale of chemical drugs, and the last reveal tensions in the religious community between the established authorities and Paracelsian medical practitioners and lay preachers of Paracelsian religion, without ever discussing medical theory or therapeutics. Seldom does the surviving documentation of one person's career reveal all the characteristics enumerated above; rather, one must be content to find several indicators and sometimes only one, and then weigh one's judgment accordingly as to the individual's commitment to a Paracelsian worldview. Several examples will illustrate this.

Two Danish students, Petrus Severinus and Johannes Pratensis, intentionally brought Paracelsian ideas home to Denmark in the 1570s from their studies abroad and began to implement them in their teaching and practice of medicine, but they left no specific indications of their religious views. Yet they certainly defended Paracelsian philosophy and deprecated blind adherence to Galen and Aristotle. They may be called Paracelsians.

Tycho Brahe, their friend and correspondent, espoused an occasional idea that is not compatible with Aristotelian philosophy but would seem to be of Paracelsian origin; moreover, he dabbled in Paracelsian medical pharmacy. He is known to have resisted orthodox Lutheranism, but the only indication that he had any

---

5. See Jole Shackelford, "Paracelsianism and Patronage in Early Modern Denmark," in *Patronage and Institutions: Science, Technology, and Medicine at the European Court 1500–1750*, ed. Bruce T. Moran (Woodbridge, Suffolk: Boydell, 1991); and idem, "Paracelsianism in Denmark and Norway in the 16th and 17th Centuries" (Ph.D. diss., University of Wisconsin, 1989), chaps. 1–4.

interest in Paracelsian religion per se is a secondhand report that he owned a Bible commentary by Paracelsus. Apparently, he had strong sympathies with Paracelsian ideology; he may even guardedly be called a Paracelsian.[6]

Brahe's student, Kort Aslakssøn, fused Paracelsian chemical philosophy with religious ideas that departed from Lutheran orthodoxy, the legal religious norm of the Danish state. Nevertheless, such views, which were associated with Danish and North German followers of Philipp Melanchthon, enjoyed some degree of free expression in late-sixteenth- and early-seventeenth-century Denmark, and Aslakssøn took them with him into the theology faculty at the University of Copenhagen.[7] He may be ranked with Tycho Brahe.

Iatrochemistry, probably of a Paracelsian sort, also found a foothold at the University of Copenhagen with the professor of natural philosophy Ander Krag, and later at the royal castle of Rosenborg in Copenhagen, with the establishment of a royal laboratory that was staffed by Peter Payngk. Too little is known about Krag's philosophical and religious views to identify him as a Paracelsian.[8] Payngk, however, was a spagyric physician from Holstein who prepared Paracelsian iatrochemical drugs, cosmetics, and confections. The enthusiasm he expressed for Paracelsus in his treatise *Rhapsody to the Life of Theophrastus Paracelsus* leaves no doubt that he regarded himself as a Paracelsian.[9]

Krag excepted, every one of these Danes exhibits two or more of the characteristics of Paracelsianism itemized above, and although they were eclectic in their natural philosophy and medicine, regarding them as Paracelsians highlights an intellectual commonality; it is historically useful. With them and in their work,

---

6. On Tycho Brahe's Paracelsianism and research program see Jole Shackelford, "Tycho Brahe, Laboratory Design, and the Aim of Science: Reading Plans in Context," *Isis* 84 (1993): 211–30; Carl-Johan Clemedson, "Något om Tycho Brahe och hans medicinska verksamhet," in *Sydsvenska medicinhistoriska sällskapets årsskrift, 1972* (Lund: Sydsvenska medicinhistoriska sällskapet, 1972), 38–59; Victor Thoren, *The Lord of Uraniborg: A Biography of Tycho Brahe* (Cambridge: Cambridge University Press, 1990), 53–54, 59, 210–13; and John Christianson, *On Tycho's Island: Tycho Brahe and His Assistants 1570–1601* (Cambridge: Cambridge University Press, 2000), 47–52, 83–124.

7. On Aslakssøn, see Oskar Garstein, *Cort Aslakssøn: Studier over dansk-norsk universitets og lærdomshistorie omkring år 1600* (Oslo: Lutherstiftelse, 1953); and Jole Shackelford, "Unification and the Chemistry of the Reformation," in *Infinite Boundaries: Order, Disorder, and Reorder in Early Modern German Culture*, ed. Max Reinhart, Sixteenth Century Essays and Studies 40 (Kirksville, Mo.: Sixteenth Century Journal Publishers, 1998).

8. On Krag's iatrochemistry, see Shackelford, "Paracelsianism and Patronage," 105–6. Holger Rørdam, *Kjøbenhavns Universitets Historie fra 1537 til 1621* (Copenhagen, 1877), 3:517–19, noted that Krag had a reputation as a Paracelsian.

9. See August Fjeldstrup, *Dr. Peter Payngk: Kong Kristian IV's Hofkemiker* (Copenhagen: A. Giese, 1911); and Shackelford, "Paracelsianism and Patronage," 106–8.

Paracelsianism had a presence in Danish academic medicine and was a part of academic philosophical and theological discussion in and around the University of Copenhagen in the last quarter of the sixteenth century and first decade of the seventeenth.

However, in the second decade the situation in Denmark changed. Historians agree that there was an orthodox Lutheran reaction to what they perceived as Calvinist-leaning tendencies in the late work of Philipp Melanchthon, and that this reaction was decisive for establishing the course of Danish religion for the rest of the century.[10] One indication of the consequences that orthodoxy had for the intellectual climate of Denmark is the official denouncement of the Rosicrucian brotherhood, in 1619, by a spokesman for the University of Copenhagen, the physician, professor, and current dean of the philosophical faculty, Ole Worm.[11] The Rosicrucians maintained theological ideas that orthodox Lutherans considered threatening, and some of these ideas were closely linked to Paracelsian religious and philosophical expression. Indeed, some scholars regard the Rosicrucians as a loose collective identity representing Paracelsian hopes and ambitions for a reform of religion, politics, and education.[12] Therefore, Worm's rejection of Rosicrucian

---

10. Svend Ellehøj, *Danmarks Historie*, ed. John Danstrup and Hal Kock, 2d ed. (Copenhagen: Politikens Forlag, 1970), 7:163–69; Helge Gamrath and E. Ladewig Petersen, *Danmarks Historie*, ed. Aksel E. Christensen et al. (Copenhagen: Gyldendal, 1980), 2:486–88; and Shackelford, "Unification and the Chemistry of the Reformation," 309–11. Bjørn Kornerup, *Den Danske Kirkes Historie*, ed. Bjørn Kornerup and Hal Koch (Copenhagen: Gyldendal, 1959), 4:197–220, discusses Resen's victory over the Philippists and succinctly summarizes (4:221): "I en lille Menneskealder efter dette Tidspunkt var Resen ubestridt den ledende Personlighed, og det kunde derfor være fristende at kalde hele Perioden efter ham 'det Resenske Tidsrum.' Rigtigere er det dog sikkert at tale om den lutherske Ortodoksis første Periode (indtil 1660)" (For almost a generation after this point in time (1614), Resen was indisputably the leading personality, and one is tempted to name the entire period after him "The Resen Period." But it is surely more accurate to speak about it as the first period of Lutheran orthodoxy (until 1660)); Kornerup in fact called the title of that chapter of the volume "Orthodoxy's Period of Greatness (1615–1660)." On Tycho Brahe and Philippism, see Thoren, *Lord of Uraniborg*, 100–01, 118–19, 372–73; and Christianson, *On Tycho's Island*, 200–1, 204, 242. Unless otherwise indicated, all translations into English from Danish and Latin sources are my own.

11. Jole Shackelford, "Rosicrucianism, Lutheran Orthodoxy, and the Rejection of Paracelsianism in Early Seventeenth-Century Denmark," *Bulletin of the History of Medicine* 70 (1996): 181–204, at 192. Although Ole Grell, "The Reception of Paracelsianism in Early Modern Lutheran Denmark: From Peter Severinus, the Dane, to Ole Worm," *Medical History* 39 (1995):78–94, at 92, does not specifically mention Worm's 1619 denouncement of the Rosicrucians, he does acknowledge that Worm had rejected Rosicrucianism by 1620.

12. For example, the overlap of Rosicrucianism and Paracelsianism is evident in John T. Young, *Faith, Medical Alchemy and Natural Philosophy: Johann Moriaen, Reformed Intelligencer, and the Hartlib Circle* (Aldershot, Hampshire: Ashgate, 1998), 15–25, 165–74 et passim, and is made explicit in Hugh Trevor-Roper, "The Paracelsian Movement," in *Renaissance Essays* (London: Secker & Warburg, 1985), 182–84. Siegfried Wollgast, "Zur Wirkungsgeschichte des Paracelsus im 16. und 17. Jahrhundert," in

heresy *de facto* extended to the radical Paracelsian religious ideas that the Rosicrucians embraced and also impugned Paracelsian theory in general and discouraged Danish scholars from defending Paracelsian medical philosophy.[13] However, some disagreement has arisen over the nature of the connection between Paracelsian religion, Paracelsian medicine, and Paracelsian natural philosophy, and therefore whether Ole Worm's denunciation of the Rosicrucians had any effect at all on natural philosophy at Copenhagen.

The various parts of the Paracelsian whole were intimately associated in a complex way that also linked academic natural philosophy to the underlying religious policies of the Danish state. The conclusions that I have reached are that acceptance, or even toleration, of Paracelsianism by academics at Copenhagen diminished sharply after the onset of Lutheran orthodoxy in the second decade.[14] However, this analysis has been challenged by the Danish historian Ole Grell. Grell argues that the leading natural philosophers, physicians, and even theologians at the university in this period not only continued to view favorably Paracelsian philosophy and medicine, but were able to separate them from sectarian religious ideas.[15] Clearly our difference of opinion depends in part on what constitutes Paracelsianism, and therefore affects whether the concept has any utility for assessing the intellectual history of the period. It is the useful definition of Paracelsianism and the survival of Paracelsianism in Denmark that this paper will examine, leaving aside for another occasion the more general problem of whether Danish academics actually separated religious and scientific views.

Resultate und Desiderate der Paracelsus-Forschung, ed. Peter Dilg and Hartmut Rudolph, Sudhoffs Archiv Beihefte 31 (Stuttgart: Franz Steiner, 1993), 113–44, points out that Nicolaus Hunnius and other contemporaries of Bartholin and Worm readily identified the Weigelians and Rosicrucians as Paracelsians.

13. Shackelford, "Rosicrucianism," 204.

14. Ibid., 192.

15. Ole Grell, "The Acceptable Face of Paracelsianism: The Legacy of *Idea medicinæ* and the Introduction of Paracelsianism into Early Modern Denmark," in *Paracelsus*, ed. Grell, 259, argues that Hans P. Resen did not oppose Cort Aslakssøn's "Paracelsian leanings" but only his "perceived crypto-Calvinism," that is, that Resen could separate the two. Likewise, he claims that "Worm's rejection of Rosicrucianism... served to disassociate 'moderate' and eclectic Paracelsianism from the danger of being connected with religious heterodoxy" (ibid., 264). In "The Reception of Paracelsianism," 88, he writes that "[Holger] Rosenkrantz was undoubtedly a supporter of some form of moderate Paracelsianism," and in "Introduction: The Enigma of Paracelsus," in *Paracelsus* (15) he extends this claim that Copenhagen's academicians could readily separate Paracelsian religious and medical ideas: "The subsequent professors of medicine at the University of Copenhagen, Caspar Bartholin and Ole Worm, in the main continued to espouse and teach a Severinian type of Paracelsianism and Worm made sure that it remained untainted by the dangers of religious heterodoxy inherent in early-seventeenth-century Rosicrucianism."

Paracelsianism and Natural Philosophy at the University of Copenhagen During the Early Years of Lutheran Orthodoxy

Pressure on Danes to abandon all elements of Reformed theology and subscribe to orthodox Lutheranism accelerated in the first decade of the seventeenth century as a generation of German-trained strict Lutherans, or Gnesiolutherans, began to exterminate Calvinist ideas from the Danish church. This program reached a decisive phase with the 1615 appointment of Hans P. Resen, who spearheaded the assault, to the prime bishopric of Sjælland, subordinate only to the king in matters of religion. During this period and the decade after, Danish theology as it was preached and taught in Denmark was dominated by Resen and his supporters, chiefly Jesper Brochmand, Holger Rosenkrantz, and Christian Friis. Brochmand followed Resen as bishop primate and extended the grip of orthodoxy on the church into the middle of the century, and beyond.[16] Rosenkrantz eventually dissented, but during the crucial first two decades, he was publicly a firm supporter of Resen and Lutheran orthodoxy. Rosenkrantz was a powerful nobleman and member of the Council of the Realm. He was widely respected and exerted a potent influence on Danish intellectuals. Christian Friis was not a theologian, but in his capacity as chancellor, his support for Resen and orthodoxy must be regarded as important.

Natural philosophy and medicine at the University of Copenhagen in this period was largely shaped by the senior professor of medicine, Thomas Fincke, and by Caspar Bartholin and Ole Worm, who were married to his daughters. Consideration of the specific theological attitudes and ideas of these leaders of Danish education lies beyond the scope of the present investigation, but it is safe to assume that these men shared at least a public commitment to the king's insistence that Denmark adhere to strict Lutheran interpretations of Scripture and church discipline.

The breadth of the conservative political reaction that accompanied or perhaps fueled the imposition of Lutheran orthodoxy is evident in Denmark's laws and foreign policy from the period. Ole Grell remarked the stark change in his 1996 article "Exile and Tolerance," noting that while prospective Dutch immigrants to Denmark were promised religious freedom in 1607, such tolerance was

---

16. Danish church historians define 1615 to 1660 as the first period of orthodoxy; see Ellehøj, *Danmarks Historie*, vol. 2; Shackelford, "Unification and the Chemistry of the Reformation"; Kornerup, *Den Danskes Kirkes Historie*.

explicitly forbidden by statute in 1617.[17] This is a fine example of how deeply conformity to strict Lutheranism, which became firmly entrenched in this ten-year period, discouraged toleration of foreign religions and confessional diversity. The repressive environment also demanded religious conformity from the intellectuals of Copenhagen and discouraged the incorporation of Paracelsian theory in both natural philosophy and theology.

Caspar Bartholin and Ole Worm were not wholly closed to the benefits of studying Paracelsian medicine, especially during their student years, before Resen succeeded in suppressing Philippism and enforcing Lutheran uniformity in the Danish church and state, and consequently at the university.[18] But can their youthful quest for chemical knowledge be construed as support for Paracelsianism or even sympathy for Paracelsian cosmology or medicine, in particular after this sea change in the intellectual and political climate? Grell believes so:

> From his appointment as professor of medicine in Copenhagen in 1613 until his death in 1629, Caspar Bartholin was keenly interested in Paracelsian medicine. His early interest is evident in the first medical disputations he presided over in 1613 and 1614: *De philosophiæ in medicina usu et necessitate* and *Exercitatio disputationis secundæ ordinariæ*. Throughout his life, even after he had left the medical faculty to become professor of theology, Bartholin continued to promote an eclectic Paracelsianism close, if not identical, to that introduced in Copenhagen forty years earlier by Severinus and Pratensis.[19]

What is the nature of Bartholin's natural philosophy and what kind of evidence does Grell evince to support his contention that Bartholin remained a Paracelsian throughout his life?

Any claims for Bartholin's sympathy for Paracelsian philosophy must be evaluated in light of his public commitment to Aristotelian philosophy and Galenic medicine, as is evident in both his teaching and his published textbooks. This is not too surprising, since Aristotelian philosophy was part of Resen's and Brochmand's program for orthodoxy in Danish teaching, and Aristotelian natural philosophy was well suited to Galenism. In other words, Bartholin fit right in with the

---

17. Ole Grell, "Exile and Tolerance," in *Tolerance and Intolerance in the European Reformation*, ed. Ole Peter Grell and Robert Scribner (Cambridge: Cambridge University Press, 1996), 173.

18. Shackelford, "Rosicrucianism," 183–88.

19. Grell, "Reception of Paracelsianism," 91.

intellectual demands of Resen's pedagogical leadership.[20] As Bjørn Kornerup notes, although Bartholin's textbooks were more widely read than Resen's and "were an even stronger expression of the Aristotelianism that the times promoted," nevertheless, it was Resen who headed up the Aristotelian initiative and must be regarded as Bartholin's predecessor in this matter.[21] When Bartholin joined the medical faculty in 1613, it was as an even stronger proponent of Aristotelianism than Resen.[22] In what way, then, can the staunch Aristotelian Caspar Bartholin be seen as an advocate of Paracelsian medicine or natural philosophy? Grell's argument is based on Bartholin's suggestions for reform of the medical curriculum in 1621 and his 1626 recommendation of a course of study for his sons and the grandson of Severinus. Both of these documents bear closer examination.

Improvements in both the financial and regulatory structure of the University of Copenhagen began during the reign of Frederik II (1559–1588), but despite his support and the efforts of reform-minded scholars such as Severinus, Pratensis, and Tycho Brahe, medical education lagged far behind what was available at more southerly universities, a fact that was clear to Danish students returning from study at Basel, Montpellier, and especially Padua. Severinus and Pratensis had introduced chemical medicine to Denmark, but it was not made a part of the officially sanctioned curriculum. Likewise, Anders Christensen attempted to improve anatomical education by introducing human dissection, but this, too, was not instituted, and social pressures eventually discouraged him from conducting dissections. So it is no wonder that when the possibility for statutory reform presented itself in the early 1620s, reform-minded university professors would be eager to bring both chemical medicine and anatomical demonstration into the

---

20. Bjørn Kornerup, *Biskop Hans Poulsen Resen: Studier over Kirke- og Skolehistorie i det 16. og 17. Aarhundrede, I (1561–1615)* (Copenhagen: Gad, 1928), 302: "Af langt større Betydning end alle disse Mænd var dog for Resen Ansættelsen af Jesper Brochmand 1610...stod han Side om Side med Resen som hans betydeligste og mest trofaste Forbundsfælle i Kampen mod Kryptokalvinismen" (However, of far greater significance than all these men, for Resen, was the appointment of Jesper Brochmand in 1610...he stood by Resen's side as his most significant and most faithful ally in the struggle against Cryptocalvinism).

21. Ibid., 178: "[Bartholin's textbooks] var endnu strengere Udtryk for den Aristotelisme, som Tiden fordrede." Bartholin names Resen as a source in the preface to *Enchiridion logicum ex Aristotele et opt. ejus intepretum monumentis* (Strasbourg, 1608), according to 178 n. 3.

22. Ibid., 302: "Blandt den ny Tids Mænd, som i Pagt med Resen førte an i den kommende teologiske og filosofiske Reaktion, maa dog især nævnes Caspar Bartholin, som 1613 blev Professor i Medicin, og som...giver Udtryk for en endnu stivere Aristotelisme end Resens" (Among the men of the new era who, in league with Resen, were the leaders in the coming theological and philosophical reaction, one must especially identify Caspar Bartholin, who in 1613 became professor of medicine and who...expresses an Aristotelianism even stricter than Resen's).

curriculum as well as to improve botanical teaching, three areas at which Europe's best medical schools excelled.

Although the New Constitutions (Novellæ Constitutiones) that were signed into law 18 May 1621 failed to enact major changes, perhaps for want of money (as Grell argues),[23] a surviving draft written by Jesper Brochmand and emended and supplemented by Caspar Bartholin reveals what reforms were actually sought by the movers and shakers of the theological and medical faculties.[24] While Grell does not claim outright that Caspar Bartholin's sympathy with Severinus's Paracelsianism is reflected in his additions to the draft of the New Constitutions, he does imply some connection when he states that "Caspar Bartholin was in agreement with Severinus and Pratensis about the eschatological urgency of medical as well as religious reform. This is in evidence in all the textbooks and manuals he produced in the 1620s and in the draft for new Statutes of the university he coauthored in 1621, which included plans for a major reform of medicine."[25] It is not amiss, then, to examine this draft for hints of Bartholin's sympathy to Paracelsian medicine as part of his vision of reform.

Brochmand's draft includes thirty-nine points, four of which deal explicitly with medical education (nos. 5–8), and another nine with literary, linguistic, and philosophical studies that might be regarded as propaedeutic to medicine.[26] The first group specify that the senior medical professor teach praxis during the winter and medical botany during the summer, and have an adjunct pharmacist as an assistant to teach pharmacy and chemistry. The junior professor is to lecture on theory and conduct one annual anatomical demonstration each winter, using either a cadaver or a skeleton. He is also to have an adjunct assistant, a surgeon who serves as the prosector during the demonstrations and also lectures as professor of surgery, covering in the course of his lectures what a barber ought to know. These four men will constitute the medical faculty and have at their disposal a special medical auditorium and anatomical theater, where the surgical and pharmaceutical lectures are to be held.

---

23. Ole Peter Grell, "Caspar Bartholin and the Education of the Pious Physician," in *Medicine and the Reformation*, ed. Ole Peter Grell and Andrew Cunningham (London: Routledge, 1993), 94.

24. Thomas Fincke was the senior professor of medicine, but seems to have had little interest in medical reform throughout his long tenure. He published little and left little record of a public voice in university matters when compared to Caspar Bartholin and Ole Worm.

25. Grell, "Reception of Paracelsianism," 91.

26. The draft and final versions of the New Constitutions are reproduced in Holger Rørdam, "Aktstykker til Universitetets Historie i Tidsrummet 1621–60," *Danske Magazine*, 5th ser., 1 (1887–88): 36–47.

Notable features of these reforms are that they aim to teach both learned students and professional "empirics" and that lectures are to be held in Danish. These provisions were to be coupled with medical regulation of the professions, which stipulated that physicians, pharmacists, and barber-surgeons either have studied at the university or have been approved by the faculty there. The final point stipulates that any medical arcana (*medicamenta rariora*) that are proffered by itinerants (*circumforanei*) must be tested and approved before being used.

Two observations need to be made about these suggested reforms. First, they do not represent a break with the extant medical faculty structure, which already specified a senior and junior professor, who taught theory and practice.[27] What is new here are the adjuncts and the institutionalization of chemical, anatomical, surgical, and pharmaceutical training. Second, the institutionalization of regular dissection and the establishment of a designated medical auditorium and anatomical theater would have been a great step toward building a medical school of the sort that trained the best anatomists of the age, such as Padua was and Leiden was then becoming. But for want of funding or whatever reason, these parts of the plan were not carried out in Denmark until after 1644, when the anatomical house (*domus anatomica*) was erected. Notably, nothing here resembles Paracelsian or Rosicrucian reforms, unless one construes as Paracelsian the demand to teach in the vernacular and the inclusion of chemical pharmacy in the curriculum, both of which were arguably not limited to the Paracelsians by 1621.

Caspar Bartholin made few emendations to these basic provisions, merely pointing out that there was not yet a botanical garden suitable for teaching pharmaceutical botany, that "itinerants" were already forbidden by a recent royal order, and other minor clarifications. His additions regarding the medical professors are short and specific: the *primus medicus* (senior chair), who is to teach praxis and medical botany, should spend half the year lecturing on plants and metals and minerals, and the *medicus secundus* should spend half of the year lecturing on anatomy and "*Historiam Animaliam*," that is, on Aristotle's treatise.[28] If there is room for a Paracelsian here, it is certainly not in the second chair, since Paracelsus and his followers generally had little use for either anatomy, in the normal sense of the word, or Aristotle's zoology. Whatever medical theory was to be taught at the University of Copenhagen, according to these statutes, was securely

---

27. P.L. Panum, "Vort medicinske Fakultets Oprindelse og Barnedom: et Bidrag til Kundskab om Lægevidenskabens og Naturvidenskabernes Udvikling i Danmark," in *Festskrifter udgivne af Det Lægevidenskabelige Fakultet ved Kjøbenhavns Universitet i Anledning af Universitetets Firehundredaarsfest Juni 1879* (Copenhagen: Gyldendal, 1879), 26.

28. Rørdam, "Aktstykker til Universitetets Historie," 43–44.

Aristotelian and Galenic. This perception is strengthened when one considers the propaedeutic subjects covered in Brochmand's draft: Greek is to be taught first of all from Aristotle's ethical treatises and then later from Plato and other authors; Eloquence begins with Aristotle's *Rhetorica*; Metaphysics relies on Aristotle's *Metaphysica*; Physics is to be learned from the traditional Aristotelian texts *On the Heavens, On the World, On Generation and Corruption, Meteors*, etc. Even the mathematics professor is responsible for teaching Aristotelian treatises on subjects like respiration and local motion, in addition to more strictly mathematical tracts. The professor of logic, another subject demeaned by many Paracelsians, is to teach Aristotelian logic, using the traditional Aristotelian texts (*Prior and Posterior Analytics, Categories, Topics, The Organon*, etc.). Moreover, he is expected to use the two extraordinary days demonstrating Aristotelian logic using theological examples, which suggests that the close connection between Aristotelian teaching and Lutheran orthodoxy was being institutionalized in the philosophy curriculum.[29]

None of this sounds at all Paracelsian, with the possible exception of the expectation that the senior medical professor lecture on minerals and metals and that chemical pharmacy should be taught. Both of these fall within praxis and, it will be argued below, were by this time in history no longer necessarily connected to Paracelsian theory, but were accommodated to Galenic theory and practice. If Bartholin had any sympathy or interest in Paracelsianism per se, it is not evident here; quite the opposite—he endorses a thoroughly Aristotelian pedagogy as far as natural philosophy and medical theory are concerned. The medical innovations suggested by him and Brochmand were mainly intended to introduce the kind of educational curriculum and practical surgical, botanical, and chemical methods that were found at Padua, Basel, and Montpellier, where Bartholin, Worm, and other Danes had themselves learned what academic medicine entailed.

Now to the second point of the argument. Grell notes that Bartholin "warmly recommended iatrochemistry as essential to the aspiring physician" in *De studio medico*, a treatise that he wrote as advice to his sons and Severinus's grandson.[30] He concludes from this document that "For an Aristotelian, Caspar Bartholin was surprisingly positively inclined towards iatrochemistry. He warmly recommended the works and arcana of Paracelsians such as Joseph du Chesne, Petrus Severinus,

---

29. Ibid., 38–39.

30. Grell, "Reception of Paracelsianism," 91. Caspar Bartholin, *De studio medico inchoando, continuando, et absolvendo, pro accurato & supra vulgus futuro medico consilium breve atque extemporaneum*, is dated 1626 but was published by Geogius Hantzch at Copenhagen in 1628 and later included in Hermann Conring, *In universam artem medicam singulasque ejus partes introductio* (Helmstadt: Hammius, 1687). The treatise is highlighted in Panum, "Vort medicinske Fakultets Oprindelse," 58–61.

John Bannister and Johannes Hartmann to his students, referring them to his own considerable alchemical library."[31] Whether this translates into enthusiasm for Paracelsian theory depends on what is meant by iatrochemistry and how "warmly" Bartholin recommended Paracelsian authors, and why.

Bartholin's recommendations are dated 1626, two years after he left the medical faculty for the theology faculty, so the fact that he begins the tract with an admonition to read from the Bible daily comes as no surprise. Neither do occasional laudatory references to Petrus Severinus, since the work was in the first instance addressed to Severinus's grandson Peter Charisius, son of Jonas Charisius and Anna Severinus. Jonas had been a high-ranking diplomat and trusted emissary in Christian IV's government. Anna was Bartholin's relative, by marriage, as was Ole Worm. After Jonas's death in 1619, Caspar Bartholin looked after Peter's education and, judging by this treatise, regarded him fondly and as a son. Whatever Caspar's opinion of Severinus as a Paracelsian, he certainly had nothing to gain by deprecating the former royal physician, relative by marriage, and father of his friend.

Caspar Bartholin's recommendations in many ways support the curriculum suggested in the New Constitutions. The aspiring physician should be well grounded in Latin and Greek—after all, he asks, how can one properly learn what Hippocrates and Galen have written without a knowledge of Greek idioms?[32] Furthermore, if one can go directly to the Greek sources, one need not master Arabic. Medical Latin should be learned from Celsus, Fernel, and Heurnius. Thus far, Bartholin's recommendations mark him as a medical humanist of the sixteenth-century academic sort. He further recommended that the aspiring physician should also be schooled in rhetoric, logic, practical philosophy, and metaphysics. Although he does not mention specific authors for these subjects, beyond pitching his own very Aristotelian *Enchiridion metaphysicum*, one can surmise from the provisions of the draft of the New Constitutions that these were meant to be mainly Aristotelian subjects, taught primarily from Aristotelian texts. For arithmetic, geometry, optics, astronomy, and astrology, Bartholin cites a variety of compendia and recent authors, including his own countrymen Tycho Brahe and Longomontanus. For physics, which is especially important for the physician ("the healer begins where the natural philosopher leaves off"),[33] Bartholin commends

---

31. Grell, "Caspar Bartholin," 95.

32. Bartholin, *De studio medico* (1687 edition), 9: "Quid ex Hippocrate vel Galeno bene vel male dictum æstimabis, Græcanici ignarus idiomatis?"

33. Ibid., 10: "Ubi enim desinit Physicus, ibi incipit Medicus." Of course, we incorporate this sentiment today in our term *physician* and requirement that medical doctors be scientifically trained.

Johannes Magirus and Sennert, and his own *Enchiridion physicum.* For a knowledge of minerals, metals, and stones, he recommends Cesalpino, Encelius (Christoph Entzel), Libavius, and Anselm Boetius de Boodt; for botany, Lobelius, Tabernamontanus, Tragus (Hieronymus Bock), Mathiolus Bauhin, and Clusius; for *historia animalium,* Konrad Gesner and Aldrovandi. A host of Europe's best anatomists are cited for study of anatomy, beginning with Vesalius.

For the art of medicine itself, the core of medical professional training, Bartholin recommends Hippocrates, Galen, Fernel, and Sennert, followed by a large number of neoteric authors, who are recommended for specific aspects of medical training. To make a long story shorter, one finds Paracelsian names only under the rubric "spagyric pharmacy and pyrotechnics," with the exception of Quercetanus, who also is recommended for ordinary (i.e. nonspagyric) pharmacy and dietetics—hardly Paracelsian subjects. Therefore, this attempt to illuminate Caspar Bartholin's interest in or sympathy for Paracelsian medicine is practically limited to his recommendations under "spagyric pharmacy."[34]

Noted Paracelsian physicians appear prominently in this list, including Quercetanus, Croll, Basil Valentine, Penotus, Hartmann, and Petrus Severinus. But so do names of noted chemists such as Jean Beguin, Angelus Sala, and Della Porta, who are less clearly Paracelsians, and even the violently anti-Paracelsian Libavius, who nevertheless was as famous as the others for the preparation of chemical medicines. Libavius is a prime example of the disentanglement of iatrochemistry from Paracelsianism in the early seventeenth century, a doctrinal autonomy that he strenuously promoted for the chemical art. Tycho Brahe and Severinus also figure into this list. Why? Tycho Brahe did not write on medicine per se, though he was clearly sympathetic to certain Paracelsian ideas. However, he was probably well known among Danish academic physicians and aristocrats for his chemical drug recipes, one of which, *Elixir Tychonis Brahei,* was still stocked in the nineteenth-century Danish pharmacy. Likewise, although Severinus was best known internationally as a theoretician, he was known in Denmark for his use of chemical remedies. This is why Bartholin makes a point of saying that young Peter Charisius can surely find much of use in his grandfather's bookshelves, whose books he can obtain from his uncle, namely Severinus's son Frederik Severinus.[35] The point here is that with the exception of the final sentence of this section, these recommendations appear to be oriented toward the use of chemical drugs regardless of Paracelsian medical theory, a conclusion that explains why Hartmann and his

---

34. Ibid., 14–15, "Pharmacopæa Spagyrica seu Pyrotechnia."

35. Ibid., 15: "Non dubito, mi Petri, Avum tuum maternum D. Petrum Severini, plurima reliquisse in pluteis latentia, quæ facile ab avunculo tuo impetraveris."

severe critic Libavius appear as bedfellows. The final sentence, on which Bartholin's sympathy for Paracelsian medicine here depends, reads, "One may learn the theories of the Paracelsians from Petrus Severinus, Petræus's *Nosologia*, Warenius, and others."[36] Can this comment be construed as strong support for Paracelsianism? Severinus was without a doubt a Paracelsian, and his theory was well grounded in Paracelsus's ideas. But then again his appearance on this list must be seen in context—he was a powerful man in Danish academic medicine as the century opened, and he was an internationally known royal physician, besides being Peter's grandfather and related to Bartholin by marriage. Petræus's *Nosologia* is less Paracelsian. In fact, Petræus's goal was to accommodate Paracelsian chemical medicine to Galenic therapeutics, that is, to adapt iatrochemistry to Galenic theory.[37] Considering Bartholin's recommendations in their entirety, his apparent advocacy of Paracelsian authors must be severely qualified. He truly was an eclectic in the spirit of Danish medical practice, but like his contemporaries, he remained fundamentally an Aristotelian and a Galenist, who sought to apply apparently efficacious chemical remedies within the framework of traditional academic medicine. Caspar Bartholin's use of chemical drugs was a Danish version of the Elizabethan compromise that Allen Debus attributed to Bartholin's English contemporaries.[38]

Beyond Caspar Bartholin's curricular recommendations, one might well inquire what light his published theoretical tracts—academic dissertations and such—can shed on his attitude toward Paracelsianism. To begin with are the two treatises that Ole Grell claims indicate Bartholin's early interest in Paracelsian medicine, *De philosophiæ in medicinæ usu et necessitate* (1613) and *Exercitatio disputationis secundæ ordinariæ* (1614).[39] Neither of these tracts discusses the merits of Paracelsian medicine explicitly, but there are some hints to Bartholin's attitude. In the former, Bartholin criticizes the Paracelsian homunculus theory and explicitly denies the three Paracelsian principles—salt, sulphur, and mercury.[40] He clarified

---

36. Ibid.: "Theorias Paracelsistarum haurire licet ex Petro Severini, Petrei Nosologia, Warenio & aliis."

37. Heinrich Petræus, *Nosologia harmonica* (Marburg: P. Egenolph, 1615–16). Petræus had earlier defended the dissertation *Contradictionum in medicina dogmata & Hermetica apparentium conciliatio: Inauguralis* under the Paracelsian Johann Hartmann in Johann Hartmann, "Disputationes chymico medicæ," pt. 7 of *Opera omnia medico-chymica* (Frankfurt: Balthasar Wustius, Jr., 1684), no. 8.

38. Allen Debus, *The English Paracelsians* (London: Oldbourne, 1965), chap. 2.

39. Grell, "Reception of Paracelsianism," 91.

40. Caspar Bartholin, *Illustrium quæstionum philosoph. et medic. miscell. prodromus de philosophiæ in medicina usu et necessitate ad disputandum publice propositus 17. Novemb. Anni MDCXIII: Præside Casp. Bartholino: Respondente Olao Slangendorpffio Joh. f.* (Hafniæ: Ex Calcographia Waldkirchiana,

this position somewhat in his 1618 treatise *Physica generalis præcepta*, where he stated that these three are sensible, but they are not primary. The actual primary principles are matter, form, and privation.[41] In that tract, which introduces the fundamentals of *physics*, that is, natural philosophy, Bartholin went on to define the four Aristotelian causes and the four primary qualities—two active (hot and cold) and two passive (wet and dry).[42]

In a 1617 treatise, Bartholin explicitly denies that there is life in the human cadaver, but recognizes that it has the power to generate life, following Aristotle's teaching about the generation of worms and flies from rotting carcasses. He specifically denies that inanimate bodies are alive, contradicting Aristotle's teaching that what moves, lives. While these claims are not directed against the Paracelsians here, they are in part used to support the notion that the human resurrection promised in the Bible demands that human cadavers be truly dead until they are resurrected, implicitly opposing the Paracelsian teaching that life remains in the bodies of creatures who have died prematurely, an idea that sanctioned the use of bodies of executed criminals as a source of vital balsam (a Paracelsian vital agency that could be used as medicine).[43]

These treatises contain no evidence to support the idea that Bartholin was a Paracelsian. Certainly he was aware of the Paracelsians very early in his career, though, since he specifically discussed the "Paracelsian school" in his 1611 treatise *Problematum philosophicorum*.

> But in the constitution of medicine, indeed for all philosophy, Paracelsus invented a new school, rejecting the principles and causes of all the old ones. He devised new principles, causes, and relationships, having taken the primary basis from the harmony and the likeness of the parts and

---

[1613]), A3r, question 17: "Ad orcum autem releganda est Paracelsia superstitiosa de Homunculo vegetabili; quomodo Homo sit fabricandus per putrefactiones chymicas ex pane & vino aut semine, in furnulo arcanorum: consimilique Bombasti sordes impietatis & mendacij plenissimæ," and B4r, question 24: "An 3 illa Chymicorum principia, sal, sulphur, & [mercurius] sint prima corporum naturalium omnium principia? Impossibile."

41. Caspar Bartholin, *Physicæ generalis præcepta: Quæ thesium loco in disputationibus esse possunt, Decerpta ex prælectionibus physiologicis publicis Casp. Bartholini Pro juniorum memoria* (Rostock: Joannem Hallervordeum, 1618), 3: "Spagyricorum tria principia, Sal Sulphur, & Mercurium, ut sensibilem admittimus, non ut prima."

42. Ibid., 23, 44–45.

43. Caspar Bartholin, *Illust. quæstionum philosoph. et medic. miscell. exercitatio disputationi quintæ ordinariæ & anniversariæ in academia regia hafniensi 6. Kal. Decembr. Anni MDCXVII: Publice subjecta: Præside Casp. Bartholino: Respondente Christiano Laurentii Thverstedio* (Copenhagen: Henrik Waldkirch, [1617]), B2r–B4r, question 2: "An in hominis demortui cadavere vitæ adhuc sint reliquiæ? N."

their changes in the macrocosm and microcosm. The Paracelsian doctrine leans on this as its greatest foundation, and he takes it for granted always and everywhere that the same reasoning applies to the world at large and to man. But we shall intentionally refrain from explaining all these things more abundantly.[44]

By 1614, when Bartholin published the second of the two treatises that Grell cites, he had decided to treat this Paracelsian foundation in greater detail, in question 3, "Can the analogy of the microcosm and the macrocosm be a basis for all philosophy and medicine?"[45] After discussing the geographical mapping of the greater world onto the human body, Bartholin concluded: "It follows from all these considerations that no solid and indubitable basis for philosophy and medicine can be derived from the analogy of the microcosm and the macrocosm."[46] Given that he regarded such an analogy as the foundation for Paracelsian theory, it is hard to reconcile his position with the claim that he was sympathetic to Paracelsianism. Certainly it is clear that he did not regard himself as a Paracelsian.

Other treatises of Bartholin indicate that he was mainly a Galenist and Aristotelian in his views of man and nature. Certainly this is true of his popular textbook on anatomy, which was republished by his son Thomas Bartholin,[47] and his didactic treatises on Aristotelian physics and metaphysics.[48] In his topical treatises

---

44. Caspar Bartholin, *Problematum Philosophicorum & Medicorum nobiliorum & rariorum miscell. Exercitationes: Ad disputandum in illustri Albiaca propositæ* (Wittenberg, 1611), G3r: "At in medicinæ constitutione, imo Philosophiæ totius, Paracelsus novam excogitavit sectam, reicens veterum omnium Principia & Causas. Nova principia excogitavit, Causas & affectiones, sumpto fundamento. 1. ex harmonia & similitudine partium & vicissitudinum in [Greek: macro & microcosmo]. Cui ceu maximo fundamento innititur doctrina Paracelsistica, proque confesso & concesso sumit eandem semper & ubique esse rationem in toto mundo & in homine. Sed hæc omnia uberius declarare, consilio supersedemus."

45. Caspar Bartholin, *Illust. quæstionum philosoph. et medic. miscell. exercitatio disputationi secundæ ordinariæ & anniversariæ in academia regia hafniensi 7. Kal. Novembr. Anni MDCXIV: Publice subjecta: Præside Casp. Bartholino: Respondente Christiano Stubæo Fridericiburgense* (Copenhagen: Henrik Waldkirch, [1614]), A4r: "Quæstio III. An analogia macrocosmi & microcosmi universæ philosophiæ & medicinæ fundamentum esse queat?"

46. Ibid., B1r: "Quibus omnibus conficitur, ex analogia Macrocosmi & Microcosmi solidum & indubitatum Philosophiæ atque Medicinæ fundamentum peti nullum posse."

47. The 1651 edition, *Anatomia, ex Caspar Bartholini...Reformata* (Leiden: Franciscum Hackium, 1651), for example, is quite in the anatomical tradition of Galen and Vesalius, although Thomas Bartholin has appended Johannes Walæus's letters on chyle and the circulation of the blood, defending Harvey's discovery. Paracelsus is mentioned as one of those who judge that it is possible for a fetus to be generated outside the uterus, but Bartholin (168) notes that nobody has tested this by experiment and doubts that it is possible to create a homunculus in vitro.

48. Caspar Bartholin, *Enchiridion metaphysicum ex Philosophorum Coriphæi, Aristotelis, Optimorumque eius interpretum monumentis adornatum opera et vigiliis Caspari Bartholini Malmogii Dani*

De mixtione and *De aquis* (both 1617) he shows himself to be a conventional university-educated humanist physician. In the former, for example, he discusses differences between Aristotle and Galen, but does not discuss Paracelsus. In *De aquis* he defines water as cold and wet, lighter than earth, but heavier than air—a suitably Peripatetic description—and takes a jab at Paracelsus's quarrelsomeness.[49] Like many Peripatetics after Aristotle, Bartholin doubted that fire was really an element, but he does not follow the Paracelsians in associating cosmic firmament with terrestrial fire; rather, fire for him is something burning.[50] This agrees with his account of the heat of thermal springs, which Paracelsian authors attributed to intrinsic heat generated by spiritual agencies—Paracelsian *semina* in spa water.[51] Unlike the Paracelsians, though, Bartholin argues, as an Aristotelian would, that the water of hot springs cannot be naturally hot, since water is by nature cold, and that therefore its heat must come from an extrinsic source.

Ole Grell is quite correct in claiming that Caspar Bartholin recommended chemical drugs, but it is not clear that he viewed these as Paracelsian remedies, and therefore that we should ascribe his pharmacology and empiricism to Paracelsian precedents.[52] In fact, in his 1613 treatise mentioned above, he argued

---

(1623), which is appended to an edition of Johannes Magirus's *Physiologiæ Peripateticæ libri sex* (Frankfurt: Johan Berner, 1624) and the revised version, *Metaphysica major, ab ipso autore plurimum aucta, & scholiis illustrata*, which is appended to a 1642 edition of Magirus's work (Cambridge: R. Daniel, 1642); I have not seen his *Enchiridion physicum ex priscis & recentioribus philosophis* (Strasbourg, 1625), but I would be greatly surprised if it were Paracelsian.

49. Caspar Bartholin, *De aquis libri II: Naturam et accidentia aquæ tam in genere quam in specie; Maris, fontium, fluminum, thermarum, aliarumque aquarum mineralium succincte explicantes: Cum earundem per varia artificia tum vulgaria tum chymica probationibus* (Copenhagen: Sartorius, 1617), G3v: In discussing authors' opinions that mineral waters are mixtures of water and various kinds of metals, minerals, stones, salts, etc., Bartholin notes that "Omnes videtur carpere Paracelsus; post quem Tabernæmontanus paucis pepercit" (Paracelsus seems to slander them all, and following him, Tabernæmontanus spared few).

50. Caspar Bartholin, *Physicæ specialis præcepta,quæ thesium loco in disputationibus esse possunt, decerpt. ex prælectionibus physiologicis publicis Casp. Bartholini pro juniorum memoria: Pars prima De corporibus naturalibus simplicibus, eorumque mixtione, & mixtionem consequentibus* (Rostock: Johannem Hallervordeum, 1619), chap. 10, "De igni": "Nam Ignis *nil aliud* est, *quam* corpus aliquod incensum."

51. Edward Jorden, *Discourse of Naturall Bathes, and Minerall Waters* (1631; repr., Amsterdam: Theatrum Orbis Terrarum, 1971), for example, adapted Petrus Severinus's theory about *semina* to thermal springs.

52. Grell, "Reception of Paracelsianism," 91, says about Caspar Bartholin's recommendation of the use of chemical drugs and his acknowledgment of "the need for personal and practical experience in the field," "the importance of experimenting with alchemy," and the importance of gleaning knowledge "from old wives and barbers about their remedies": "These were opinions which resembled those held by Severinus and Pratensis and they demonstrate that a moderate Paracelsianism remained important within the medical faculty of the University of Copenhagen."

that such drugs were useful, but were conformable to Galenic medical theory. In the absence of evidence that Bartholin adopted any sort of Paracelsian therapeutic theory or chemical philosophy, it must supposed that he was a Galenist using iatrochemical preparations, which was not unusual for the times, as is discussed below in the case of Ole Worm's medicine.

The possibility that Bartholin's empiricism stems from Paracelsian demands for direct observation of nature warrants careful consideration, but one must proceed cautiously. It is certainly true that expressions of fidelity to firsthand knowledge and to the value of such "maker's knowledge," when it is gotten from the unlettered worker and used by the scholar to build doctrine, are not unknown in Danish medicine in the seventeenth century, nor are variations on the credo "Aristotle is my friend, Plato is my friend, but I love neither as much as the truth." Indeed, Petrus Severinus is perhaps best known to modern historians of science for his plea to abandon one's books and worldly possessions and take to the field in search of natural knowledge, to seek knowledge through chemical experiment, and to glean information from any who have it, regardless of their level of education or social rank.[53] Such claims for the scientific value of personal experience and practical knowledge may be a Severinian and Paracelsian legacy, but one must recognize that this empirical sentiment was not limited to Paracelsians, at least not by the turn of the century. Taken by itself, the exaltation of maker's knowledge and empiricism is no more Paracelsian than it is Baconian.[54] The question is whether these points are sufficient in Bartholin's case to sustain the claim that Grell makes for them regarding his "moderate Paracelsianism." There is no specific evidence from Bartholin's life or work that speaks to this issue, and further discussion of the relation between Danish empiricism and Paracelsian theory will be deferred to consideration of Ole Worm's methodology, below.

Caspar Bartholin was clearly interested in learning about a variety of approaches to medicine while he was a student, including Paracelsian ideas and methods. However, nothing about his teaching or publication as a professor at the University of Copenhagen suggests a sympathy for the philosophy or medicine of Paracelsus or his followers, beyond what could be readily subsumed under an Aristotelian and Galenic worldview. Indeed, where Paracelsian ideas could not be

---

53. Petrus Severinus, *Idea medicinæ philosophicæ* (Basel: Henricpetri, 1571), 73, quoted in Debus, *English Paracelsians*, 20.

54. However, it should be noted that Graham Rees, *Francis Bacon's Natural Philosophy: A New Source* (Chalfont St. Giles: British Society for the History of Science, 1975), has shown some similarity between some of Bacon's ideas and Paracelsian concepts.

reconciled with peripatetic philosophy or were in conflict with Lutheran orthodoxy, he rejected them.

OLE WORM AND PARACELSIANISM

Caspar Bartholin was an important influence on teaching at the University of Copenhagen, both in his own right, as a professor of natural philosophy, medicine, and theology, and as a close colleague of Ole Worm. But Worm, who succeeded Bartholin as junior professor of medicine when the latter moved to the theological faculty, was even more important than Caspar Bartholin in determining how natural philosophy and medicine were taught at the university in the first half of the century. Worm was not a highly visible reformer or an exceedingly original thinker, but in the course of his long career he quietly introduced changes in research methodology and pedagogy, was a vigorous correspondent and an influential teacher and mentor, and published books on many topics. But was Paracelsian chemical medicine part of the natural philosophy and medicine that he taught and practiced?

Before his appointment to the University of Copenhagen in 1613, Ole Worm was an eager student of Paracelsian medicine and traveled extensively to learn iatrochemical methods, but he grew increasingly wary of the religious implications of Paracelsianism and, specifically, of the claims of the Rosicrucians.[55] Ole Grell and I are in agreement about Worm's early interest in chemical medicine—both theory and practice—and about Worm's rejection of the Rosicrucians in 1619. But we differ in our assessment of his earlier sympathy for the Rosicrucians and of his longer-term opinion of Paracelsianism.[56] Whereas I have claimed that Worm did not support Paracelsian theory after 1619,[57] Grell has argued that Worm retained his interest in Paracelsianism after 1624, that he continued to be

---

55. Shackelford, "Rosicrucianism," 183–88, 192–99.

56. Consideration of Ole Worm's public attitude toward religion lies beyond the scope of the present study. However, according to H. D. Schepelern, *Museum Wormianum, dets forudsætninger og tilblivelse* (Aarhus: Wormianum, 1971), 44, 54–55, Ole Worm must have fallen in with an orthodox crowd while a student, since he was among the group of strict orthodox Lutherans who were studying at Marburg when Moritz decided to make the university Calvinist and, with them, followed the Lutheran faculty to the academy at Giessen, which Ludwig of Hessen-Darmstadt elevated to the status of Lutheran university in 1607. During these years Worm studied under Balthasar Mentzer, a strict Lutheran, but soon thereafter he decided to leave off study of theology and take up medicine. A thorough examination of Worm's letters and treatises, published and manuscript, may yet reveal something about his religious convictions after this period, but in the meantime there is no reason to suppose that he departed from the Lutheran orthodoxy of his friends and colleagues.

57. Shackelford, "Rosicrucianism," 198, 204.

sympathetic to Paracelsian philosophy even though he rejected the radical *religio Paracelsica*.[58]

Grell's argument that Worm continued to propound some form of Paracelsianism after his denunciation of the Rosicrucians in 1619 depends on five supporting propositions: (1) that Worm sought to publish the manuscript treatises left behind by Petrus Severinus, who was an internationally known Paracelsian, which he would presumably not have done if he had felt any qualms about their Paracelsian contents; (2) that various disputations that Worm wrote for his students to respond to bear witness to his continued subscription to Paracelsian ideas; (3) that Worm late into his career sought information about palingenesis, which he presumably would not have done if he had not still believed in it; (4) that Worm made, recommended, and administered Paracelsian chemical remedies; and (5) that his professing of empiricism is a vestige of Severinus's Paracelsian legacy. These propositions will be examined in some detail.

First, in the year following his condemnation of the Rosicrucians, Ole Worm wrote to Frederik Severinus, Petrus Severinus's only surviving son, to inform him that the king's right-hand man and administrative head of the university, Chancellor Christian Friis, wished to have the elder Severinus's manuscripts published. Ole Grell has interpreted this correspondence, two letters of which have survived, as an indication that Worm was still interested in Paracelsianism in 1620.[59] In this assessment he differs from H.D. Schepelern, the foremost Worm scholar, who thought that Worm's letters expressed a polite reluctance to publish and therefore were evidence that Worm *did not* want to be associated with Paracelsianism.[60] Schepelern's interpretation is better suited to the evidence.

In the first place, Worm said that he was writing to Frederik as a direct consequence of a recent conversation with chancellor Friis.[61] Worm understood that the Chancellor had gotten a description of these manuscripts from Frederik's sister,

---

58. Grell, "Acceptable Face of Paracelsianism," 263 n. 57, and 264.

59. Grell, "Reception of Paracelsianism," 92. Grell persists in this assessment in "The Acceptable Face of Paracelsianism," 265.

60. Schepelern, *Museum Wormianum*, 128, attributes Worm's reluctance to publish Severinus's manuscripts to his desire to avoid commitment to any particular philosophical school of thought and speculation in general. Grell, "Caspar Bartholin," 98 n. 38, disagrees with Schepelern: "That Ole Worm should have been unwilling to edit the unpublished works by Severinus in order not to be associated with Paracelsianism, as claimed by Schepelern…, is a misinterpretation of Worm's politeness towards Severinus's son, the physician, Frederik Sørensen. Worm, in fact, ends his letter to Sørensen by declaring his willingness to undertake the task if Frederik Sørensen would agree to it."

61. H.D. Schepelern, ed., *Breve fra og til Ole Worm*, (Copenhagen: Det danske sprog- og litteraturselskab, 1965–68), 1:46, letter no. 73; and Hans Gram, ed., *Olai Wormii et ad eum doctorum virorum epistolæ* (Copenhagen, 1751), 52, epistola 60. Schepelern's edition comprises Danish translations

who was married to Jonas Charisius and therefore was in the social circle of leading members of the Danish government, and he attributed the chancellor's choice of him to contact Frederik about this matter to the fact that he and Frederik were friends and were related by marriage.[62]

However, despite the fact that Worm and Frederik Severinus were on familiar terms, Worm's letter shows a lack of any personal initiative in the matter of publishing the manuscripts, beyond his obligation to fulfill the chancellor's request. He clearly stated that it was the chancellor's wish to have them published,[63] and that he, Worm, was to encourage Frederik *in the chancellor's name* to revise them and publish them at the first opportunity.[64] For this Frederik could expect both the chancellor's and the king's favor. Worm did not express any gratitude of his own.

As it happened, Worm already possessed Severinus's incomplete commentary on Aristotle's *Physics*, which he must have gotten from Frederik earlier, and when the chancellor learned of this, he personally asked him to organize, edit, and publish the text in order to ease Frederik's burden and to encourage him to publish the remaining treatises. But, despite the chancellor's wishes, Worm begged off, writing to Frederik that he really had too much of his own work to do and that he did not want to get involved in anybody else's. The result was that Worm asked Frederik to edit and publish the *Physics* commentary, too, instead of following the chancellor's specific request. However, he wrote that if Frederik really wanted him to publish the *Physics* commentary, he would do so out of friendship and in memory of the old Danish Paracelsian, but he would add a preface to the reader explaining that the edition originated *with Frederik*.[65] One could interpret Worm's words as a

---

of Worm's correspondence, some of which was published in its original Latin. I have cited the Latin letters, too, in cases where I have consulted them.

62. Ibid. Worm writes about Friis, "mihi tecum haud vulgarem intercedere familiaritatem intelligeret, dixit se percepisse" (*Epistolæ*, 52). In fact, Worm and Frederik Severinus were probably friends and were related through Worm's marriage into the family of Thomas Fincke. Ole Worm's wife and Caspar Bartholin's wife were sisters; both had a great-grandfather in common with Frederik Severinus and Anna Severinus, who was Peter Charisius's mother. For an explanation of the interrelatedness of Denmark's leading physicians at this time, see Eyvind Bastholm and Hans Skov, trans., *Petrus Severinus og hans Idea medicinae philosophicae: 'En dansk paracelsist* (Odense: Odense Universitetsforlag, 1979), 33. Frederik Severinus later moved to Copenhagen, and Worm lodged at least one of his students, Ambrosius Rhodius, at Frederik's house, suggesting a continuing friendship. Rhodius married Frederik's daughter, and Ole Worm later helped her look after her economic affairs in Copenhagen.

63. Gram, *Epistolæ*, 52: "quæ divulgata sumopere cuperet."

64. Ibid.: "quocirca jussit, tibi ut scriberem ejusque nomine adhortarer, ut iis revidendis incumberes, ac quædam ex prima occasione, typis mandares."

65. Ibid.: "a te hæc profecta esse, in epistola ad Lectorem indicabo" (I will indicate in a letter to the reader that this was carried out by you).

very polite reluctance to take any credit for the publication. Or they may indicate that although he was at one time sufficiently *interested* in Severinus's comments on Aristotle's natural philosophy to borrow the manuscript, he was unwilling to have any part in *the publication* of Severinus's manuscripts, beyond what might be specifically required of him by his superiors or obligation to his friend.

In a second letter to Frederik Severinus, Worm requested that Frederik send whatever manuscripts he had in his possession to Chancellor Friis, who he said was affected by a unique benevolence for the memory of Petrus Severinus and wanted to diligently study his manuscripts and promote his reputation. With the exception of the reconstructed list of Severinus's manuscripts that was eventually printed in Thomas Bartholin's *Cista medica hafniensis* (1662), which may well have been based on the description that Frederik's sister had provided to the chancellor, this is the last we hear of Severinus's unpublished works, so we do not know if they ever actually came under Friis's scrutiny.[66] It is hard to discern in this history any particular enthusiasm for Severinus's Paracelsian reading of Aristotle on the part of either Worm or Chancellor Friis.

As to the second argument, if Ole Worm were publicly sympathetic to Paracelsian ideas, it stands to reason that they would have left some mark on his exchanges with friends and colleagues and on his published treatises, including those that he used in teaching. Ole Grell claims that Worm's April 1621 inaugural lecture as professor of natural philosophy "confirmed his continued belief in the Severinian form of Paracelsianism,"[67] and furthermore, that "he remained an advocate of a Severinian type of Paracelsianism, as can be seen from a number of the annual disputations he presided over as professor of medicine."[68] It is worthwhile to revisit some of these documents in order to assess Worm's adherence to Paracelsianism.

In his 1621 inaugural lecture as professor of natural philosophy, Worm repeated a litany that is now, since the work of D.P. Walker and Frances Yates, quite familiar to early modern scholars as a fundamental tenet of Renaissance Neoplatonism, specifically Hermeticism: Prior to his fall and expulsion from Eden, Adam possessed perfect knowledge of nature, the *prisca sapientia*, which was embodied in the very language which he passed down to the early sages. From them, it was transmitted through an unbroken line of wise men, from Solomon to the sixteenth-century cabalists, Neopythagoreans, Neoplatonists, and

---

66. Perusal of the titles given in Thomas Bartholin, *Cista medica Hafniensis* (Copenhagen: Haubold, 1662), 130–32, suggests that at least some of the manuscripts did not treat Paracelsian topics.
67. Grell, "Reception of Paracelsianism," 93.
68. Grell, "Acceptable Face of Paracelsianism," 265.

Hermeticists.[69] Grell distills from Worm's account evidence for his "continued belief in the Severinian form of Paracelsianism."

It is true that Severinus regarded Paracelsus as a possessor of *prisca sapientia*, but it should be observed that the idea of an original wisdom that has been handed down through the ages was not limited to the Paracelsians.[70] Much more interesting is Worm's tracing of its lineage: Adam's prelapsarian knowledge passed down to Plato and Aristotle, who, Worm wrote,

> succeeded in collecting and deducing the hidden knowledge and presenting it in a comprehensible form. *Although they were heathens, we are permitted to seek the knowledge from them, which God originally gave to Man, and that is why natural philosophy helps us to understand God's omnipotence through his Creation.*[71]

Clearly, Worm has co-opted the Hermetic doctrine of the *prisca sapientia* and *used it to legitimate Aristotelian natural philosophy*, not Paracelsian chemical philosophy. Inasmuch as the Paracelsians' rejection of the authority of Aristotle's philosophy was based on its unchristian nature, this cannot be regarded as evidence that Worm was at that time sympathetic to Paracelsianism—quite the opposite; Worm was making Aristotle a source of divine knowledge!

From 1624 to 1652—spanning most of his thirty-year career as professor of medicine—Ole Worm produced a series of disputations called *Controversiarum medicarum exercitationes*. Grell is correct that Worm mentions Severinus in several of these, identifying him as a leading iatrochemist,[72] citing him for his *semina*

---

69. Part of Ole Worm's inaugural lecture is quoted, in English translation, by Grell, "Reception of Paracelsianism," 93. I have not seen the original manuscript and rely here on Grell's rendering.

70. It is clear from Petrus Severinus's treatise *Epistola Scripta Theophrasto Paracelso* (Basel, 1570?) that he regarded Paracelsus as an heir to the *prisca sapientia*. He did not have in mind a literary tradition, since he specifically lauds Paracelsus for having "received a treasury of these secrets from living and flourishing streams, and not, as we are accustomed, from dead and silent literature" (Nec dubito quin ex viventibus vigentibusque rivulis, non ex mortuis et silentibus literis, ut nos solemus, horum arcanorum thesaurum receperis) (fol. A2v), but in other respects his account is similar to the lineage that Frances Yates attributed to the *prisci theologi* in her book *Giordano Bruno and the Hermetic Tradition* (Chicago: University of Chicago Press, 1964), where she established that Renaissance Platonists and Rosicrucians from Marsilio Ficino to Robert Fludd regarded the ancient wisdom of the *prisci theologi*, principally Moses, Orpheus, and Hermes Trismegistus, as having been handed down secretly to the late Renaissance. Paracelsus fits into her narrative of the Hermetic tradition of sages and interpreters of nature extending down through time, but in general the Paracelsians do not figure as prominently as Ficino, Pico della Mirandola, Cornelius Agrippa von Nettesheim, and others whom we cannot consider to have been Paracelsians.

71. Grell, "Reception of Paracelsianism," 93 (Grell's translation; emphasis added).

72. Ole Worm, *Controversiarum medicarum exercitatio II* (Copenhagen, 1626), Controversia 2.

doctrine,[73] and listing him among those who equated innate moisture with spirit.[74] However, these few references cannot be construed as support for Severinus's brand of Paracelsianism.

Worm did laud Severinus as an *iatrochemist*, that is, as a physician who used chemically prepared drugs, but without legitimizing the theoretical basis for his drug therapy. If one considers Worm's profession of natural philosophy and medical theory at large, it is readily evident that these scattered references are embedded in a Galenic and Aristotelian context. For example, Worm cited Severinus's *semina* theory of disease causation in the first controversy of the 1644 *exercitatio*, titled "Is disease a hindrance against nature?" The title itself indicates that the question to be disputed is framed in the traditional Galenic way. Although Worm quoted Severinus's realist position, that disease categories should be created to describe the diseases—the diseases should not be forced to conform to *a priori* categories (which Severinus meant as an attack on scholastic Galenism)—Worm concluded the controversy by affirming the Galenic definition of disease as contranatural and as a hindered function. The very point of Severinus's *semina* theory was to define disease as having a positive ontological existence, which cannot be reconciled with Worm's conclusion that disease is a deficiency or privation.

Worm more directly criticized Paracelsian natural philosophy in a treatise composed just three years after his denunciation of the Rosicrucians.[75] In question 3 of this treatise, Worm rejected vital philosophy, mentioning Severinus specifically as a source. Vital philosophy is the very soul of Severinus's doctrine and was the particular subject of a treatise drafted by Johann Hartmann, who had readily acknowledged the influence of Severinus on his ideas. Worm had studied this treatise very carefully, and it probably did not escape his notice that Andreas Libavius had done so, too. Libavius wrote a tract specifically attacking it, thereby ranking Hartmann and Severinus among other Paracelsians, Rosicrucians, and empirics whose ideas were skewered by the vigorous anti-Paracelsian critic. Part of Libavius's agenda was to protect academic education from the neoteric, heretical currents that were threatening to undermine Aristotelian philosophy. Ole Worm had a similar commitment to academic tradition and weighed in on the side of Aristotelianism and

---

73. Ole Worm, *Controversiarum medicarum exercitatio XII* (Copenhagen, 1644), Controversia 1. Summarized in Ejnar Hovesen, *Lægen Ole Worm 1588–1654: En medicinhistorisk undersøgelse og vurdering* (Aarhus: Aarhus Universitetsforlag, 1987), 185.

74. Ole Worm, *Controversiarum medicarum exercitatio IV* (Copenhagen, 1630), Controversia 1. Summarized in Hovesen, *Lægen Ole Worm*, 165.

75. For my analysis of questions 3, 4, and 9 of this treatise, see Shackelford, "Rosicrucianism," 195–98.

traditional pedagogy after Libavius and like-minded critics had polarized the debate.

Question 9 of Worm's 1622 treatise reveals his handling of Severinus's ideas. Here he seems at first glance to adopt the Paracelsian *tria prima*, the idea that three spiritual principles (salt, sulphur, mercury) underlie material reality. This would be an important indication of Worm's sympathy for Paracelsian philosophy, were it not for the fact that he focused on a concessionist passage from Severinus's *Idea medicinæ*, where Severinus compares the *tria prima* to Aristotle's three principles—form, matter, and privation—and Plato's three principles—god, exemplar, and matter. Clearly this is not what Paracelsus or the Paracelsians had in mind. Even Severinus had rejected privation as a basis for ontology. Worm's analysis effectively removed the *tria prima* from chemical medicine and reduced it to an insignificant adjunct to Peripatetic philosophy. He was willing to entertain the Paracelsian idea only insofar as it could be subjected to Aristotle.

Something similar is found in the *exercitatio* for 1626, where Worm quotes Severinus to establish the chemists' view of the four terrestrial elements as "wombs" or passive material locations where bodies are constructed by the active spiritual principles. But Worm concludes that elements are in fact simple homogenous bodies, from which sensible bodies are composed, and into which such bodies are ultimately resolved. This is not Paracelsian matter theory.[76]

Casting a wider net and sampling Worm's literary production more generally, he is found presenting and affirming Galenic humoral pathology and Galenic therapeutics and lauding the advances of modern anatomists, while criticizing Paracelsian diagnosis by means of the "anatomy" of the urine.[77] Where he was critical of medieval Aristotelian theory, he did not adopt Paracelsian principles, but rather adopted Hippocratic and Galenic correctives and, increasingly, the methodological reforms associated with natural history, collection, and experimentation.

Worm's attitude toward Paracelsian theory as revealed in his published works is corroborated by his surviving letters. Scattered references to chemical study and chemical drugs in his voluminous correspondence affirm that Worm's interest in chemical drugs continued throughout his career. These references indicate that

---

76. Worm, *Controversiarum medicarum exercitatio II*, Controversia 2, section 3, identifies Severinus as a leader (*agmen*) among the *chymici* and introduces Severinus's theory of the elements as matrices or wombs, but then goes on to criticize this Paracelsian view. In the end, Worm rejects the Paracelsian notion: "Elementa sunt corpora simplicia homogenea, ex quibus corpus humanum primum componitur, & in quæ ultimo resolvitur" (Elements are simple homogenous bodies, from which the human body is initially constructed, and into which it is, in the end, resolved).

77. Ole Worm, *Controversiarum medicarum exercitatio XV* (Copenhagen, 1648), found Paracelsian distillation of urine to be humbug. See Hovesen, *Lægen Ole Worm*, 196.

Worm was familiar with standard laboratory procedures, such as distillation and fusion, and was conversant with major written sources of such knowledge.[78] Apparently Worm's friends also understood him to command these skills, if a letter from Arngrim Jonsson in Iceland is any indication. Arngrim sent him a hard mineral of some sort with the expectation that he could analyze it chemically if needed.[79] But none of this is clear evidence that Worm himself continued to work at the furnace after his return to Copenhagen. He indicated in two letters from his later years that he had *in the past* made chemical preparations,[80] and there are references to specific substances—cream of tartar, which he recommended to a colleague for constipation in 1643, and Johann Hartmann's oil of tartar, which he said did not work.[81] But there is nothing to indicate that he continued to prepare drugs or dabble in chemistry in Copenhagen, and nothing pointing to anything Paracelsian, except possibly one drug recommendation, considered below.

Worm certainly advised his students to study iatrochemistry, but these references pose another interpretive problem. Why would he urge these students to study chemistry in Paris, which he commended to the son of a colleague as the best place to learn chemical methods, if he or anyone else were teaching it in Copenhagen?[82] In a letter to his own son Willum Worm he criticized the boy for wasting his time in Leyden studying Greek and Hebrew and reading Plautus—subjects easily learned at home—when he could be studying chemistry, botany, and medicine, which was the purpose of his travel abroad in the first place.[83] Taken together, these references in Worm's treatises and correspondence do support, albeit weakly, Worm's lifelong interest in chemical medicine, but not the conclusion that this was Paracelsian in nature.

---

78. For example, in 1653 he commented in a letter to Johann Daniel Horst that the latter's report of a mineral that produced exhalations capable of melting gold did not strike him as unusual, since he knew that exhalations of even common salt could do that (Schepelern, *Breve*, 3:482, letter 1733). Several years earlier, in a report to a French correspondent about that delightful Baltic gem, amber, Worm noted that chemical analysis supported the hypothesis that it was like pitch (ibid., 3:153, letter 1385). In a 1641 letter to his countryman Steffen Hansen Stephanius, Worm cited Martin Ruland's *Lexicon Alchymiæ* and even Geber on how to prepare copper in a crucible (ibid., 2:307, letter 987). Worm's summary of a letter he sent to Marcus Christensen Humble in 1641 (ibid., 2:285, letter 961) indicates that he had borrowed and was now returning a book on chemical experiments. This shows that Worm was at least interested in keeping abreast of the literature on laboratory methods.
79. Ibid., 2:198–200, letter 859.
80. Ibid., 3:238, letter 1483, and 3:321, letter 1589.
81. Ibid., 2:480, letter 1161, and 1:38, letter 55.
82. Ibid., 3:259, letter 1518.
83. Ibid., 3:489, letter 1745.

The third point that Grell brings to support his argument is Ole Worm's interest in palingenesis, which is suggested by certain letters written toward the end of his career. Grell quotes Worm's letter to Rasmus Bartholin (1648), in which Worm explains that he is looking for an ash or salt that Quercetanus once reported secondhand, which produced visible but spiritual plants with lifelike colors when heated, but returned to ash when cooled again. Grell interprets this as evidence that "[Worm's] own research continued to incorporate Paracelsian elements."[84] While palingenesis may have been of interest to Paracelsians and might be a Paracelsian concept, inasmuch as it presupposes that real objects have a corresponding spiritual or astral existence that dwells within their elemental matter, it is not at all clear from this letter that Worm was convinced of the truth of Quercetanus's account or that he interpreted the alleged phenomenon within a Paracelsian theoretical framework.

Looking at how this curiosity fits into Worm's correspondence more generally, it seems that he was looking for a *specimen* of this spiritual plant to add to his museum collection, and that he and his correspondents had failed to prepare it from the recipes he had.[85] He told one correspondent, whom he advised to try all three recipes concurrently, that he was especially interested in obtaining the spiritual plant because he was engaged in an argument with a friend, who denied the possibility.[86] This puts Worm in the position of trying to test a phenomenon reported both in published books and in personal correspondence; unlike his friend, he was not willing to dismiss what was reported by credible colleagues, but yet he required evidence to confirm what he had read. This pattern is repeated during Worm's career. It is a methodology that is perhaps implicated in the rise of experimental science, but also one that is historically located in medical

---

84. Grell, " Reception of Paracelsianism," 93.

85. Rasmus Bartholin knew that Worm was looking for a specimen of this spiritual plant, and he understood that Joel Langlot, physician to the duke of Gottorp, knew its secret. Rasmus had asked Langlot to send a specimen to Worm for his museum (Schepelern, *Breve*, 3:309, letter 1572). In his response to Rasmus, Worm said that he already had two recipes for preparing such a salt but that they had not proven successful. He wanted to know if Rasmus had actually seen Langlot demonstrate the palingenesis described by Quercetanus. He specified that he did not want a description of the recipe, but wanted to know if Rasmus had seen the plant (ibid., 3:310, letter 1575). Meanwhile, he was inquiring about it from others.

86. Ibid., 3:313, letter 1578. Two more letters, both from 1653, the year before Worm died, attest to his persistence in trying to locate a specimen of the ashes in order to produce the spiritual plant by palingenesis (ibid., 3:483, letter 1734, and 3:490, letter 1747). Still trying to track down a certain Mr. de Clave, who was reported to demonstrate the spiritual plant whenever anyone asked, and still searching the literature for descriptions and procedures, Worm advised Joel Langlot to try simultaneously at least three of the recipes, each in its own vessel.

empiricism.[87] Although placing one's faith in observation is often credited to the Paracelsians, and rightly so, it is also a feature of humanist medicine, which succeeded in finding Galen's own empiricism from original texts that did not figure prominently in medieval scholastic medicine. In Worm's case, Galen appears to have been a more cogent influence than Paracelsus.

Methodology aside, there is still the matter of whether Worm's interest in palingenesis belies an ongoing adherence to Paracelsian theory. After all, palingenesis was a demonstration of Paracelsian theories about the existence of spiritual bodies. However, the phenomenon, beyond being a curiosity in its own right, and therefore attractive to a collector, can also be viewed in an Aristotelian light, as a demonstration of the Aristotelian idea that forms inhere in matter independent of their corporeal expressions,[88] which is an explanation for resurrection. Lacking any clear intellectual context for Worm's curiosity about palingenesis, the conclusion that it is an expression of Paracelsianism is not justified.

A fourth point of Ole Grell's argument is that Worm remained sympathetic to Paracelsianism because he "kept a manuscript book" of Paracelsian chemical medicine, recommended that his students study under Paracelsian chemists, and "continued to prescribe Paracelsian drugs throughout his life."[89] It is clear that Worm accepted the medical efficacy of chemically prepared medicines and incorporated them into his practice, and it only makes sense that he would have conveyed this to his students, just as Caspar Bartholin had to his sons and Peter Charisius. However, the evidence is not strong or unambiguous and does not prove that Worm's interest in iatrochemistry after he returned to Copenhagen and took a position at the university extended to Paracelsian chemistry, since it was not uncommon for seventeenth-century Galenists to accommodate chemical medicaments into a thoroughly Galenic theory of health, disease, and therapeutics.

Indeed, Worm kept chemical manuscripts which were later bound into two volumes.[90] These have yet to be carefully studied and offer several problems for historical interpretation. In the first place, although they are mostly written in Ole Worm's hand, the identity of the texts has not been hitherto determined, with the exception of one labeled as Duncan Burnett's *De præparatione et compositione medicamentorum chymicorum artificiosa Duncani Bornetti Scoti* (On the

---

87. I have argued this point in Jole Shackelford, "Documenting the Factual and the Artifactual: Ole Worm and Public Knowledge," *Endeavour* 23 (1999): 65–71, at 69.

88. This is the interpretation given by Schepelern, *Museum Wormianum*, 189.

89. Grell, "Acceptable Face of Paracelsianism," 266.

90. MS. Rostgaard 33 8° and 43 8°, Det Kongelige Bibliothek, Copenhagen.

artificial preparation and compounding of chemical medicines).[91] Nor is it clear when they were written down, where, or in what order. Two other texts were originally written by Johann Hartmann, suggesting that Worm may have copied them out when he was Hartmann's student in Marburg. However, one of the texts is a lengthy *criticism* of "Hermetic medicine," suggesting that Worm did not uncritically follow Hartmann when it came to theory. The manuscripts as a group are at best an ambiguous indication of the nature of Worm's iatrochemical interests and have not been fixed to any time or place. It seems that all or nearly all of them date to Worm's student years, but this has yet to be proven.

Likewise, Worm's recommendation of study under the Paracelsian chemist Guy de la Brosse is a weak indicator of Paracelsianism.[92] Guy de la Brosse was indeed an iatrochemist and leaned decidedly toward Paracelsianism, incorporating Severinus's theory into his published work. He was, moreover, by the time of his death in 1641, probably the best-known teacher of laboratory chemistry in France, perhaps in Europe, having at his disposal the facilities of the Jardin des Plantes in Paris and the support of the French crown. But Worm's recommendation must be evaluated in light of another recommendation to a student not to waste his money learning chemistry from the Paracelsian Hartmann, when he could learn what he needed—that is, the practical methods and not the Paracelsian theory—from any good pharmacist who was skilled at preparing chemical drugs.[93] Worm's recommendation of Guy de la Brosse, seen in this light, pertains to iatrochemistry generally, and not necessarily to Paracelsian iatrochemistry. In order to try to bridge this gap, Grell points to Worm's personal prescription of a chemical drug that he attributes to Severinus as evidence that Worm continued to practice the Dane's Paracelsian medicine. Leaving aside the fact that Severinus was reported to have practiced a mixture of Paracelsian and Galenic therapy,[94] Grell's evidence also has a problem that needs to be ironed out before it can testify to Worm's iatrochemical practice. He cites a letter that Worm wrote to Jens Dinesen Jersin with regard to a drug that he was sending for the treatment of Jersin's

---

91. For a list of the titles of the sections of Rostgaard 33, see Shackelford, "Rosicrucianism," 185 n.14.

92. Grell, "Acceptable Face of Paracelsianism," 266.

93. On Guy de la Brosse and Severinus's *Idea medicinæ*, see Shackelford, "Paracelsianism in Denmark and Norway," 151–59. In 1616, Worm wrote to Niels Christensen Foss advising him not to study with Hartmann. Schepelern, *Breve*, 1:9, letter 16.

94. Johannes Paludanus wrote to Henricus Smetius about Severinus, 3 June 1605: "Medicamentis Paracelsicis non semper est usus, verum & compositionibus Galenicis sæpe: Sed extremis morbis extrema adhibebat remedia" (He did not always use Paracelsian medicines, and in fact often [used] the Galenic compounds. But in extreme diseases he applied extreme remedies). This letter was published in Bartholin, *Cista medica*, 127–29.

wife, which Grell has identified as "highly likely" the same prescription as an arcanum used by Severinus.[95] The Latin original for this letter was not available, but the Danish translation, which Grell cites, does not support his conclusion that Worm prepared the drugs himself.[96] Nor is there any other evidence available that Worm had the necessary laboratory equipment to undertake elaborate chemical procedures. Without such evidence, it is safer to suppose that he obtained whatever spagyric drugs he needed from one of several Danish apothecaries, perhaps from the court apothecary, or even from the court Paracelsian chemist, Peter Payngk. One may believe that Worm, like Bartholin, recommended and used iatrochemical medicines, but there is as yet no persuasive evidence that he used them according to Paracelsian therapeutic principles.

This leads to the fifth and final point. Probably the most modern-sounding aspects of Ole Worm's practice and teaching are his perseverance in collecting natural and artificial objects, his subjecting of reported claims to the test of experience, his use of natural objects in teaching, and even his occasional recourse to experiment. This empirical methodology is striking and has rightly attracted historians' attention as a step in the direction of establishing a new basis for scientific research. It is also true that Paracelsians, Severinus included, have long been recognized as important agents in calling for empiricism in philosophy and medicine. Perhaps this juxtaposition encouraged Ole Grell to draw the conclusion that "Ole Worm continued the empirical tradition begun by Peter Severinus,"[97] but it does not necessarily warrant the conclusion that Worm's empiricism was derived from Paracelsian sources.

Worm's empiricism was not mainly inspired by the Paracelsian tradition, but by academic natural history.[98] The systematic collection, preservation, and classification of natural objects was well under way in the nascent museums of Italy that Worm visited and in the research at Basel by professors under whom Worm and his colleague Caspar Bartholin had studied, notably Felix Platter and Caspar Bauhin. The Paracelsians were experimenters and sometimes incorporated the reports of laymen, but they were not systematic collectors. Worm's empiricism was accumulative and vested a certain amount of authority in the collected object, instead of

---

95. Grell, "Acceptable Face of Paracelsianism," 266.

96. Schepelern, *Breve*, 1:91, letter 166, is a very short abstract of the lost original. It indicates that Worm is sending pills for Jersin's wife's uterine problems—these are chemically prepared. But unless there is an error in the Danish translation, Grell's conclusion is wrong: Worm says that the pills are made by "din egen Kemi," indicating *Jersin's chemistry*, and not "min egen Kemi," his own.

97. Grell, "Acceptable Face of Paracelsianism," 265.

98. Shackelford, "Documenting the Factual and Artifactual," 68.

focusing only on the macrocosmic-microcosmic relationship or harmony that it might illustrate. His empiricism was in this sense more traditional, inasmuch as he sought to accumulate simple facts about what worked and what did not, what was objectively real and what was mere fable. In the preface to his last work, which presented his collection to the learned world, Worm hinted at the changes that his pedagogy had undergone during his years at the university. It was becoming increasingly empirical. He cited two authorities, the ancient Galen and the neoteric Gassendi, not Paracelsus or a Paracelsian.[99] Whatever he thought about the transmission of *prisca sapientia* down through the ages, it is clear that for natural history, he believed one should dissect and observe nature and learn from the objects themselves, as Galen had done.

■ ■ ■

How, then, shall Paracelsianism in seventeenth-century Denmark be defined in a way that has any useful meaning? Grell's argument for the continuing interest in Paracelsianism at the University of Copenhagen is tenable only if Paracelsianism is defined in terms that exclude theoretical aspects that were antithetical to Aristotelian philosophy or had heterodox religious implications. Such a Paracelsianism would be characterized by (1) a natural philosophy that was divested of sectarian religious implications and sufficiently altered to be compatible with Aristotelian philosophy or (2) an iatrochemical practice largely stripped of Paracelsian theoretical basis and adapted to Galenic theory.

Caspar Bartholin and Ole Worm both entertained Paracelsian concepts only when they were integrated into an Aristotelian worldview, subordinated to Aristotelian philosophical principles, and compatible with Galenic theory and practice. Can a Paracelsianism that was so denatured as to be fully compatible with medical and philosophical schools that were antithetical to Paracelsus and to those of his followers who identified themselves with his teachings (such as Severinus and Pratensis) be considered really Paracelsian? It seems not; to define Paracelsianism in such a way would render the term useless as a tool for historical understanding. Even the "acceptable face" that Severinus had put on Paracelsus's theory could not be stretched indefinitely, to accommodate the needs of the Aristotelian Bartholin and Worm. Only bits remained—such as aspects of the Paracelsian *semina* theory—and these were alienated from their original Paracelsian theoretical context and made to serve Peripatetic purposes.

---

99. See Shackelford, "Documenting the Factual and the Artifactual," 69.

In order to give Paracelsianism a clear identity, one useful for understanding the dynamics of early-seventeenth-century social and intellectual issues and concerns, which sometimes embraced the ideas and terminology of sixteenth-century Paracelsian medical philosophy, it must be defined as an ideology. Using various theoretical and practical expressions as indicators of this ideology, as outlined above, this paper argues that Paracelsianism, defined as something more than the use and recommendation of chemical drugs, was indeed rejected by Caspar Bartholin and Ole Worm after the critical years around the beginning of the Thirty Years' War. Finally, as stated above, Paracelsianism had a presence at the court of Christian IV, but it was eschewed, even actively denied in some cases, at the University of Copenhagen, especially by Ole Worm. Furthermore, arguments that Worm and Bartholin had a voice in attempting to reshape academic education there on the basis of Paracelsianism are unconvincing because of their denial of characteristic Paracelsian ideas and because the direction of the reforms they were supporting can be better accounted for by other academic stimuli, particularly from neo-Aristotelian, Galenic, and natural-historical schools as well as from the philosophical criticisms of Gassendi.

Finally, to address a challenge posed by Grell: How can one define Severinus, whose "acceptable face" of Paracelsianism was more or less devoid of radical religious theory, as a Paracelsian, and yet draw attention to the essential connections between Paracelsian natural philosophy and Paracelsian religion in the early seventeenth century? Any definition of "Paracelsian" must have a consistent ideological core content and yet still be flexible enough to apply to the evolving tapestry of intellectual history. Ideological concerns shift with time and political circumstance, and different issues emerge as focal points of debate and touchstones of belief. As Carlos Gilly points out, the very early Paracelsians, which would include Severinus, tended to avoid discussing theology and therefore the theological consequences of Paracelsus's theory, or else to distance themselves from Paracelsian theology by emphasizing Platonic or Neoplatonic interpretations.[100] Severinus fits both of these categories, inasmuch as he did not tread far onto theological ground in the *Idea medicinæ* and definitely emphasized a Neoplatonic intepretation of Paracelsus's metaphysics. Nevertheless, at one point Severinus's doctrine did touch on the theology of creation and the origins of disease, and for this he was attacked as a Manichaean by Thomas Erastus, and probably others.[101] Moreover,

---

100. Carlos Gilly, "'*Theophrastia Sancta*'—Paracelsianism as a Religion, in Conflict with the Established Churches," in *Paracelsus*, ed. Grell, 157–58.

101. Jole Shackelford, "Early Reception of Paracelsian Theory: Severinus and Erastus," *Sixteenth Century Journal* 26 (1995): 123–35.

this concern with the theological implications of Severinus's Paracelsian account of the historical origin of disease and his theory of the ontology of disease as a metaphysical form or class outlived the author and was still sufficiently cogent for Ambrosius Rhodius to defend it in a 1643 treatise.[102] So Gilly's observation must not be interpreted to mean that early Paracelsianism was inherently devoid of religious consequences, which it obviously was not, but rather that early Paracelsian medical writers, such as Severinus, did not see any need or at least did not venture to develop those consequences in print. They were not ready to go to press with a unified Paracelsian natural philosophy and religion of the sort found in the "Rosicrucian" period of the next generation.[103] It was, however, apparently very important to later writers to make such a linkage, to merge Paracelsian philosophy with Paracelsian religion into a coherent worldview. The definition of Paracelsian must be sufficiently nuanced to span this shift in historical circumstances, if the term is to be used as a tool for understanding the continuities and discontinuities that give the period meaning today.

The reception and interpretation of Severinus's Paracelsianism is a case in point. Certainly some longer-term echos of the ideas of Paracelsus and Severinus bear witness to the lasting if limited influence of the *Idea medicinæ*'s "acceptable face," specifically Severinus's formulation of *semina* theory. But it is possible, even likely, that some of these influences came from outside Danish academic circles. Clearly the *Idea medicinæ* itself, as a historically detached text, continued to influence Danish medical scholarship into the early nineteenth century.[104] However, when Ambrosius Rhodius, already educated in medicine under Daniel Sennert and others, came under the tutelage of Ole Worm in the 1630s, perhaps aspiring to an M.D. in Copenhagen, his new mentor apparently did not recommend that he study Severinus. It was only after Rhodius came under the roof of Frederik Severinus and married his daughter that his attention turned toward the great Danish

---

102. Ambrosius Rhodius, *Disputationes supra Ideam medicinæ philosophicæ Petri Severini Dani Philosophi & Friderici II Daniæ & Septentrionalis Regis Achiatri olim felicissimi, quibus loca illius libri obscura & difficilia illustrantur, adversarij refutantur, & multi discursus ex intimis naturæ adytis deprompti moventur* (Copenhagen: Sartorius, 1643). I have discussed Rhodius's Paracelsianism in Shackelford, "Paracelsianism in Denmark and Norway," 295–336; and idem, "A Reappraisal of Anna Rhodius: Religious Enthusiasm and Social Unrest in Seventeenth-Century Christiania, Norway," *Scandinavian Studies* 65 (1993): 349–89.

103. I have argued that such a desire for a unified ideology lay behind Kort Aslakssøn's work in Shackelford, "Unification and the Chemistry of the Reformation."

104. For example, we see the longevity of Severinus's ideas in Mauritius Petri Køning, *Dissertatio de rerum principiis et mechanica seminum liturgia* (Copenhagen: Godicchenius, 1663); and Andreas Fredericus Bremer, *Dissertationis de vita et opinionibus Theophrasti Paracelsi particula posterior* (Copenhagen: Trier, 1836).

Paracelsian's book. When Rhodius finally queried Worm about Severinus's ideas—after he had left Copenhagen for Norway—Worm was dismissive, referring his former protégé to the less Paracelsian Horst.[105] Arguably, Severinus's theory was not kept alive and passed on in this instance by Worm and other members of the medical faculty. They had long ago rejected any parts of its "acceptable face" that contradicted Aristotle, Galen, and the Book of Concord and had plucked other parts, such as Severinus's theory of seminal contagion, from their Paracelsian matrix.

Acknowledgments: I wish to thank Frankie Shackelford and Charles Gunnoe for reading and commenting on drafts of this paper.

---

105. See Shackelford, "Rosicrucianism," 184 n. 11.

"A Spedie Reformation"

Barber-Surgeons, Anatomization, and the Reformation of Medicine in Tudor London

Lynda Payne

During the late sixteenth century two generations of ambitious and articulate surgeons came to exercise increasing influence within the newly combined Barber-Surgeons' Company of London (1540). Against the shifting background of the Tudor Reformation and widespread anxiety about health in the capital, learned surgeons called for their own reformation in the "decayed and ruined" art of medicine.[1]

The foundations upon which the leading surgeons of the United Company of Barber-Surgeons erected their medical reformation were classical learning and Christian conduct. In anatomy they modeled themselves upon Galen of Pergamum (ca. 130–200 C.E.), whose surgical works had been largely recovered and published by the 1540s, while in surgical practice they principally followed the ethical guidelines of Aulus Cornelius Celsus (ca. 14–37 C.E.). The nascent community of learned surgeons cultivated and publicized an image of themselves as godly in their manners and bearing. They sought to instill a Christian reverence for the body during dissections at the Barber-Surgeons' Hall and promoted themselves as protectors of the gullible public from quacks and unskillful surgeons alike. Unlike some English Paracelsians such as Richard Bostocke who viewed their reformation of medicine as entailing the necessary overthrow of the ancient pagan medical authorities, the newly incorporated Company of Barber-Surgeons looked to the same authoritative sources as the university-trained physicians.

In practice the medical reformation was carried out on three fronts. First, prominent surgeons in the Company enacted strict regulations concerning the

---

1. John Read, *A Most Excellent and Compendious Method of Curing Woundes in the Head* (London, 1588), epistle 3.

frequency and contents of the anatomical and surgical demonstrations to be staged for the benefit of the apprentices and freemen. Second, learned surgeon-authors publicized their desire to reintegrate physic and surgery in numerous vernacular works and poured scorn upon the university-trained physicians for opposing them in this mission. They characterized their detractors as divorced from the realities of contemporary medical practices. Finally, armed with specialized knowledge from their extensive anatomical training, learned surgeons described their unique abilities to unmask and punish quacks.

As anatomical training entailed the formation of values as well as skills, the process of attempting to reform medicine affected learned surgeons' relations with their patients. Throughout their writings they scattered numerous claims to an anatomical ability, derived from their learning and constant experience of the workings of the body, literally to see through a patient to the root cause of their sufferings. Equipped with this (in)sight, or *autopsia*, London surgeons demonstrated an increasing desire and confidence to intervene in their patients' bodies.[2]

In 1540, through an Act of Parliament and the Royal Assent of Henry VIII (32 Henry viii. cap. 12), the unincorporated fraternity of surgeons in London united with that of the Company of Barbers and thus became the largest livery company in England—the United Company of Barber-Surgeons. The factors leading to this formal amalgamation have been variously interpreted by historians of medicine.[3] Both the Company of Barbers and the fraternity of surgeons had originated prior to the fourteenth century and had at various times claimed the right to regulate the practice of surgery in the city of London. However, of the two bodies united in 1540, the fraternity of surgeons was by far the smaller, consisting of less than twenty members. Typically these were surgeons attached to the households of the higher nobility, with significant military experience from having accompanied their patrons into battle during the Hundred Years War and the Wars of the Roses.

---

2. *Autopsia* was literally seeing for oneself with one's own eyes. I acknowledge my debt here to Andrew Wear for drawing my attention to the importance of autopsia as both experience and observation. See "William Harvey and the 'Way of the Anatomists,'" *History of Science* 21 (September 1983): 223–49. Also Anthony Pagden, *European Encounters with the New World* (New Haven: Yale University Press, 1993), 91; and Shigehisa Kuriyama, *The Expressiveness of the Body and the Divergence of Greek and Chinese Medicine* (New York: Zone Books, 1999).

3. See Jesse Dobson and R. Milnes Walker, *Barbers and Barber-Surgeons of London* (Oxford: Alden Press, 1979), particularly chap. 3; D'Arcy Power, *Selected Writings* (Oxford: Clarendon Press, 1931), particularly chap. 4; and D'Arcy Power, ed., *Memorials of the Craft of Surgery in England* (London: Cassell, 1886). Also see Margaret Pelling, "Appearance and Reality: Barber-Surgeons, the Body and Disease," in *London 1500–1700: The Making of a City*, ed. A. L. Beier and Roger Finlay (London: Longmans, 1985), 82–112; and Zachary Cope, *History of the Royal College of Surgeons* (London: Anthony Blond, 1959), 2–3.

The fraternity enjoyed certain privileges from the city of London, which, to a large extent, put the surgeons on par with the physicians of the capital. Although not university graduates, they were entitled to wear the academic cap and gown, be addressed as "Master," and identify and report unskillful surgeons within seven miles of the capital to the Court of Aldermen. These were significant distinctions because, while a physician could practice surgery, a surgeon by tradition could not act as a physician. Given their apprenticeship training and lack of university education, surgeons were viewed as being capable of dealing only with external or clearly visible injuries and diseases. Surgeons, therefore, performed amputations and cutting for the stone, applied ointments to the skin, and set broken bones. In contrast, all internal bodily matters were claimed as the province of the physician. Education in Aristotelian natural philosophy and knowledge of diseases from the writings of Hippocrates and Galen allowed him to make visible the invisible through the art of diagnosis, and treat and prescribe for internal disorders.

While the fraternity of surgeons was treated well by the Mayor and Corporation of London, the Company of Barbers supplied their economic support. Most likely this was due to the Company's large membership of eight hundred or so, and the involvement of the barber-surgeons in the life of the capital. In particular, as superintendents of baths they were ideally situated to report carriers of contagious diseases to the municipal authorities.[4]

The College of Physicians, consisting largely of medical humanists and royal physicians, was chartered in 1518 to promote medical learning, much along the lines of similar continental organizations. In its early years it appears to have met at the home of its president, Thomas Linacre. The College had been formed during a major outbreak of plague, which saw the first issuance of crown plague orders by an English government. Margaret Pelling has pointed to the overall sense of crisis surrounding the chartering of medical organizations, arguing that the Tudor period saw an escalation of urban decay and a decline in public health:[5]

> Above all London represented mobility, change and instability.... Urban living raised an already high level of anxiety about health to the point of obsession, and prompted an increasing resort to medical practitioners. In essence a sense of crisis in respect of health in this period lay behind the introduction of the ecclesiastical licensing system in 1511, the foundation

---

4. As Carole Rawcliffe points out in *Medicine and Society in Later Mediaeval England* (Stroud: Sutton, 1995), 133–35, this rivalry was not typical of other cities in England. On this point also see Pelling, "Appearance and Reality," 94–95.

5. See Harold J. Cook, *The Decline of the Old Medical Regime in Stuart London* (Ithaca: Cornell University Press, 1986), particularly chap. 2.

of the College of Physicians in 1518, the uniting of the Barbers' and Surgeons' Companies in 1540, the refoundation of a few hospitals following the dissolution of the monasteries, and the emergence of the Apothecaries' Society in 1617.[6]

That the United Company of Barber-Surgeons, specifically, may have viewed its founding as a response to a need for urgent action in the face of medical decay and ruin is supported by the analysis of a painting commissioned by the Company to celebrate the union of the barbers and surgeons in 1540.

Hans Holbein the Younger portrayed two court physicians, John Chambre and William Butts, kneeling with the royal apothecary, Thomas Alsop, on the right before the king. On the king's left are a large group of supplicants composed of the royal surgeons, Thomas Vicary, Sir John Ayliffe, and Richard Ferris, and the king's barbers, Nicholas Simpson, John Pen, and Edmund Harman, behind them. Also represented is a miscellaneous group of surgeons and barber-surgeons, including Nicholas Alcocke and James Monforde. The king is shown seated, holding the sword of state in his right hand and the Charter of Incorporation in his left.[7]

Traditionally the painting has been analyzed only from the viewpoints of what it suggests regarding the attitude of the College of Physicians to the union of the barbers and surgeons in 1540, and the presentation of Henry VIII. The inference is that the presence of Chambre and Butts in the portrait demonstrated that the College of Physicians not only approved of the union but may even have instigated it.[8] However, scant attention has been paid to what the painting as a whole tells of what the leading surgeons of the newly combined Company saw as their mission.

While the medical community of physicians, surgeons, barbers, and apothecaries is shown united under the imposing figure of the king, the two physicians and the sole apothecary, with their downcast expressions, appear marginalized. Their gazes are distant, in contrast to those of the surgeons and barbers, who look eagerly toward the king as he hands the Charter to Thomas Vicary, much as God created Adam to enact his mission on earth. This sense of being given the task of healing the decaying and ruined art of medicine as their monarch had undertaken the task of purifying and healing the religion of his kingdom is reinforced by the

---

6. Pelling, "Appearance and Reality," 82.

7. For further details on the painting, see Bertram Cohen, "King Henry VIII and the Barber Surgeons: The Story of the Holbein Cartoon and A Tale of Two Paintings," Vicary Lecture (London: Royal College of Surgeons, 1980).

8. See George Clark, *A History of the Royal College of Physicians of London*, 2 vols. (Oxford: Clarendon Press, 1964–1966), 1:200–3.

King Henry VIII and the Barber-Surgeons at the time of the Union (1540) between the Company of Barbers and the Guild of Surgeons. Painting by Hans Holbein the Younger (1497–1543), courtesy of the Worshipful Company of Barbers, London, where it currently hangs in their Great Hall. Oak panels. 305 cm x 185 cm.

Latin dedication inscribed on the cartouche, glorifying the king as "defender of the Faith, and next to Christ, Supreme Head of the Church of England and Ireland." Below, verses in praise of the king cast him in the role of "Christus-medicus," or Christ the Physician:

> Sadder than ever had the plague profaned the land of the English, harassing men's minds and besetting their bodies; god, then from on high pitifully regarding so notable a mortality, bade thee undertake the office of a good Physician. The light of the Gospel flies around about thee on glowing wings; that will be a remedy for the mind diseased, and by thy counsel men study the monuments of Galen; and every disease is expelled by speedy aid.[9]

The cartouche highlights the Company's right to study, as physicians did, the newly translated writings or "monuments of Galen," and to apply this learning to

---

9. Jonathon Sawday, *The Body Emblazoned: Dissection and the Human Body in Renaissance Culture* (London: Routledge, 1995), 190. Sawday, too, focuses only upon the presence of the physicians in the portrait and the stance of the king, rather than the picture as a whole.

conquer disease. The surgeons' equal standing with physicians is further emphasized by the fact that the two groups wear identical gowns. Ranked like an army of scholars behind their sword-wielding king-physician, the barber-surgeons pay homage and prepare to do battle in his name. The religious imagery is striking; clearly the leading members of the new Company of the Barber-Surgeons saw themselves as uniquely qualified to effect a reformation in medicine such as their king had in religion.

In 1540, the newly formed United Company of Barbers and Surgeons was large, with four masters and a Court of Assistants made up of between twenty and thirty members. This hierarchical structure was repeated in the yeomanry or freemen. The founding act confirmed all the Company's previous privileges relating to the exemption of surgeons from bearing armor, making up the watches, or attending inquests. It also spelled out the right of barbers and surgeons to keep apprentices, or "servants," and clearly delineated the separate tasks of barbers and surgeons. Barbers were to attend to the hair and shaving of their clientele and practice some dentistry, while surgeons were to bleed and apply instruments to their patients:[10]

> No manner of person within the City of London, or the suburbs thereof, or within one mile compass of the same, after the feast of the Nativity of our Lord God next coming, using barbery or shaving or that, shall hereafter use barbery or shaving . . . neither he nor they nor none other shall occupy any surgery, letting of blood, or any other thing belonging to surgery, drawing of teeth only except. And furthermore, in like manner, whosoever useth the mystery or craft of surgery, shall in no wise occupy nor exercise the feat or craft of barbery or shaving, neither by himself nor by none other for him to his or their use.[11]

In this new organization quickly emerged a small group of learned surgeons primarily concerned with reforming the decayed and ruined art of medicine. Who were these men? Like the members of the medieval fraternity of surgeons, they were socially mobile as a result of attachments to royalty and noble families during the many wars against France and Spain and their subsequent appointments to large hospitals in the city of London. Knowledgeable in Latin and French, they were eager to construct a new identity for their profession. Over the course of two generations, they formulated and passed regulations concerning the entry, education, and licensing of the capital's surgical apprentices. They also

---

10. J.J. Keevil, *Medicine and the Navy: 1200–1900*, 3 vols. (Edinburgh: E. & S. Livingstone, 1957), 1:126.

11. Power, *Memorials*, 100.

translated continental works on surgery and wrote books condemning empirical practitioners of their art. These publications reflected the surgeons' careers, demonstrating their interests in anatomy and their exposure to the treatment of war injuries, particularly those caused by the new weapon of gunpowder. In addition, the learned surgeons claimed expertise in treating syphilis and skin diseases such as struma or scrofula. The three most influential figures of this first generation of surgeons in the Company were Thomas Vicary, Thomas Gale, and John Halle.

Thomas Vicary (1495–1561) had been a surgical practitioner in Maidstone, Kent. He reportedly came to the monarch's attention after being called in to cure Henry VIII of his "sore leg" following a riding accident. Rewarded with twenty pounds, Vicary moved to the capital and was appointed the King's Sergeant Surgeon in charge of mustering the barber-surgeons in times of war. By 1530 he was Master of the Company of Barbers, and he became the first Master of the United Company of Barber-Surgeons in 1540. Eight years later, as Reader in Anatomy to the Company's freemen and apprentices, he published *A Profitable Treatise of the Anatomie of Mans Body*.[12] In the frontispiece Vicary addressed the need for more reading materials for the new Company's thousand or so apprentices.

Thomas Gale (1507–87) acquired surgical experience with Henry VIII's army in the battle of Montreuil in 1544 and with Philip II's navy at St. Quentin in 1557. He was Master of the Company in 1561 and surgeon to the hospitals of St. Thomas's and St. Bartholomew's. Gale showed some knowledge of Latin in his publication, *Certaine Workes of Chirurgerie, newly compiled and published* (1563), which mainly dealt with the treatment of war wounds. He also supported the revival of Greek medicine in the vernacular by publishing books 3 through 6 of Galen's *De methodo medendi* as *The institucion of chyrurgerie* in 1567.

John Halle (ca.1529–?), like Vicary, was a practicing surgeon in Maidstone, Kent. His exposé of quacks, *An Historiall Expostulation: Against the beastlye Abusers, both of Chyrurgerie and Physyke in oure tyme* (1565), was widely quoted by his fellow surgeons in the Company.

Through their works this first generation of learned surgeons strove to give meaning and status to the craft of surgery as a skill that could be employed only by those trained in the appropriate method and experienced in practice. Atypically for medical authors of the sixteenth century, Vicary, Gale, and Halle wrote in the vernacular. They built their arguments for the worthiness of surgery around case histories of patients which served both to show their expertise and condemn the

---

12. The earliest extant edition of this work is from 1577. References below are to a later edition: Thomas Vicary, *The Englishemans Treasure, or Treasor for Englishmen: With the True Anatomye of Mans Body* (London: J. Perin, 1586).

fatal ignorance of quacks. The publications of this first generation of learned surgeons set the stage for the men who succeeded them in the Company of Barber-Surgeons.

The key players in the second generation of learned surgeons included William Clowes (ca. 1540–1604), who came from Warwickshire gentry and attended the armies of both the earls of Warwick and Leicester. Clowes later became chief surgeon at St. Bartholomew's Hospital and visiting surgeon to Christ's Children's Hospital. Again, as in the case of Gale, the experience of warfare influenced both his choice of subject and his values in practicing surgery. In 1588, Clowes was in charge of all the surgeons in the English Fleet as it faced the Spanish Armada. In 1575, he translated the seventh book of the French physician Jean Fernel's *Pathologia* on surgical procedures, and in 1591 he published *A Prooved Practice for all Young Chirurgians, Concerning Burnings with Gunpowder, and Woundes made with Gun Shot, Sword, Halbard, Pyke, Launce, or Such Other*.[13] As Clowes aged, his publications reflected his changing interests, from binding wounds as a youthful military surgeon to caring for tumors and skin complaints as an older, respected London hospital surgeon. In 1602, he published his final work, *A Right Frutefull and Approved Treatise for the Artificiall cure of that malady called in Latin struma, and in English, the evill, cured by Kinges and Queenes of England*.[14]

John Banister, or Banester (1533–1610), had accompanied Leicester during his campaigns to the East Indies in the 1550s. In 1575, while practicing as a surgeon and physician in Nottingham on the strength of a license from Oxford University, he issued *A Needefull, New, and Necessarie Treatise of Chyrurgerie*. This was largely on the topic of skin ulcers. In the 1580s, as Reader or Lecturer in Anatomy to the Company, Banister published *The Historie of Man, Sucked from the Sappe of the Most Approved Anathomistes, in This Present Age*. In his dedication, addressed to the Company of Barber-Surgeons, Banister discussed what led him to compile *The Historie of Man*. While calling for a "spedie reformation" in surgery, Banister reminisced:

> I called to remembrance, that the greatest want that raigneth in Chirurgians at this Day, is ignoraunce in the subject of their worke…then did I clearly see, how that to write Methodes or meanes to cure the affected

---

13. William Clowes, *A Prooved Practice for all Young Chirurgians, Concerning Burnings with Gunpowder, and Woundes made with Gun Shot, Sword, Halbard, Pyke, Launce, or Such Other* (London: T. Orwyn, 1591).

14. William Clowes, *A Right Frutefull and Approved Treatise for the Artificiall cure of that malady called in Latin struma, and in English, the evill, cured by Kinges and Queenes of England* (London: E. Allde, 1602).

partes of the body, the partes themselves beyng altogether unknowen, or falsely imagined of, might rather be a meane to indurate the cataract of inscience, then to eate it through, or take it away.[15]

Not to know a part of the body was, for Banister, equivalent to being functionally blind. He literally saw an anatomical apprenticeship as the only way to acquire true insight or *autopsia* as a medical practitioner. Only a true anatomist could imagine the architecture of the body and discern the purpose behind the arrangement of the parts and their functions. The skill needed to cut into the body became, in its own right, an instrument of knowledge and, as surgical practices were translated into print, a further propaganda tool for the reformed surgeon. It was Banister's father-in-law, the Gloucester surgeon John Read, who translated from Spanish a work by Francisco Arceus, *A Most Excellent and Compendious Method of Curing Woundes in the Head* (1565). Read opened his translation with a blistering attack on the traditional division of surgery and physic.

George Baker (1540–1600) was a learned surgeon with eclectic interests. While attached to the household of the earl of Oxford, he published a translation of *Evonymus* by the Swiss naturalist Konrad Gesner, entitled *The Newe Jewell of Health* (1576). Baker also wrote a small iatromedical treatise, *The Nature and Propertie of Quicksilver*, which was attached to William Clowes's *Booke of Observations* (1596), and supplied the introduction to John Gerard's *Herball* (1597). Though Baker and Clowes seemed to be good friends, Clowes was censured and then pardoned in 1577 by the Company's Court of Assistants for repeatedly fighting with Baker in the fields outside the Barber-Surgeons' Hall.[16]

In 1540, as the first part of the reformation in medicine that these learned reform-minded surgeons aspired to bring about, they turned to the question of how the apprentices and freemen of their new Company should be trained. Weekly surgical demonstrations and anatomies were instituted for the first time within the confines of the Barber-Surgeons' Hall. Above all, the intention was that anatomy would not only provide the basis for the standardization and improvement of surgery but would be the very foundation upon which rested the surgeons' code of Christian behavior. By reforming the method of teaching anatomy, the learned surgeons would also reform the conduct of their colleagues and so make anatomy a profession in its own right.

---

15. John Banister, *The Historie of Man, Sucked from the Sappe of the Most Approved Anathomistes, in This Present Age* (London, 1578), dedication.

16. See Allen Debus, *The English Paracelsians* (New York: F. Watts, 1965), esp. chap. 2, for a discussion of Baker's sympathy toward chemical medicines but rejection of Paracelsus's theories.

Having obtained an annual formal grant as part of the Act of Union in 1540, "to make incision of...four persons condemned, adjudged, and put to death for felony," the Company of Barbers and Surgeons arranged a series of public dissections for each Winter period.[17] The Masters decreed that each anatomy session was to last six days in total and to be followed by a feast second in size and expense only to that held on election days, thus demonstrating the ceremonial importance of anatomizing to the image of the new Company.

Demonstrations of physical anatomy were arranged systematically, beginning with viscera, followed by muscles and arteries, and ending with bones, ligaments, and joints. The apprentices were given access to the body, whose anatomy, as well as the nature of surgical operations and the use of the appropriate apparatus and instruments, was explained to them. In 1566, recognizing the increasing interest in dissection, the officers of the Company stepped in once again and reminded the anatomists of their Christian duty regarding the treatment of bodies: "[I]t was ordered that all private anatomies taking place in the Hall would henceforth be reverently buried as public anatomies...any skelliton to be made onelye excepted."[18]

From 1587, individual members were held responsible for any costs resulting when a body revived during a public or private dissection. Such a concern sprang from an incident in February of that year where:

> A man hanged for felonie . . . being begged by the Chirurgions of London, to have made of him an Anatomie, after hee was dead to all mens thinking, cut downe, stripped of his apparell, laide naked in a chest, throwne in a carre, and so brought from the place of execution through the Borough of Southwarke over the bridge, and through the Citie of London to the Chirurgions Hall nere unto Cripelgate: The chest being there opened, and the weather extreeme cold hee was found to be alive, and lived till the three and twentie of Februarie, and then died.[19]

Despite the right in law to take certain criminal bodies for dissection, the use of the term *begged* indicates how difficult it was to obtain them.[20] The problem

---

17. The dissections were public events in the sense that the body anatomized was a public one, meaning it was the body of an executed felon which had arrived at the Barber-Surgeons' Hall under statutory provision for the supply of corpses from the hangman.

18. Sidney Young, ed., *Annals of the Barber-Surgeons of London* (London: Blades, East & Blades, 1890), 180.

19. Power, *Memorials*, 126. The clothes of the hanged were often auctioned off by the hangman in a local public house.

20. Approximately 560 people were hanged annually at Tyburn alone from ca.1536 to 1553, but after that the figures start to fall dramatically. During the subsequent reigns of Mary I (1553–58) there

may well have been that responsibility for requesting and then removing a hanged body rested with the Company's servants rather than the civil authorities. Occasionally, faced with the protests of family and friends at the gibbet, they had to use force in order to make off with their prize. The anonymous writer of the above passage outlines the indignity to which the body was put before it arrived at the "Chirurgions Hall." It was stripped by the hangman, who took the clothes as part of his fee, thrown naked into a pauper's coffin in a cart, and trundled hurriedly across half of London.

However, once the ill-treated corpse was deposited in the Barber-Surgeons' Hall by the Stewards, no effort was spared to stress the dignity and gravity of the dissection. According to the eleventh article of the ordinances made by the Company in 1556, the two Stewards were to pay increasing attention to the comfort and professional appearance of the Reader of Anatomy. They were to make sure

> that there be every yere a matte about the harthe in the hall that Mr Doctor [be] made not to take colde upon his feete…there be ij fyne white rodds appointed for the Doctor to touche the bodye when it shall please him and a waxe candell to loke in the bodye and that there be always…two aprons…and two payr of Sleaves.[21]

In 1556, additional regulations were passed by the Company regarding the duties of the spectators at public anatomies. Every licensed barber-surgeon, or freeman, of the Company, along with the one thousand or so apprentices, now had to attend these weekly lectures under penalty of a fine.[22] Indeed, such was the concern of the learned surgeons that their less educated brethren should grasp the noble art of anatomy in the appropriate manner, that they laid down rules of etiquette concerning attendance at the dissections. For the honor of the Company, freemen were enjoined to come properly dressed, remain the whole time, and not ask questions until the end of the lecture, when they might courteously point out anything that they thought had been misrepresented.[23] By "inviting" freemen to come, watch, learn, and, above all, behave themselves at anatomical

---

were 280, Elizabeth I (1558–1603) 140, and James I (1603–25) 140. It seems unlikely that in the earlier period the removal of four bodies annually would have made dissection that visible. The College of Physicians did obtain a grant of four bodies annually in 1564/5. See Sawday, *Body Emblazoned*, 56–58.

21. Power, *Memorials*, 136.

22. By ordinance of 1572 a sliding scale of fines was established for "freman, fforeyn, or alian straunger," who failed to turn up and/or stay for the "forenoone" and "afternoone." See Power, *Memorials*, 142.

23. Ibid., 86.

demonstrations, learned surgeons were imitating the behavior of physicians toward surgeons themselves. Intending to bring about a reformation in medicine, the learned surgeons undoubtedly made a significant effort to reform the manners of and teach civility to future practitioners of their art. Such concern over surgeons' manners directly reflected the new links between civility and education promoted during the Reformation by humanists such as Erasmus and Thomas More.[24]

An entry dated 1577 in the Company's minute book tersely points to the enormity of the task the learned surgeons had undertaken when they set about reforming their colleagues' behavior:

> Yt is agreed and condescended that no person or persons of this Companie do presume at anie tyme or tymes hereafter of Anathomies to take and carrie awaie or cause to be taken or carried or conveyed awaie any parte of the skynn of any bodie which shall at any tyme be... cause to be tanned like lether.[25]

Flaying was part of the anatomist's art, and skin was probably readily available during the dissections. But why would it have been carried away and, moreover, tanned? Human body parts were commonly bought and stolen as souvenirs and health aids by executioners and midwives.[26] However, the concern in this passage is focused on members of the Company who have engaged in such a disrespectful, irreverent act. The theft of the skin was perhaps not so shocking as its having been treated and tanned as if it were an animal hide.[27] This association of the art of dissection with butchery, and the unpleasant reputation of tanners and executioners, were precisely what the learned surgeons were so earnestly fighting to dispel with their new rules concerning courteous behavior.

The Company decided in the same year to conceal the body behind specially erected curtains:

---

24. See Norbert Elias, *The Civilizing Process: The Development of Manners*, trans. E. Jephcott (New York: Pantheon, 1971); and Marvin B. Becker, *Civility and Society in Western Europe* (Bloomington: Indiana University Press, 1988).

25. Power, *Memorials*, 88.

26. See Sawday, *Body Emblazoned*, 79–84; and Katherine Stuart, "The Executioner's Healing Touch," in *Infinite Boundaries: Order, Disorder, and Reorder in Early Modern German Culture*, ed. Max Reinhart (Kirksville, Missouri: Truman State University Press, 1998), 349–79, for discussions of the ambivalent status of the executioner in early modern society.

27. On the historical twinning of surgery and butchery see Andrea Carlino, *Books of the Body: Anatomical Ritual and Renaissance Learning* (Chicago: University of Chicago Press, 1999).

The Anatomical Table of John Banister (1581). Part of a series commissioned by Banister, probably executed by Nicholas Hilliard. Reproduced with permission of the University Library, Special Collections, Glasgow. Boadsheet, painting on paper.

And also ther shalbe pyllers and Rod of Iron made to beare and drawe Courteynes upon & aboute the frame where within the Anathomy doth lye and is wrought upon, for bycawse that no persone or persones shall beholde the desections or incysyngs of the body, but that all maye be made cleane and covered with fayer clothes untyll the Docter shall com and take his place to reade and declare upon the partes desected.[28]

As in a play, the performance would not begin until the Stewards of Anatomy whisked open the curtains to reveal the body covered with yet more "fayer" cloths. Finally, the Reader would ceremoniously enter the Hall and expose the cadaver.

Many of the reforms instituted in the teaching of anatomy and surgery at the Barber-Surgeons' Hall in the sixteenth century, and the aspirations underlying them, are reflected in an anonymous portrait of John Banister lecturing on anatomy in 1581, shown above.

28. Young, *Annals*, 315.

Banister was Reader of Anatomy at the Barber-Surgeons' Hall during the late 1570s and early 1580s. He is shown in the 1581 portrait pointing with one of the "fyne white rodds" to the lower rib cage of the Company's skeleton while his bejeweled hand rests on the intestines of a cadaver which appears to be sleeping peacefully on a table surrounded by surgical instruments. No blood or gore stains the white cloth, despite the opened thorax and abdomen, giving the anatomical scene the appearance of orderly accepted protocol. The calm and dignified representation of Banister, the corpse, the spectators, and the "science" of anatomy stands in stark contrast to the description of the Stewards' having to "beg" for the body at Tyburn and, amidst the emotional chaos and confusion of the place of execution, make off hurriedly with the naked cadaver. The portrait conveys no suggestion of infamy in the dissector touching the corpse, although he deals with it in much the same way as the hangman did and the corpse is that of a murderer. Instead, the tableau is designed to assert the religious significance and medical authority of anatomizing the body by comparing it to any (body's) final state, namely, the skeleton.

The choice to present an opened cavity suggests that Banister is delivering one of the visceral lectures. Wearing the traditional cap and gown of a physician along with one of the "payr of Sleaves," Banister stands with the two Masters of Anatomy by his side. They hold the traditional instruments of surgery and dissection, the probe and scalpel. The Masters were elected annually and were primarily responsible for guarding the keys to the Company's library and surgical instruments. In the foreground stand the two Stewards of Anatomy, one holding a probe and perhaps handing it to the other. To the right and slightly to the back of Banister an open book sits upon a reading desk. This has been identified as the most recent edition of Realdo Columbo's *Anatomy*, published in 1572 in Paris. Perhaps for the benefit of his students, Banister is using the cheaper *vade mecum* or portable pocketbook edition of Columbo.[29] The apprentices' eyes, however, are not focused on the book but rather follow the line of Banister's rod to the skeleton, which stands supported and crowned with the blue and white of the Barber-Surgeon's arms. At the head of the corpse, a bearded old man, wearing the elaborate furred gown of a philosopher, with one hand points to the skull of the dissected body and with the other touches his own heart, possibly alluding to Galen's admonition to practice anatomy with both head and heart.

Andrea Carlino has suggested that in early modern anatomical portrayals the book signifies the relationship between physician and theory while the corpse

---

29. Power, *Selected Writings*, 78.

signifies the relationship between physician and his practice.[30] The painting of Banister, then, suggests that knowledge of surgical practice rather than theory was the primary emphasis of his demonstration. Banister's very stance implies that anatomical and surgical knowledge come, not from books of theory, but from demonstrations of the workings of the interior of the body and the skeleton. Calmly touching both book and corpse, Banister acts as the conduit for anatomical knowledge, carefully directing his students' gaze to the final stage of anatomy, the skeleton. This, rather than the book, was the basis upon which all medical knowledge rested.

The portrait powerfully displays the two pillars of the new reformed medicine. The first, reflecting the discovery of Galen's texts and the humanistic context within which they were placed, is the classical tradition of anatomy. The second is found in the emphasis on the Renaissance Christian belief that dissecting a body was akin to worshipping God, the body being a microcosm of the divine macrocosm. Above the book in the portrait is the text "Anatomia sciential dux et aditumque ad dei agnitionem praebet" (Experience of the structure of the human body leads one to a better understanding of God).

The new social and intellectual mobility of the learned surgeon was supported and his identity as an expert in anatomy constructed and, most important, publicized through the medium of print. The language was English. Sixteenth-century surgeons used print to argue against physicians and untrained practitioners who lacked the surgeon's anatomical expertise and professional confidence. Moreover, writing for the members and apprentices of their Company, who did not know Latin, learned surgeons translated a series of continental medical works into English and so began to develop a vernacular surgical tradition.

In 1548, Thomas Vicary, Sergeant-Surgeon to Henry VIII, hospital surgeon at St. Bartholomew's, and possibly the first Reader of Anatomy at the Company, published *A Profitable Treatise of the Anatomie of Mans Body*. The treatise was based upon the mediaeval writings of the surgeon-anatomists Lanfranc (d. 1315) and Henri de Mondeville (1260–1320) and may have been the text which Vicary used to lecture on anatomy. In the preface, Vicary states that the "two principal rootes of phisicke and Surgery" were reason and experience.[31] Some physicians had neglected experience—which Vicary took to mean the practice of anatomy—at the risk of losing their reputations and their patients' lives.

---

30. Andrea Carlino, "The Book, the Body, the Scalpel: Six Engraved Title Pages for Anatomical Treatises of the First Half of the Sixteenth Century," *RES* 16 (Autumn 1988): 31–49.

31. Vicary, *The Englishemans Treasure*, 11.

Vicary's complaint was amplified by one of his surgical contemporaries in the Company, John Halle, who in 1565 published a new translation of Lanfranc's *Chirurgia Parva*, or "Little Surgery," with a "compendious worke of Anatomie appended." Halle took issue with the idea of what constituted learning for his fellows in two ways. First, while demonstrating his own knowledge of Latin, Halle supported the use of English for surgeons who were not Latin-literate:

> For as phisiciens thynke their learning sufficient, without practice or experience: so the chirurgien for the most parte havyng experience or practice, thinketh it unnedefull to have any learnyng at all…a good chirurgien (that will avoide wicked crafts and abuses) should first learne, and then worke and use experience.[32]

This allowed Halle to apologize to physicians for his use of English while insisting that learned surgeons, such as himself, were more than mere empirics. Halle's idea of learning differed significantly from that taught at the universities. In fact, it initially appears close to Paracelsian notions of the importance of empirics. Quoting Socrates, Halle claims that "Lerning ought not to be written in bokes, but rather in mennes mindes."[33] Yet, with a reference to Galen, Halle makes clear what he meant. He explains that Galen had advised surgeons to rely on "oure propre eyes: which are to be trusted above all other authores, ye before Hippocrates and Galen…the eyes are judges bothe true and certaine."[34]

Like Banister, Halle accuses the physicians of having insight. By blindly relying upon learning and, above all, by not following the instructions of the great anatomist Galen to practice and experience the workings of the body, physicians had neglected to cultivate their "propre eyes." Halle, therefore, concludes that "I woulde have no man thinke him selfe lerned, otherwise than chiefly by experience: for learning in chirurgery and physic, commeth not in speculation only, nor in practice only, but in speculation well practiced by experience."[35]

Halle's attempts to find a way out of the dilemma that learned surgeons were not really learned (i.e., university trained), and therefore, by Reformation standards, could not exercise medical authority, were echoed by other members of the Barber-Surgeons' Company during the late sixteenth century. Thomas Gale, John Read, and his son-in-law John Banister all argued that the learned surgeon ought

---

32. John Halle, *An Historiall Expostulation: Against the beastlye Abusers, both of Chyrurgerie and Physyke in oure tyme* (London, 1565), fol. 18v.
33. Ibid., fol. 14v.
34. Ibid.
35. Ibid., fol. 18v.

to be not only knowledgeable in physic but allowed to practice it too. Physic was defined very narrowly as "prescribing inward medicines and convenient die[t]."[36] The point was being made that an anatomically skilled surgeon had sufficient learning to act as a physician. This belief may reflect the military and hospital backgrounds of the learned surgeons and the sense of crisis and despair those experiences had instilled in them. Thomas Gale, a Master of the Company in 1561, recalled with horror his experience with quacks on the battlefield at Montreuil and in the London hospitals of St. Thomas's and St. Bartholomew's. At Montreuil he had witnessed "a great rabblement there, that took upon them to be surgeons. Some were sow-gelders, and some horse-gelders, with tinkers and cobblers."[37] Gale reported that of the three hundred or so patients who were waiting for or who underwent amputations at the hospitals, he worked in, nine tenths were there because of quackery.

The argument that surgeons were perfectly capable of performing as physicians is also part of a wider debate concerning the historical division of surgery and physic. Banister and Read made the strongest appeal for the reformation and reunification of surgery and physic, basing their arguments on pagan authors, especially Hippocrates and Galen, who had defined physic and surgery as essentially indistinguishable. Moreover, Banister and Read argued that given that the body had to be cured from the inside out, restricting the surgeon to performing only topical cures made little sense. John Read poetically exclaimed:

> How can the Surgeon well dissolve,
>   the thing contained in his cure,
> Except he do evacuate,
>   and purge the same that is unpure,
> Or is he able for to cure,
>   all woundes and ulcers redlie,
> Without the administration,
>   of divers medicine inwardlie....[38]

For John Read, physic and surgery were both anatomical endeavors. Because physicians lacked experience with the body, surgeons, with their training in the precise workings of the body, ought to be allowed to prescribe internal remedies.

In contrast to Halle's self-consciousness about the use of the vernacular in medical writings, Read's unapologetic stance regarding his use of English demonstrated

---

36. John Read, *A Most Excellent and Compendious Method*, epistle 1.
37. Power, *Memorials*, 187. Gelders = castrators.
38. Read, *A Most Excellent and Compendious Method*, 29.

how the identity of the learned surgeon evolved in the first decades of vernacular surgical publications. In his 1574 edition of *A Most Excellent and Compendious Method of Curing Woundes in the Heade*, Read explicitly referred to the "good menne" in the past, especially "Master William Clowes," who had publicized the twin virtues of surgery: its anatomical basis and Christian virtues. Specifically addressing physicians, and citing the necessity to spread God's gift of this "godly science" for the relief, comfort, succor, and health of all, Read exclaimed: "Why grutch they Chirurgerie should come forth in English? Would they have no man to know but onely they?...what reason is it, that we should huther muther heere among a few, the thing that was made common to all?"[39]

Read called on his "good and loving friends, Banester, Clowes, and all Masters in Chirurgerie," to lead a general reformation in surgery:

> For you are thay, which is most delighted in chirurgerie, you are they, by whom chirurgerie being decayed may hope for reformacion, you are they which for your singular skill heerein, are able to judge of all imperfections as lurke among those bufardlie empiricks. You are they that are able to defend the true and sincere chirurgerie against the false and corrupt...by your good assistance, learned professours may be appointed in all convenient places, to publish and set foorth the same.[40]

John Read's appeal for learned surgeons to use their anatomical knowledge to expose false practitioners of medicine reflects contemporary transformations of medicine from the 1560s on.

In 1565, John Halle issued *An Historiall Expostulation against the beastly abusers, both of Chyrurgerie and Phisicke in our tyme*, where he looked back to a golden age of medicine:

> For where as in the tyme paste there were fewe Chirurgians, & they very cunning, learned, and experte: whereby they were accepted as precious jewels or honourable treasures of the common weale: Ther are nowe many, and the moste parte ignorant...yea very caterpyliers to the publique orders.[41]

Drawing upon his long experience as a practicing surgeon and Justice of the Peace in Maidstone, Kent, Halle vividly describes quacks he claimed to have unmasked in his region through interrogation of their medical knowledge:

39. Ibid., 5.
40. Ibid., 3.
41. Halle, *An Historiall Expostulation*, fol. 1v.

> One Robert Nicol, a false deceiver, and most ignoraunt beaste, hath in tymes passed boasted him selfe to have been the servaunt of Master Vicary, late Sargeant Chyrurgien to the Queenes highnes. Among whose wicked and prodigious doynges…one very notable chanced in…1564 the 26 of September. He poured in a purgation to an honest woman of good fame…whiche within three or foure houres at the moste, purged the lyfe out of her body.…[42]

Charged with murder by the magistrates of Maidstone, Nicol was questioned by Halle, among others, regarding his "knowledge and doings" of physic and chirurgerie. While Halle did not record the prisoner's examiners by name or occupation, he did give the exact questions and responses. Interestingly, Halle strongly implies that it was only when the interrogation turned to Nicol's lack of knowledge regarding human anatomy that his quackery was fully revealed: "It was asked him what the splene was, and he answered, that it was a disease in the syde, baked harde lyke a bisket: denying that there was any thing called the spleen but the disease (he saith) so called."[43] After further questioning regarding the nature of wounds, ulcers, and cancers, etc., Halle declared that while Nicol could avoid detection when examined about the speculative and imprecise art of physic, he was easily proven to be an ignorant, beastly quack when interrogated in the practical art of anatomy.

Thus, the absence of anatomical knowledge constituted an indictment of medical practitioners who did not measure up to the standards of a learned surgeon. This is apparent in the case from the 1560s of "one William a Shoemaker, [who] came into Kente, pretending to be very cunning, in curing diseases of the eyes." Asked by the friends of a potential patient to examine the man's expertise, Halle questioned William as to his credentials and anatomical knowledge:

> I asked him, whether he were a surgien, or a phisitien, and he answered no, he was a shomaker. But he coulde heale all maner of sore eyes.… Well sayde I, seyng that you can heale sore eyes: what is an eye, whereof is it made, of what members or partes is it composed and he sayde he knewe not that. Then I asked hym if he were worthy to be a shoemaker, or so to be called, that knew not howe, or wherof a shoe was made. He answered no, he was not worthy. Then sayde I, howe dare you woorke upon suche a precious, and intricate member of man, as is the eye.… Well sayde he, I

42. Ibid., 14v.
43. Ibid.

perceive you doe but skorne me, and flunge out of the doores in a great fume.[44]

Halle's emphasis on the practical knowledge of surgeons and the anatomical basis of their learning was echoed by William Clowes. In 1585, he issued a small treatise on syphilis entitled *De Morbo Gallico*, in which he describes the tricks quacks played upon their hapless patients. He mentions several deceitful doctors, including Valentine Rawsworme (or Rosswurm) of Smalcade, "a stranger borne, who was not only a deceitful stone cutter but falsely taken to be a wise alchemist."[45]

> Be it known that the third day of April 1574 Valentine Rawsworme did take upon him deceitfully to cure for the stone in a bladder one Helen wife of M. Curraunce, musician dwelling in London, in the presence of divers honest persons did attempt with his instruments to take out of the bladder a stone. But finding none there privily he took a stone out of the pocket of his hose and conveyed it into a sponge holding it for a space in a baisin of hot water and subtily and slyly forced it in pudendo, yet presently he was espied and charged there withall that they did plainly perceive and see him take that stone out of the pocket of his hose and did put it in the sponge, etc.[46]

Rawsworme was a Paracelsian and perhaps a practicing alchemist. Certainly his manipulation of the stone, including warming it in water prior to "subtily" pushing it inside the patient and then removing it, is suggestive of alchemical practices. His willingness to promote himself as capable of cutting the bladder stone appears to have been what offended Clowes, rather than his Paracelsian beliefs. Elsewhere, Clowes distinguishes between false and true Paracelsians on the basis of their use of visual spectacle and proud words versus quiet charitable actions. False ones prated and true ones did good works.[47] He admits to having "practised certaine of

---

44. Ibid., 11v.

45. For information on Rawsworme see Robert Jutte, "Valentin Rosswurm: Zur Sozialgeschichte des Paracelsismus im 16. Jahrhundert," in *Resultate un Desiderate der Paracelsus-Forschung*, ed. Peter Dilg and Hartmut Rudolph, Sudhoffs Archiv Beihefte 31 (Stuttgart: Franz Steiner, 1993), 99–112. See 107–8, where Jutte describes Rawsworme's interactions with Clowes. I am grateful to my colleague Dr. Andrew Bergerson for translating the relevant section into English. Deborah Harkness is currently preparing an article, "Paracelsian Therapeutics in Elizabethan London: The Case of Valentine Russwurin of Schmalkald," for publication. Unfortunately, I have not had the opportunity to consult it prior to completing this essay.

46. William Clowes, *De Morbo Gallico* (London, 1585), fol. 10v. Debus, *English Paracelsians*, 70–71, briefly discusses the encounter between Clowes and Rawsworme.

47. Clowes, *De Morbo Gallico*, fol. 59v.

his [Paracelsus's] inventions Chirurgicall, the which I have found to be singular good, and worthy of great commendations."[48] He was willing, he continues, to take anything he learned "eyther by reason or experience" that might be to the good of his patients from any source, "be it Galen or Paracelsus; yea, Turke, Jewe, or any other infidell."[49] In this he supported Paracelsus in his promotion of the twin bases of reformed medicine as experience and curiosity.

Rawsworme denied the charges and took his fee of ten pounds from the husband, giving him the stone in return. However, he was soon called to treat the patient again, her pain having returned. Rawsworme prescribed a powder: "[A]fter the receipt of this powder she could never void any urine. And moreover the powder did so blister her mouth, her nose and face and likewise the inward parts of her body that she never afterwards received any sustenance but died most pityfully by his wicked dealing."[50]

The story was not over, for as anatomy had enabled Clowes to identify Rawsworme as a quack, dissection allowed him to reveal the extent of his medical malpractice:

> Then she was opened where it did most manifestly appear that she never had a stone in her bladder nor any matter wherof the stone is engendered; neither any offence in the bladder or parts thereabouts that then could be conjectured saving that himself most vilainously had committed and done, but only in her kidneys and there was all the whole trouble of her sickness.[51]

The moral of this tale is a harsh one. Patients must accept that they do not know and cannot see the workings of their own bodies—Helen Curraunce could not even tell that the quack had inserted the stone into her vagina. Only Clowes's *autopsia* could reveal to all that Rawsworme was a wicked deceiver of this gullible woman and her family and friends. They could not protect her; only a learned surgeon like Clowes could do that, but only if she was prepared to entrust her ailing body to his capable hands.

To this end, as part of the campaign to effect a reformation in medicine, in 1566 the Court of Assistants issued an order whereby all patients in peril of "mayme" or death were to be presented and their case made known to the Masters within three

---

48. Clowes, *A Right Frutefull and Approoved Treatise*, "Epistle to the Reader." See Debus, *English Paracelsians*, 70.
49. Clowes, *A Right Frutefull and Approoved Treatise*, "Epistle to the Reader."
50. Ibid., fol. 11v.
51. Ibid.

days.[52] Three learned surgeons were to be appointed by the Court to see the patient and "assist" in the cure. Injured patients and their families also appeared before the Court in the hope of obtaining a public confession and apology from the surgeon concerned. The Court issued fines and had the power to imprison and to permanently remove surgical licenses from repeat offenders.[53] The Court was another public forum where the learned surgeons could seek to discipline those among them who had not been fully reformed and could participate in restoring the decayed and ruined art of medicine to its previous glory.

During the later sixteenth century, learned surgeons within the newly combined Barber-Surgeons' Company set about effecting a reformation in medicine such as their king had in religion. Changes were instituted in the frequency and teaching of anatomy and surgical operations to the apprentices and freemen; ethical guidelines were promulgated regarding the "right" behavior of spectators at anatomies. The abilities of surgeons to unmask quacks were publicized in vernacular works. All of these reforms came to influence how surgeons interacted with their patients. In numerous works, leading surgeons publicized their belief that anatomy brought spirituality and an elevated status to their skills. The crises in health care that learned surgeons witnessed on the battlefields of Europe and in the hospitals of London reinforced their belief in the necessity of applying the special skills of *autopsia* with greater frequency. The sense of religious urgency evoked by this mission was palpable.

As the dissections in the Barber-Surgeons' Hall became more and more formalized during the sixteenth century, young barber-surgeons were being trained to perceive, experience, and so imagine the bodies of their future patients in new ways. By tightly controlling their rights to dissect, barber-surgeons made anatomy an accepted part of being a surgeon, transparently obvious and necessary to surgeons' ability to advance medical knowledge and help their patients. The body had been concealed behind curtains and draped in cloth; now surgeons could open its messy, smelly interior and see within. Following in the footsteps of Galen, leading surgeons of the Company such as Banister, Gale, and Clowes espoused anatomy as a religious service. Establishing a dignified ceremony of dissection, they sought to develop the new conception of the body not as a patient per se but as a temple attesting to the wisdom of God that must be explored for a greater good.

---

52. Young, *Annals*, 201.

53. For example, strict penalties attended Hewe Placket, who took "upon him to heale a pacient who ys deade and comaundement geven that he shall medle no more in surgerie." Ibid., 139.

SEEING "MICROCOSMA"
Paracelsus's Gendered Epistemology

Hildegard Elisabeth Keller

> *Licence my roaving hands, and let them go,*
> *Before, behind, between, above, below.*
> *O my America! my new-found-land,*
>   *My kingdome, safeliest when with one man man'd,*
>   *My Myne of precious stones, My Emperie,*
>   *How blest am I in this discovering thee!*
>   —John Donne (1572–1631), Elegie 19,
>   "To His Mistress Going to Bed"

The two words "microcosma" and "seeing" are intended to represent fundamental pillars in the cosmology of the Swiss doctor and lay theologian Theophrast von Hohenheim (1493/4–1541), otherwise known as Paracelsus. *Microcosma* is a term resulting from Paracelsus's conception of the world, humanity, and the sexes.[1] The first creation, the work of the first five days of Genesis, is the large (first) world or macrocosm. Its child is the male human being (Adam, created from the primal element, *limbus*), who is thus referred to as the small (second) world, or microcosm. Since the woman, or matrix, the smallest (third, or last) world, derives not from the earth but from this "lebendigen fleisch" (living flesh), she is thought to be ruled by an "andere monarchei" (a different monarchy): "this necessitates producing a new theoretical work about woman and dealing with her separately in a separate work of physiology."[2] This idea underlies Paracelsus's vehement argument for a new doctrine of seeing in contemporary medicine, which to his view is blind. The female

---

1. The expression is found repeatedly in Paracelsus, *Opus Paramirum*, bk. IV, in I, 9, 197. All my quotations of Paracelsus's works are from Paracelsus, *Sämtliche Werke, I. Abteilung: Medizinische, naturwissenschaftliche und philosophische Schriften*, ed. Karl Sudhoff, 14 vols. (Munich: Otto Wilhelm Barth, 1922–33), 197; *II. Abteilung: Theologische und religionsphilosophische Schriften*, ed. Kurt Goldammer (Wiesbaden: Franz Steiner, 1955–). Paracelsus's works are cited here by Roman numerals to denote the parts of the collected edition of the complete works of Paracelsus (I or II), followed by the volume and page numbers.

2. "[D]as ursacht ein neue theoricam zu machen von der frauen und sie zu besöndern in ein sondere physicam." Paracelsus, *De matrice*, I, 9:201.

cosmos should be considered by itself, quite separately from men and their illnesses, since women have women's brains, women's hearts, and women's hair and, if they are hungry, hunger in a woman's way.[3] Here he is not simply emphasizing the necessity for gynecological specialization. Quite the contrary: the partial differentiation of the woman, or rather of her lower abdomen, would represent the exact opposite of his fundamental principle. The whole woman constitutes a world of her own. Paracelsus proposes an unconventional, gendered medical approach, as comprehensive as possible, which should treat all illnesses as gender-specific: men's ailments and women's ailments.[4] Consequently, he not only speaks of the "sunderen welt" (separate world) of women, but also gives it a new name: "microcosma."

In contrast to earlier as well as contemporary male medical authors who try to advertise their texts by seductively titling them *secreta mulierum* (women's secrets),[5] Paracelsus is not appealing only to his male professional colleagues. In *De Caduco Matricis* he explicitly exhorts women to know themselves:

> Thus, o women, take notice of yourselves and the nature of your illnesses.... This is why I describe them..., so that you may recognize that you carry a double microcosm within yourselves, since your body is

---

3. Cf. ibid., 179 ff.

4. E.g., ibid., 200. Cf. Karl-Heinz Weimann, "Die deutsche medizinische Fachsprache des Paracelsus" (doctoral diss., University of Erlangen, 1951), 86.

5. See Margaret Schleissner, "A Fifteenth-Century Physician's Attitude toward Sexuality: Dr. Johann Hartlieb's Secreta mulierum Translation," in *Sex in the Middle Ages: A Book of Essays*, ed. Joyce E. Salisbury, Garland Medieval Casebooks 3 (New York: Garland, 1991), 110–25; idem, "Pseudo-Albertus Magnus, 'Secreta mulierum': Ein spätmittelalterlicher Prosatraktat über die Entwicklungs- und Geburtslehre und die Natur der Frauen," *Würzburger medizinhistorische Mitteilungen* 9 (1991): 115–24; Karma Lochrie, *Covert Operations: The Medieval Uses of Secrecy*, The Middle Ages Series (Philadelphia: University of Pennsylvania Press, 1999), 118 ff.; Britta-Juliane Kruse, *Verborgene Heilkünste: Geschichte der Frauenmedizin im Spätmittelalter*, Quellen und Forschungen zur Literatur- und Kulturgeschichte 5 (239) (Berlin: De Gruyter, 1996), 18 ff.; and Monica H. Green, "From 'Diseases of Women' to 'Secrets of Women': The Transformation of Gynecological Literature in the Later Middle Ages," *Journal of Medieval and Early Modern Studies* 30 (2000): 5–39. More generally for the wide range of scientific or pseudoscientific texts called *secreta* (*naturae*) cf. William Eamon, *Science and the Secrets of Nature: Books of Secrets in Medieval and Earl Modern Culture* (Princeton, N.J.: Princeton University Press, 1994). A complementary view on the discourses of "women's secrets" (based upon the literary motif of eavesdropping) is presented by Ann Marie Rasmussen, "Gendered Knowledge and Eavesdropping in the Late-Medieval *Minnerede*," *Speculum* 77 (2002): 1168–94. As Gérard Genette has pointed out, a text's title affects its reception in significant ways. Cf. Gérard Genette, *Paratexts: Thresholds of Interpretation*, trans. Jane E. Lewin (Cambridge: Cambridge University Press, 1997); and Hildegard Elisabeth Keller, "Totentanz im Pardies: Titelblätter und Widmungsvorreden in Geburtshilfebüchern des 16. Jahrhunderts," in *L'art macabre*, Jahrbuch der Europäischen Totentanz-Vereinigung 4 (Düsseldorf: Europäische Totentanz-Vereinigung, forthcoming).

the same as that of men and you are [simultaneously] the world of human birth.[6]

Paracelsus follows Genesis when he holds that a woman is perhaps not a higher being of increased subtlety (this is true only of Mary, the mother of God, whom Paracelsus does not regard as a generic woman), but instead is an entity apart, who should be considered separately from the other sex. She embodies, in the inimitable Paracelsian formula, "zwo welt in einer haut" (two worlds under one skin).[7]

But the male and female microcosms should not only be considered separately. They should both be regarded in relation to the macrocosm. And what this means for Paracelsus in concrete terms is that creation, both visible and invisible, can help human beings to know and understand themselves. This concept influences Paracelsus's entire practice of medicine, so that these two aspects—both the idea of micro- and macrocosm and that of the separateness of the two sexes—are an essential part of his epistemology. Paracelsus presents this view in two gynecological treatises: first in *De Caduco Matricis* (the second part of his work on epilepsy of 1530, in which he gives a new definition and description of the *suffocatio matricis*),[8] and again in *De Matrice* (the fourth, gynecological, book of his *Opus Paramirum* [1531], about the three principles governing the cosmos).[9] Paracelsus's doctrine of seeing reveals itself as a particularly independent worldview when considered in the context of its time and of the voyages of discovery.

This paper will show how much Paracelsus's concept of knowledge and his training of the physician's eye are informed by his thoughts about the two sexes and their genesis. The first section of this article will explain some basic points about Paracelsus's view of the eye as an organ; the second section will deal with the physician's gaze, which should penetrate into the visible and the invisible;[10]

---

6. "Also haben acht ir frauen auf euch selbst, wie eure krankheiten geschaffen sind.... [D]arum beschreib ichs..., damit zu erkennen wie das ir ein doppeln microcosmum in euch traget, den leib gleich den mannen und seind die welt der geberung der menschen." Paracelsus, *De Caduco Matricis*, I, 8:345.

7. Cf. Katharina Biegger, *"De invocatione Beatae Mariae Virginis": Paracelsus und die Marienverehrung*, Kosmosophie 6 (Stuttgart: Franz Steiner, 1990), 201ff; and Paracelsus, *De Caduco Matricis*, I, 8:356. See below for the Paracelsian interpretation of the account of the first transformation of (earthly) matter into gender-differentiated (bodily) matter in Genesis.

8. Paracelsus, *De Caduco Matricis*, I, 8:261–308. See n. 17.

9. Paracelsus, *De Matrice*, I, 9:177–230.

10. 10.Cf. also Gunhild Pörksen, "Paracelsus und der Augenschein—Notizen zum ärztlichen Blick," in *Paracelsus und seine internationale Rezeption in der Frühen Neuzeit*, ed. Heinz Schott and Ilana Zinguer, Brill's Studies in Intellectual History 86 (Leiden: Brill, 1998), 1–12; and Alois M. Haas, "Wahrnehmung im 'Licht der Natur': Magie als ein Schlüssel zur Natur," in *Nova Acta Paracelsica*, Jahrbuch der Schweizerischen Paracelsus-Gesellschaft 7 (Bern: Peter Lang, 1993), 3–10.

and the third section will show in concrete terms how Paracelsus differentiates the roles of man and woman.

"Concordanz" of the Eye

For Paracelsus it is an irrefutable fact that the eye is the "central organ through which we register the world"[11] and that seeing is the true vocation of human beings: "For the human being's duty is to experience things and not to be blind to them; for this is why he was created: to talk of the wondrous deeds of God and to hold them up [for admiration]."[12] Part of his doctrine of seeing, which is aimed especially at the physician, is a sort of dual vision. This constitutes a peculiar epistemological situation, for the physician should look outwards and inwards simultaneously: "Because the human being learns from the large world and not from the human being. The concordance makes the physician complete, so that he knows the world, and from the world [he] also knows the human being, as these are one thing and not two as I know from experience."[13] What makes a good physician, then, is not only that he sees the microcosm in the macrocosm and vice versa. It is rather that he should also—as the etymology of "concordance" suggests—see with his heart: "The heart makes the physician, who proceeds from God; he is [born] from natural light and experience."[14] Even at this point it becomes clear that the two spheres of Paracelsus's writings—on medicine and on theology and natural philosophy—should not be viewed in isolation. They constantly interact,

---

11. Uwe Pörksen, "War Paracelsus ein schlechter Schriftsteller?" in *Nova Acta Paracelsica: Jahrbuch der Schweizerischen Paracelsus-Gesellschaft* 9 (Bern: Peter Lang, 1995), 25–46, at 41.

12. Paracelsus, *Das Buch von den Nymphen, Sylphen, Pygmaeen, Salamandern und den übrigen Geistern: Faksimile der Ausgabe Basel 1590*, ed. Gunhild Pörksen (Marburg an der Lahn: Basilisken, 1996), 8 (hereafter *Liber de Nymphis*). See also the English translation of the tract in Paracelsus, *Four Treatises of Theophrastus von Hohenheim*, trans. C. Lilian Temkin et al., ed. Henry E. Sigerist (1941; repr., Baltimore: Johns Hopkins University Press, 1996), 213–53. This theologically founded predilection for the eye is characteristic of Paracelsus; cf. Alois M. Haas, "Unsichtbares sichtbar machen: Feindschaft und Liebe zum Bild in der Geschichte der Mystik," in *Konstruktionen Sichtbarkeiten: Interventionen*, ed. Jörg Huber and Martin Heller, Interventionen 8 (Vienna: Springer, 1999), 265–86. Of course, there is a cultural context for such convictions; see *La visione e lo sguardo nel Medio Evo* (View and Vision in the Middle Ages), ed. Véronique Pasche, vols. 1 and 2, Micrologus 5 and 6 (Turnhout: Brepols, 1997); and Carl Havelange, *De l'oeil et du monde: Une histoire du regard au seuil de la modernité* (Paris: Gallimard, 1998).

13. "[D]an der mensch wird erlernt von der großen welt und nit aus dem menschen: Das ist die concordanz die den arzt ganz macht, so er die welt erkent und aus ir den menschen auch, welche gleich ein ding sind und nit zwei, daz ich der erfarung heimsez." Paracelsus, *Opus Paramirum*, I, 9:45.

14. "Im herzen wechst der arzt, aus got get er, des natürlichen liechts ist er, der erfarenheit." Paracelsus, *De Caduco Matricis*, I, 8:321. Cf. also "concordieren" in Paracelsus, *De Matrice*, I, 9:210.

complementing and sometimes contradicting one another: "Paracelsus was and remained a physician, though in due course (from 1530 at the latest) he extended his efforts to heal people in the spiritual dimension."[15] His doctrine of seeing inextricably interweaves the two worlds with one another. But what does Paracelsus mean when he says that the physician "grows in the heart"?[16]

Even in purely physical terms, the eye is connected with the heart, for the eye moves in tandem with all the movements of the heart. This echo of the stirrings of the heart in the senses (the ear can also be affected) is referred to by Paracelsus as "anatomei des herzens" (the anatomy of the heart).[17] And the connection should be obvious in professional practice, in that it is the physician's eye which takes in the sick person, but his heart which is moved by his or her misery. Precisely when he mentions the worst of women's diseases, the "fallende[n] siechtag der mutter" (the falling-womb disease),[18] he urgently appeals to the compassionate eye:

> But look at the sick as they labor in paroxysms, and do not shy away from them and share with them your compassion in your eyes, and see their toil, see their misery, see their great need and fear. Let yourself be moved…, let that persuade you, since Christ himself says that the sick need a physician and the healthy do not, that the basis of medicine is truthful and arose from God and was given by God so that we should have our heart in God. Pray and beseech him that he may teach us and grant us this.[19]

This comprehensive sympathy via the eye is an essential part of visual competence. Paracelsus also makes clear why it is necessary to learn this way of seeing where the

---

15. Biegger, "*De invocatione Beatae Mariae Virginis*," 28.

16. See n. 14.

17. Cf. Paracelsus, *De Caduco Matricis*, I, 8:353.

18. Criticizing contemporary humoral medicine and its terminology, Paracelsus here coins a new term for hysteria, which was known by the label "suffocatio matricis and other erroneous names"(suffocatio matricis, und dergleichen auch mit andern irrigen name) (Paracelsus, *De Caduco Matricis*, I, 8:326). He rejects the classical concept of hysteria (in both the pathological and linguistic sense). For Paracelsus, this illness—instead of being the effect of a matrix wandering around in the whole body until suffocating the woman—belongs to the range of epileptic phenomena.

19. "Aber besehen die kranken so sie laboriren in paroxysmo, und scheuhen sie nit und teilen mit inen euer barmherzikeit in den augen, und secht ir arbeit, secht ir elend, secht ir groß not und angst. laßt euch bewegen…, lasst euch das bewegen dieweil iedoch Christus selbst spricht, die kranken bedürfen eins arzts und die gesunden nit, das do der grund der arznei wahrhaftig ist und aus got erstanden und geben, auf das wir in got sollen unser herz haben, bitten und suchen, damit ers uns lern und diselbig gebe." Paracelsus, *De Caduco Matricis*, I, 8:355. Cf. also the numerous calls for compassion, particularly in the books *De Caducis*.

anatomy and physiology of women are concerned.[20]

First, traditional humoral medicine had failed to recognize this "smallest world," lumping it together with the "small world" of the man. Since this does the woman an injustice, Paracelsus tirelessly polemicizes against the professional blindness of his colleagues. In concrete terms, he accuses bad doctors (those who practice humoral pathology) of considering and treating the state of the humors, which appears to him "as if Jill were ill and they gave the medicine to Jack."[21] Here, too, professional competence turns into a question of ethical attitudes. Paracelsus even views this blindness as a betrayal of one's Christian baptismal vows, since these are obviously at the root of the physician's power of sight: "Step forward, you false physician, you despairing Judas or whoever you are, since you were born in the baptism of Christ and yet fell from the power of that baptism, namely from the Holy Spirit, as though he were dead and no longer a physician."[22]

Secondly, essential truths lie hidden in the woman. Paracelsus refers not merely to the medical stereotype of the secrets of women (see below), but to yet another secret. The female body holds within it two worlds. The matrix itself is hidden,[23] and—in contrast to the wish for disclosure and discovery prevalent in his day—Paracelsus is content for it to remain obscure.[24] He introduces several arguments in support of this idea. One argument rests on the fact that it is its nature to remain hidden: "Nobody may search for anything or take anything away from the matrix; it is hidden. And what use is something hidden to the physician?"[25] Like the very first matrix of the world, the female birth organ is invisible to the observing eye. Apart from being able to see one's own mother as a whole woman, nobody has ever seen the uterus from which he was born.[26] But being by

---

20. For this special aspect of blindness see also Hildegard Elisabeth Keller, "Speculum-speculatio-augenlehr: Visualitätskonzepte in der (Frauen-) Medizin um 1500," in Akten des X. Internationalen Germanistenkongresses Wien 2000, ed. Peter Wiesinger, vol. 8: *Mediävistik und Kulturwissenschaften*, ed. Horst Wenzel, Alfred Ebenbauer, and Stephen C. Jaeger (forthcoming, Bern: Peter Lang).

21. "[A]ls wenn die Greta krank leg und man arznei Hansen [gäbe]." Paracelsus, *De Caduco Matricis*, I, 8:359.

22. "Trit herfür du falscher arzt oder wer bistu verzweifelter Judas, das du im tauf Christi geboren bist, und fellest aus den kreften des taufs, nemlich vom heiligen geist, als sei er tot und sei kein arzt mehr." Paracelsus, *De Caducis*, I, 8:270.

23. On the semantics of the matrix in the works of Paracelsus, see the study by Lucien Braun, "L'Idée de 'matrix' chez Paracelse," in *Paracelsus und seine internationale Rezeption in der Frühen Neuzeit*, ed. Heinz Schott and Ilana Zinguer, Brill's Studies in Intellectual History 86 (Leiden: Brill, 1998), 13–23.

24. See pp. 101–6 below.

25. "[A]us der matrix mag niemants nichts suchen noch nemen, sie ist verborgen. was nüzet dem arzt das verborgen ist?" Paracelsus, *De Caduco Matricis*, I, 8:343.

26. Paracelsus, *De Matrice*, I, 9:190–91, also 177.

nature invisible constitutes the active principle of the matrix: "But what it is that fabricates the human being, nobody sees.... But the matrix is invisible in its nature. That which is invisible suffers nothing, which is why we do not wish to speak of invisible things."[27] Thus, Paracelsus's representation of women's mysterious bodies does not invest them with danger and excitement. Nonetheless, he gives to them and to the whole woman (i.e., to both of the worlds she is) a special dignity: they do deserve to be known by women and men. This represents the decisive difference of Paracelsus's writings from the traditional texts on *secreta mulierum*—by women and men: "Men, by definition, are not secrets, but they must protect themselves against dangerous feminine secrets by *knowing* them."[28] As we will see more clearly later on, Paracelsus's position on secrets was different. Small wonder that his professional adversaries and intellectual enemies came from the ranks of his own gender and profession.

Bodily matter itself, which Paracelsus calls the "cadaver," does not yield any sort of knowledge which he would call anatomical ("thus anatomy does not lie in knowledge of the cadaver").[29] The matrix as an organ can only be understood fully if it is recognized as a workshop where the future child is formed and receives the stamp (Latin *matrix* = mold, cast) of its character and physical features.[30] Paracelsus shares the traditional medieval view on the shaping function of the uterus.[31] Part of the "anatomei" of the womb is its expansion—to a certain extent incorporeal—far beyond the organ itself, within the body of the woman.[32] As evidence for the presence of the matrix in the whole woman, Paracelsus coins the apparently paradoxical term "anatomei der ungreiflichen cörper" (anatomy of intangible bodies).[33] So anyone who wishes to carry out medical research into the uterus should observe the macrocosm, "since the woman's matrix is described in

---

27. "[A]ber was das sei das den Menschen fabricirt, das sicht niemandts.... [A]ber matrix ist unsichtbar in seinem wesen; was unsichtbar ist, das leidet nichts, darumb wir von den unsichtbaren dingen nicht reden wöllen." Ibid., I, 9:194.

28. Lochrie, *Covert Operations*, 121.

29. "[A]lso ligt die anatomei nicht in erkantnus des cadavers." Paracelsus, *De Caduco Matricis*, I, 8:337.

30. Cf. the passage in Paragraphus V in which Paracelsus is concerned in the first instance with the (formative and deformative) forces at work in the organs. The matrix is no exception. The forces are not identical with the organ, but are part of its "anatomy." Ibid., I, 8:346.

31. Cf. the section on the "genealogic forms" and the "matrix" in Georges Didi-Huberman, *L'Empreinte: Catalogue de l'exposition* (Paris: Éditions du centre Georges-Pompidou, 1997).

32. Cf. the comparisons with the shadow and the disembodied spirit (Paracelsus, *De Caduco Matricis*, I, 8:347).

33. Ibid., I, 8:347.

the book of the stars, their names, their dimensions, their nature and gestures, their health and their sickness."[34] With this plea for exploring the internal through the external ("for what does a human being hope to teach me in that which is not in him?"),[35] Paracelsus reveals himself as a cosmographer among physicians, but particularly among the gynecologists of his time. It can scarcely be coincidental that, particularly in his gynecological writings, Paracelsus advocates and teaches the training of the eye, his so-called doctrine of the [experienced] eye. It seems then that he hopes to solve the question of the physician's insight into the hidden matrix, as with all that is latent in creation, through training by personal experience, training one's own eye. Therefore, the female sex seems to provide favorable conditions for improving the physician's visual competence.[36] In this way, Paracelsus's theory of seeing has both epistemological and ethical implications since, from the perspective of medieval intellectual culture, knowing and loving are mutually dependent. Goldammer summarizes this as follows:

> The decisive factor is above all that human beings in antiquity and in the Middle Ages relate differently to the objects of (in our sense "scientific") knowledge and research. Even if the same vocabulary is used, their conception of knowledge is different.... Viewed descriptively, in the ideal case it is an act of loving devotion, even of dedication to the other, of union with it, which is reminiscent of the intuitive aspects of knowing and the way it is passed on in gnosis (which of course also finds expression in the use of "to know" for the act of procreation in Hebrew, and also partly in Greek *gignoskein* and Latin *cognoscere*). Even for Paracelsus, love and loving are the prerequisites for the process of cognition, which occurs in the interplay of faith, love, knowledge and understanding; and this is also true of scientific cognition (Labyrinthus medicorum errantium, c. 9).[37]

---

34. "[D]an dieweil der frauen matrix beschriben stehet im buch der gestirn, ir namen, ir lenge, ir weise und geberd, ir gesuntheit und ir krankheit." Ibid., I, 8:343.

35. "[D]an was wil mich der mensch leren in dem, das in im nit ligt?" Ibid., I, 8:363.

36. For Paracelus's ideal of the physician, in which he is a kind of cosmographer, see Hildegard Elisabeth Keller, "Zwo welt in einer Haut: Paracelsische 'Augenlehr' am Beispiel der Frau," in *Nova Acta Paracelsica: Jahrbuch der Schweizerischen Paracelsus-Gesellschaft* 14 (Bern: Peter Lang, 2000): 41–78; Eamon, *Science and the Secrets of Nature*, 161ff. An overview on the contemporary relationship between medicine and cosmography is given in Hannes Kästner, "Der Arzt und die Kosmographie: Beobachtungen über Aufnahme und Vermittlung neuer geographischer Kenntnisse in der deutschen Frührenaissance und der Reformationszeit," in *Literatur und Laienbildung im Spätmittelalter und in der Reformationszeit. Symposion Wolfenbüttel 1981*, ed. Ludger Grenzmann and Karl Stackmann, Germanistische Symposien, Berichtsbände 5 (Stuttgart: J.B. Metzlersche Verlagsbuchhandlung, 1983), 504–31.

37. Kurt Goldammer, *Der göttliche Magier und die Magierin Natur: Religion, Naturmagie und*

"ARZNEISCH AUGEN"

What constitutes this particular ability which makes the eyes of the physician into "arzneisch augen" (physician's eyes)? Paracelsus defines it as follows: "Just as outer seeing is suited to the farmer, inner seeing, which is secret seeing, is suited to the physician."[38] "Inner" here does not refer to anatomical dissection (which was taking hold in Italian universities during Paracelsus's lifetime), which penetrates the bodies of dead men and women. Rejecting the sort of insights gained by Andreas Vesalius (1514–1564) with his dissecting knife, Paracelsus takes the living as his starting point— "the foundation emerges from the living."[39] Thus, "inner" here refers to the inner, hidden, invisible side of life. The key term associated with it is "heimlich," or the synonymous noun "heimlichkeiten"—German expressions for the arcana, the secrets which have been planted in macro- and microcosmic creation.[40] Whether a physician can fathom this hidden side

---

die Anfänge der Naturwissenschaft vom Spätmittelalter bis zur Renaissance; mit Beiträgen zum Magie-Verständnis des Paracelsus, Kosmosophie 5 (Stuttgart: Franz Steiner, 1991), 8–9. Goldammer points to a very important issue in medieval spirituality, which seems to pass over, as a kind of early modern adaption, to Paracelsus's epistemology; cf. Gustav Adolf Wyneken, *Amor Dei intellectualis: Eine religionsphilosophische Studie* (Greifswald: Julius Abel Verlag, 1898); Robert Javelet, *Psychologie des auteurs spirituels du XIIe siècle* (Strasbourg: Muh-Leroux, 1959), esp. 114–73; idem, *Image et ressemblance au douzième siècle: De Saint Anselme à Alain de Lille*, 2 vols. (Strasbourg: Editions Letouzey & Aney, 1967), esp. 368–450; Gisbert Kranz, *Liebe und Erkenntnis: Ein Versuch* (Munich: Pustet, 1972); Colin Morris, *The Discovery of the Individual 1050–1200*, Medieval Academy Reprints for Teaching 19 (Toronto: University of Toronto Press, 1987). Authority in mystical literature is based fundamentally on the erotic interaction between a human being and a transcendent partner, disclosing divine mysteries to the former; see Hildegard Elisabeth Keller, *My Secret Is Mine: Studies on Religion and Eros in the German Middle Ages*, Studies in Spirituality Supplements 4 (Louvain: Peeters, 2000), esp. chaps. 1 and 2.

38. "[A]lso das eußer zusehen, ist dem pauren beschaffen, das inner zusehen, das ist das heimlich, das ist dem arzt beschaffen." Paracelsus, *Opus Paramirum*, I, 9:46. Paracelsus here relates the internal and the external to the three basic principles of the cosmos: *sal*, *sulphur*, and *mercurius*; in *Opus Paramirum* he treats of their effects in both worlds.

39. Cf. Barbara Duden, *Anatomie der guten Hoffnung: Darstellungen des Ungeborenen bis 1799* (Frankfurt am Main: Campus, forthcoming), chap. 10; idem., "Zwischen 'wahrem Wissen' und Prophetie: Konzeptionen des Ungeborenen," in *Geschichte des Ungeborenen: Zur Erfahrungs- und Wissenschaftsgeschichte der Schwangerschaft*, ed. Barbara Duden, Jürgen Schlumbohm, and Patrice Veit, Veröffentlichungen des Max-Planck-Instituts für Geschichte 170 (Göttingen: Vandenhoeck & Ruprecht, 2002), 11–48; see also idem, *Disembodying Women: Perspectives on Pregnancy and the Unborn*, trans. Lee Hoinacki (Cambridge, Mass.: Harvard University Press, 1993).

40. This terminological and semantic field is central for Paracelsus. For discussion of its use in the context of mystical literature in the late Middle Ages, see Hildegard Elisabeth Keller, "Absonderungen: Mystische Texte als literarische Inszenierung von Geheimnis," in *Deutsche Mystik im abendländischen Zusammenhang: Neu erschlossene Texte, neue methodische Ansätze, neue theoretische Konzepte: Kolloquium Kloster Fischingen*, ed. Walter Haug and Wolfram Schneider-Lastin (Tübingen: Max Niemeyer, 2000), 195–221.

depends on how experienced he is ("*Arcanum* is mainly to be fathomed through experience"),[41] and particularly how experienced his eye is, as Paracelsus emphasizes pleonastically: "sichtige erfarenheit vor den augen" (seeing experience for the eyes).[42] While the eye of the farmer is bound to the visible, the "physician's eye" penetrates the surface and "das Heimliche" (the secret).[43] This perspective (in an etymological sense) is the professional secret of the physician.

What is behind "das Heimliche" and its Latin equivalents? "Das Heimliche" is the invisible, the hidden, the latent and concealed, in modern terms also the uncanny. It represents one of the greatest challenges to medieval and early modern epistemology and science. Talking of "secrets" (of nature or of women) "organizes our view" (of nature or of women). William Eamon has pointed to the different thinking about secrets before and after the invention of printing, emphasizing their special discourses on the distribution or the divulging of knowledge.[44] In the teachings of Paracelsus, "das Heimliche" lies hidden in the darkness of life since life constitutes a mantle which cloaks all that happens within it: "[D]asselbig ist also ein solcher deckmantel der die ding verbirgt."[45] However, it is exclusively the human being, male and female, which constitutes the book, over and beyond all other books, "in which all secrets are written."[46] For the human being consists of the visible (the physical body made of the four elements) and the invisible (the astral body). Moreover, the terminology of sexual anatomy is also among the "Heimlichkeiten": Paracelsus terms the genitalia of man and woman "gemechte der heimligkeit" (organs of secrecy) or "heimliche Zeichen" (secret signs).[47] This terminology is in accordance with the linguistic usage of the

---

41. "Arcanum ist ein houptstuck durch die erfarenheit zu ergründen." Paracelsus, *Von des Bades Pfäfers Tugenden, Kräften und Wirkung, Ursprung und Herkommen, Regiment und Ordnung. Pfäfers, 31.8.1535*, I, 9:658.

42. Paracelsus, *De Matrice*, I, 9:185.

43. Ibid., I, 9:46, 196.

44. For the terminology of secrecy and its powerful use cf. Eamon, *Science and the Secrets of Nature*, 38–90.

45. Paracelsus, *Opus Paramirum*, I, 9:49.

46. "[D]arinnen alle Heimlichkeiten geschrieben stehen." *Liber Azoth sive de ligno et linea vitae*, in Johann Huser, *Opera, Bücher und Schrifften* (Strasbourg: Zetzner, 1603), 9:29, quoted in Alexandre Koyré, *Mystiques, spirituels, alchimistes du 16ième siècle allemand*, Collections idées 233 (Paris: Gallimard, 1971), 85 n. 3.

47. Paracelsus, *Liber de honestis utrisque divitiis* in *Sämtliche Werke*, II, 2:29: "[W]as ist dann, dass du aus deiner Hand [= Erwerbsfähigkeit] (sie sei im kopf, in füssen, in der zungen, in augen, in den gemechten der heimligkeit) vil gewinnest dein notdurft, darinnen dich dann gott nicht verlasst" (What is it then that you may gain all that you need by your own hand [whether it be in your head, in the feet, tongue, eyes, or the organs of secrecy], in which God does not desert you).

time. In investigating documents in the archives of Augsburg, Lyndal Roper finds examples of this expression that are used for the genitalia of both sexes, though more often for the female genitalia, probably because they are less visible and thus potentially less public. The expression is therefore also a stronger reflection of the taboo, the secret and hidden character of the corresponding part of the female anatomy.[48] When Sigmund Freud in his article "The Uncanny" (1919) explains the concept of *das Unheimliche* (literally, "unhomeliness," or sense of estrangement and alienation), he is elaborating on a concept already familiar to the Middle Ages. Medieval surgery was the first to see the female genitalia as the "archetype of the hidden."[49] Like Freud's idea of the "Unheimliche," this archetype is spatial (both bodily and architectural) but it nevertheless is different from his specifically modern idea of the "dedomesticated subject."[50] During Paracelsus's lifetime the surgeon Ambroise Paré (1509–1590) put it this way: "Women have as much hidden inside as men discover outside."[51]

It is in this context of "heimligkeit" that one can see that Paracelsus's gender-specific discourse of secrecy stands out from the male discourse on women's medicine of his day. It does not refer primarily to what, by a gynecological or obstetrical textual tradition from the Middle Ages into the early modern period, were known as the "Heimlichkeiten der Frauen" (women's secrets), the so-called *secreta mulierum*.[52] Paracelsus's view of physical anatomy and the "secrets" associated with physiological processes is very matter-of-fact: this obvious, physiological, anatomical "surface" of the woman must also be known to the farmer, he states,

48. For further examples of the expression *haimliche ort* or similar formulations in contemporary documents in the Stadtarchiv Augsburg, see Lyndal Roper, "Will and Honor: Sex, Words, and Power in Augsburg Criminal Trials," in *Oedipus and the Devil: Witchcraft, Sexuality, and Religion in Early Modern Europe* (London: Routledge, 1994), 53–78, esp. 56–59.

49. "Les secrets des femmes représentent pour ainsi dire l'archétype du caché." Marie-Christine Pouchelle, *Corps et chirurgie, à l'apogée du moyen âge: Savoir et imaginaire du corps chez Henri de Mondeville, chirurgien de Philippe le Bel* (Paris: Flammarion, 1983), 224.

50. Anthony Vidler, *The Architectural Uncanny: Essays in the Modern Unhomely* (Cambridge, Mass.: MIT Press, 1992), x. For medieval and early modern patterns how to organize (architectural and social) space and gender cf. Jan Hirschbiegel and Werner Paravicini, eds., *Das Frauenzimmer: Die Frau bei Hofe in Spätmittelalter und früher Neuzeit. 6. Symposium der Residenzen-Kommission der Akademie der Wissenschaften in Göttingen*, Residenzenforschung 11 (Stuttgart: Jan Thorbecke Verlag, 2000), esp. Peter Strohschneider, "Kemenate: Geheimnisse höfischer Frauenräume bei Ulrich von dem Türlin und Konrad von Würzburg," 29–45; Margarete Hubrath, ed., *Geschlechter-Räume: Konstruktion von "gender" in Geschichte, Literatur und Alltag*, Literatur-Kultur-Geschlecht 15 (Cologne: Böhlau, 2001); for medieval (gendered) concepts of secrecy in literature see Lochrie, *Covert Operations*.

51. "[L]es femmes ont autant de caché dedans que les hommes découvrent au-dehors." Ambroise Paré, *Des monstres et des prodiges* (Geneva: Droz, 1971), quoted in Pouchelle, *Corps et chirurgie*, 227.

52. See n. 6.

since he knows that she menstruates.[53] Certainly Paracelsus also discussed such secret subjects as menstruation, pregnancy, fetal nutrition, and lactation partly because he felt that there were so very many erroneous views in circulation about them.[54] Here Paracelsus reveals his remarkably different attitude. Paracelsus's discourse on female secrecy is free of polemics against the female sex. He polemicizes against men, against "blind" professionals looking at women's bodies. For Paracelsus, to consider only the restricted areas of women's medicine, those "classical" areas, would be the wrong way of looking. This is also true if a physician examines the uterus, "which lies in the lower body" (so unden im leib ligt), but treats the rest of the woman as if she were a man; such a physician "has a skin before his eyes" (ihm hängt ein Fell vor den Augen).[55]

A concealing mantle cloaks the nature of things; a "skin" hangs before the eyes of most physicians. The way Paracelsus expresses himself reveals his affective epistemology: the human being must find his way to fundamental sight, which is sight that knows phenomena at their very core. Though this observation may sound mundane, this sort of natural science, this way of considering the world, presupposes first that creatures have a fundamental core, and secondly that a human being must command a specific skill to be able to "read" himself and other creatures. First comes the question of human competence with regard to signs, which is twofold: it concerns the designation and the reading of signs. In Paracelsus's view it is a basic principle of human behavior that human beings designate the things around them, that is, provide them with signs. The human being is a *signator* if, for example, he gets Jews to wear "gelbe Flecklin am Rock" (yellow patches on their garments) for identification. The womb is a more involuntary designator, which may stamp "monstrosische Zeichen" (monstrous signs) onto unborn children as a result of fright and *imaginatio*; it too is a powerful signator.[56]

The fact that the human being constantly allocates signs, both with his semiotic instruments and within his own body, so that he himself becomes a signifier, shows that he too is subject to a fundamental cosmic principle. In an optimistic

---

53. Paracelsus, *De Matrice*, I, 9:181.

54. See ibid., I, 9:197–99.

55. Ibid., I, 9:194.

56. Paracelsus, *De Natura Rerum: Liber nonus: De Signatura Rerum Naturalium*, in *Avreoli Philippi Theophrasti Bombasts von Hohenheim Paracelsi... Opera, Bücher und Schrifften,... durch Ioannem Hvserum Brisgovivm in zehen unterschiedliche Theil, in Truck gegeben: Jetzt von newem mit vleiß ubersehen* (Strasbourg: Lazarus Zetzner, 1653), 908–21, at 908–10. Regarding this edition of Paracelsus's works, see Karl Sudhoff, *Versuch einer Kritik der Echtheit der Paracelsischen Schriften*, pt. 1 (Berlin: Georg Reimer, 1894), reprinted as *Bibliographia Paracelsica: Besprechung der unter Hohenheims Namen 1527–1893 erschienenen Druckschriften* (Graz: Akademische Druck- und Verlagsanstalt, 1958), no. 256.

view, this is perhaps the basis for his being able to learn to read the macro- and microcosm as a network of signs, an exercise which allows him to gain access to secrets which start on the surface of things but are not limited to that. On all that he created, God imprinted an expression of its (otherwise hidden) inner self, whether in its form, face, gestures, or reactions. The "Antlitz" (countenance) of the creature therefore provides the perceiver with a key with which he may unlock the innermost "Heimlichkeiten der Natur" (secrets of nature). *One* creature thus becomes the key to *another*—everything can be perceived and known from everything else. This chain of cognition leads to the human being and his understanding of himself. In contrast to the medieval interpretation of the world, which uses allegory to relate all created things beyond the world itself to their transcendent creator, early modern people begin to see the invisible in the creature itself. It is in terms of this medieval doctrine of signs, the so-called "Signaturenlehre," that Paracelsus views himself as the innovator of a doctrine, drawn from nature itself, of "verborgnen heimlichkeiten der natur" (the hidden secrets of nature).[57]

In this way, Paracelsus becomes an expert witness to the fact that from the Middle Ages to the early modern period there is an unbroken epistemological tradition of searching in what is latent, despite the undoubted hermeneutic differences. His confidence in the eye as the instrument of deeper knowledge, and in the revelation of the hidden in the small as well as the large cosmos, remains as important for our understanding of his person as his often noted solitude: "the lonely, wandering scholar seeking secrets of nature from empirics everywhere."[58] In the framework of his doctrine of signs, this confidence extends far beyond the things designated as "heimlichkeiten": it embraces the whole created world. Methodologically Paracelsus proceeds entirely in accordance with his time. He develops an empirical (in his sense) experience of the world by using "experientz" to get to the bottom of things. Thus he does make a reference to the newly discovered world (Paracelsus was born ca. 1493!), the "verborgenene insulnen" (hidden islands) and their mysterious inhabitants. But he does not reflect any more intensely on the "neue insulen" such as America, Japan, and China than he does on mermaids and other elemental spirits. They are all mysteries, of equal value, which are "noch verborgen" (still concealed) and must be fathomed.[59] It is in this comprehensive system of knowledge that the question of the sexes has its place.

---

57. Cf. Friedrich Ohly, *Zur Signaturenlehre der Frühen Neuzeit: Bemerkungen zur mittelalterlichen Vorgeschichte und zur Eigenart einer epochalen Denkform in Wissenschaft, Literatur und Kunst* (Stuttgart: Hirzel, 1999), 52 ff.

58. Eamon, *Science and the Secrets of Nature*, 143, 161 ff.

59. Cf. the preface to Paracelsus, *Liber de Nymphis*, 8 ff.

And this is where the issue of the otherness of microcosma becomes relevant. Precisely in microcosma does that which is latent meet the eye. Only this optimism can explain why in the fourth book of his *Opus Paramirum* Paracelsus concentrates so much on opening people's eyes to the differences between the sexes.

Microcosma's Monarchy

The fact that two sexes exist is seen by Paracelsus as rooted in the divine Trinity, particularly in God the Father. He explains the incarnation of Jesus Christ in the context of a division of the first person of the godhead: "He left his divinity whole, but divided the persons within the divinity from himself a woman, so that he and she are just one God."[60] The generation of the sexes from inside the Trinity leads, via the intermediary concept of creation in the image of God, to the creation of human beings as man and woman. Consequently, what is recounted in Genesis about the origin of the sexes, that each had its own generative nature, is also as God intended. But God also intends, and this is of central importance in the present context, that the very first "birth act" of man and woman will be repeated in human procreation. In Paracelsus's writings on Genesis, he explains how he sees human beings coming into existence in both a male and a female variety. The sexually defined human being is made of *massa*, which is "an extract from all creatures in heaven and earth."[61] This extract is what constitutes the human being as a "kleine welt, das ist microcosmus" (small world, that is, microcosm).[62] In the man it is the prime element, *limbus*; in the woman it is itself already an extracted and materially transformed substance: "But so that she may rule another monarchy, she is made subsequently from living flesh, which has been flesh, but which has made another flesh out of the flesh.... For that reason

---

60. "[E]r hat sein gottheit ganz gelassen, aber die personen in der gottheit geteilt von ihme ein weib, also daß er und sie nur ein gott seindt." Paracelsus, I, 3:244 ff., quoted from Alois M. Haas, "Paracelsus der Theologe: Die Salzburger Anfänge 1524/25," in *Paracelsus und Salzburg: Vorträge bei den internationalen Kongressen in Salzburg und Badgastein anlässlich des Paracelsus-Jahres 1993*, Mitteilungen der Gesellschaft für Salzburger Landeskunde, supp. vol. 14 (Salzburg: Gesellschaft für Salzburger Landeskunde, 1994), 369–82, at 376.

61. "[E]in auszug aus allen geschöpfen in himel und erden." Cf. the informative overview in Gerhild Scholz Williams, *Defining Dominion: The Discourses of Magic and Witchcraft in Early Modern France and Germany*, Studies in Medieval and Early Modern Civilization (Ann Arbor: University of Michigan Press, 1995), 45–65, esp. 54–57.

62. Paracelsus, I, 10, 648, quoted in Kilian Blümlein, *Naturerfahrung und Welterkenntnis: Der Beitrag des Paracelsus zur Entwicklung des neuzeitlichen, naturwissenschaftlichen Denkens* (Frankfurt am Main: Peter Lang, 1992), 147. Blümlein (148 n. 20) points out that "Auszug" can mean both "extract" and "selection."

the woman comes from the flesh of the man."[63] Thus, we learn about the metamorphosis to which the male sex and even more the female sex owe their existence. The repetition of the word "flesh" demonstrates not only the principle of extraction, but a principle of physical appropriation also active in digestion, which Paracelsus calls "transmutation": the first woman can "transmute" the flesh of Adam into woman's flesh, thus making her sex into the quintessence embodied by Adam.[64] Interpreting Genesis, Paracelsus teaches his readers a basic fact, a primary condition of mankind: both sexes live with "the fundamental fact that something can become something else."[65] This Paracelsian interpretation of the *massa* at the beginning of creation which has been transformed into men's flesh and, in a second act, into women's flesh makes evident how essentially material change and (gendered) identity are intertwined, even centuries after the time analyzed by Bynum.[66]

There is a second aspect linking Paracelsus's concept of (divine) creation and (human) procreation: his "homuncular ruminations," as William Newman calls Paracelsus's innovative conceptions. They represent a decisive historical step in answering the complex question whether it is possible to generate a homunculus by alchemical means. Newman traces the interaction between the (male) seed and the female matrix or any other place in which it develops its creative power. For Paracelsus, the female womb is crucial for the procreation of human beings, that is, *homines*, not *homunculi*:

> Paracelsus has extremely ambivalent views on the matter of generating seed.... If one does in fact generate seed, he or she must look very carefully to its ultimate resting place. Once the sperm has been produced, neither abstinence nor emission per se is acceptable, since both can result in the generation of uncontrolled and dangerous monstrosities. According to Paracelsus' *De homunculis*, the only proper destination for male sperm is the female womb, the one environment guaranteed not to produce a homunculus. The *De natura rerum*, on the other hand, whether genuine or not, has turned the pangenerative vice of human seed into a

---

63. "[D]arumb aber das sie ein ander monarchey füren sol, so ist sie nachfolgend gemacht aus eim lebendigen fleisch, das fleisch gewesen ist, und aber aus dem fleisch ein ander fleisch gemacht.... [A]lso ist die frau aus des mannes fleisch." Paracelsus, *De Matrice*, I, 9:201.

64. Cf. ibid., I, 9:189–90.

65. Caroline Walker Bynum, *Metamorphosis and Identity* (New York: Zone Books, 2001), 18. Cf. her definition of "change" at 19ff.

66. A similar conclusion could be drawn if we considered Paracelsus's reflections on the digestion and assimilation of food or his concepts on resurrection and the respective bodily conditions.

virtue. By means of the "alchemical" technique employed in incubating a flask at moderate heat, one can isolate the male seed from the female and thereby produce a transparent, "bodiless" homunculus. In this fashion, human art can generate a being unimpeded by the materiality of normal female birth, hence surpassing the artifice of nature.[67]

In this context one may place a passage from *De Matrice* where Paracelsus aims to draw attention to the "besondere monarchey" (separate monarchy) of the microcosma and the man, respectively, by relating the generative nature of man and woman to the creation of the first human couple according to Gen. 2:7.[68] In principle, the conception of a human child follows the example of the hand of God forming the earth. Paracelsus's argument begins in the Garden of Eden. Human beings should know "that the world has a hole in it, through which God's hand reaches from heaven and does whatever he wishes in the world, and that he made women for a world in which the human being would be born."[69] In this way the whole woman has become the matrix and "aus allen iren glidern ist des menschen acker genomen" (the human being's seedbed is won from all her limbs). The child growing within her draws "alle Kräfte der Welt" (all the forces of the world) from her—just like a fruit growing on a tree.[70]

The role of the man, on the other hand, "der mit der frauen handlet" (who has dealings with the woman) (i.e., has sexual relations with her), is associated with the literal "intervention" of God. Just as God "reached from his kingdom into the world of the heavens and earth and took the limbus and made the human being...the man must also have this grasp."[71] Although Paracelsus here sees the

---

67. William Newman, "The Homunculus and His Forebears: Wonders of Art and Nature," in *Natural Particulars: Nature and the Disciplines in Renaissance Europe*, ed. Anthony Grafton and Nancy Siraisi, Dibner Institute Studies in the History of Science and Technology (Cambridge, Mass.: MIT Press, 1999), 321–45, esp. 335–36. The power of male sperm plays an important role in more than the generation of proper human beings and homunculi; male sperm has another sacramental and even quasi-divine power: in sexual intercourse it can infuse a soul into an originally soulless "Naturgeist," or nymph; see Hildegard Elisabeth Keller, "*Homo medietas–homo mediator*: Eine Aufstiegsidee im paracelsischen Zoom," in *Homo Medietas: Festschrift Alois M. Haas*, ed. Claudia Brinker-von der Heyde und Niklaus Largier (Bern: Peter Lang, 1999), 207–20.

68. Paracelsus, *De Matrice*, I, 9:179, 197 ff.

69. "Das die welt ein loch hat, dadurch gottes hant aus dem himel in sie greift und macht in ir was er wil, und das er also die frauen zu einer welt gemacht hat, in der der mensch geboren sol werden." Ibid., I, 9:195.

70. Ibid., I, 9:195, 209–10.

71. "[So wie Gott] von seinem reich in die welt der himel und erden griffen hat und den limbus genomen und den menschen gemacht...muss der den griff auch haben." Ibid., I, 9:195.

man as "an der stat gottes" (in God's place) making evident the quasi-divine creative power of his sperm, he does not seem to be concerned with the traditional question of measuring the contribution of the man and the woman respectively in fertilization and conception. Rather, what fascinates him is the emergence of life in the invisible matrix—a process involving both sexes in the still ongoing creation. His crucial question is "aber was das sei, das den menschen fabriciert" (what it is that fabricates the human being). When he answers "das sicht niemants" (nobody sees this),[72] he alludes to the function of the *imago Dei* in Genesis, which affects both sexes.

These images are symbols of the invisible mystery of the beginning, which is perpetuated in the conception of all human children, "because just as God created man in his own image, the same is still happening."[73] In linking the moment of the first creation with that which is constantly repeated in human procreation, Paracelsus can switch almost at will between the different levels of the time of Genesis and his own time. In the passage immediately following he again summarizes his view of the creative "Geist des Herrn" (spirit of the Lord), which moved over the waters and created life. Paracelsus seems to see God imprinting his own image on human beings in the process of creation as mirrored in the process of procreation as well as in its central locus, the matrix, where the child receives its stamp, as described above. The difference from Genesis is revealed to be that, in place of the "large world," the "smallest world" has now become the seat of life, and man can thus intervene in the sexual transmission of life. Both sexes cooperate in this human continuation of divine creation by sexual means. This linking of Genesis with *genus/sexus* presents what is arguably the most demanding task for the visual competence of the physician and the gynecologist. Paracelsus seems to be aware of this and advocates—at the very least—that the physician should see that there *are* "gendered secrets," gendered in two senses: secrets associated with the human genus and with the sexes.

Thresholds

There is a synchronicity that seems more than coincidental. Paracelsus's demands for a doctrine of seeing informed by Christianity came at a time when, in the context of a newly legitimated curiosity, the eye reached and penetrated ever further

---

72. Ibid., I, 9:194.

73. "[D]an zu gleicher weise wie got den menschen beschuf nach seiner bildnus, derselbige tuts noch." Ibid., I, 9:194.

into the undiscovered darkness of both the large world and the small world.[74] His vehemence and his unconventional solution to a fundamental epistemological problem can be seen more clearly in this scientific and cultural context.

Long before and during Paracelsus's lifetime, the sphere called "microcosma" (in physical terms, the lower abdomen of the woman) was considered a taboo subject. The eye was excluded, shut out, from it. Even in the late Middle Ages the female genitalia were not accessible to the eyes of those responsible for medical practice, particularly obstetric practice. Obstetrics was literally in the hands of the midwives.[75] The extent of their practical and theoretical knowledge can scarcely be determined any longer, either in absolute terms or in relation to the university-trained, Latin-speaking male physicians, who, from the fifteenth century onwards, also practiced anatomical dissection. It is certain, however, that direct visual and tactile contact with the lower abdomen of the woman developed into a gendered matter of dispute and a power struggle between 1500 and 1800.[76]

> Only men were trained and permitted to perform dissections. Indeed, in the early modern period men wielding the surgeon's knife gained access to women below the skin, probed ever deeper into the mystery of the female body, mapped it and named it for themselves (e.g., the fallopian tubes were named after university professor Gabriele Falloppio [1523–62]); essentially, male anatomists in the early centuries of modern medicine colonized women's bodies.[77]

Like his colleagues in the profession who polemicized against the incompetence of midwives, Paracelsus offers his opinions on gynecological questions from a very different point of view, but also as a physician who is primarily academically

---

74. Hans Blumenberg, "Der Prozeß der theoretischen Neugierde," pt. 3 of *Die Legitimität der Neuzeit*, rev. ed. (Frankfurt am Main: Suhrkamp, 1996).

75. For the interplay of eye and hand, text and image, male professionals and female midwives in late medieval gynecological literature, see Keller, "Speculum-speculatio-augenlehr."

76. For this important dispute and its context cf. Evelyne Berriot-Salvadore, "The Discourse of Medicine and Science," in *Renaissance and Enlightenment Paradoxes*, vol. 3 of A History of Women in the West, ed. Georges Duby et al. (Cambridge, Mass.: Belknap Press, 1993), 348–88; idem, *Un corps, un destin: La femme dans la médecine de la Renaissance*, Confluences—Champion 5 (Paris: Champion, 1993); Danielle Jacquart, "La morphologie du corps féminin selon les médecins de la fin du moyen âge," Micrologus: Natura, scienze e società medievali: I discorsi dei corpi 1 (1993), 81–98; Manuel Simon, *Heilige, Hexe, Mutter: Der Wandel des Frauenbildes durch die Medizin im 16. Jahrhundert* (Berlin: D. Reimer, 1993); Monica H. Green, *Women's Healthcare in the Medieval West: Texts and Contexts* (Aldershot: Ashgate, 2000) and the literature quoted in n. 5.

77. Lynne Tatlock, "Speculum Feminarum: Gendered Perspectives on Obstetrics and Gynecology in Early Modern Germany," *Signs* 17 (1992): 723–60, 733.

educated and not involved in obstetric practice. The embittered struggles about it began around Paracelsus's day, and he was involved in them. As a radical theorist of gendered medicine, he advocated the theory that the sexes were differentiated from the very beginning: from the beginning, men and women exist as gendered entities through a fundamental change, the transmutation of their flesh.[78]

In the mid-sixteenth century the trenches were dug that characterize the history of gynecology, obstetrics, and anatomical dissection of the female genitalia in the early modern period. One of the central disputes was over the methodology of cognition: concretely, the question of whether manual palpitation by an experienced woman was inferior in its diagnostic potential to visual examination by the male physician.[79] The eye becomes the instrument of power by acquiring both perspective and insight. This can be proven not only in the exploration of women's inner sexual anatomy but also in its outer details. Charles Estienne, who (re)discovered the clitoris, shows this expansion of the male interest in the frontispiece of his *De la dissection des parties du corps humain:* "En ce protraict / t'est assez confusement remonstré ce qui appartient en partie au membre honteux de la femme / qui depend de la description de la matrice."[80] The clitoris and its function moved into the center of disputes between surgeons and anatomists in Latin and vernacular publications.[81] One of them, the Paduan anatomist Mateo Renaldo Colombo,[82] recently had his ambivalent revival in *El Anatomista* (The

---

78. Paracelsus's medico-theological writings are an ideal starting point for a differentiated revision of Thomas Laqueur's *Making Sex: Body and Gender from the Greeks to Freud* (Cambridge, Mass.: Harvard University Press, 1990). Psychoanalysis attempts to visualize the difference between the sexes in ways very different from Paracelsus's medical theology. The result, at least in Jacques Lacan's work, is that the signifier of this difference is the phallus, leaving the signification of the female sex a blank. Monika Gsell, in turn, has pointed out that visualizations of the female genitalia have a rich cultural history, to which the Middle Ages made a significant contribution. Monika Gsell, *Die Bedeutung der Baubo: Kulturgeschichtliche Studien zur Repräsentation des weiblichen Genitales*, Nexus 47 (Frankfurt am Main: Stroemfeld, 2001).

79. See Tatlock, "Speculum Feminarum," and for the feminization of the geographical space by the strategy of appropriation by naming discoveries see Monika Wehrheim-Peuker, "Die Konstruktion eines kolonialen Raumes: Die Feminisierung Amerikas," in *Geschlechter-Räume: Konstruktion von "gender" in Geschichte, Literatur und Alltag*, ed. Margarete Hubrath, Literatur-Kultur-Geschlecht 15 (Cologne: Böhlau, 2001), 163–78.

80. Charles Estienne, *De la dissection des parties du corps humain* (Paris: Simon de Colines, 1546), quoted in Katherine Park, "The Rediscovery of the Clitoris," in *The Body in Parts: Fantasies of Corporeality in Early Modern Europe*, ed. David Hillman and Carla Mazzio (New York: Routledge, 1997), 171–93, 170.

81. For the different positions, see the lucid analysis by Katherine Park, "The Rediscovery of the Clitoris," 175–79.

82. Mateo Renaldo Colombo, *De re anatomica* (Venice, 1559).

anatomist) (1997), by Federico Andahazi,[83] a "Eurotrashy period piece," as one critic, Lisa Zeidner, called it.[84]

What is interesting for our present purposes about this commercially successful and—at least in certain Argentine circles—scandalous novel (which "mocks its own historical pretensions," as Lisa Zeidner writes)is the author's use of the invasive male eye as an emblem of colonial power. The preface, entitled "The Dawn of Observation," contains a sketch that typifies the author's same-old-story approach to the epoch: "[N]ot a single map is left unchanged. The cartography of Heaven changes as well as that of Earth and that of the body. Here now are the anatomical maps that have become the new navigational charts of surgery."[85] Andahazi's plot concerns two parallel discoverers, Columbus and Colombo, devoted respectively to the exploration of the macrocosm and the microcosm. The latter (echoing *De re anatomica*) exclaims, "O my America, my new-found-land!"[86]

However one-dimensional his portrait of the first decades after Paracelsus, Andahazi does show a sensitivity to the burning issue which seems to be a key to the intellectual climate of the time: the eye, an eye, travels the world, looking both inwards and outwards, and charts this world afresh.[87] In the combination of these two factors—knowing the outer through the inner and vice versa—lies the epistemological "central concept of the epoch": "not only that which bordered on the known world in the old *mappae mundi* was viewed as 'terra incognita,' but also that which lay hidden in the inner space of the human being."[88]

Geographical conquest and self-empowerment become mutually supportive. Here the epistemological horizon of this and the following centuries opens up: the initially discovering eye is not yet necessarily the truly seeing eye. How far the one has to travel to meet the other has been demonstrated impressively by Barbara

---

83. Federico Andahazi, *El anatomista: La historia del descubrimiento que pudo revolucionar el amor* (Barcelona: Planeta, 1997); trans. by Alberto Manguel as *The Anatomist* (New York: Doubleday, 1998); subsequent references are to the English translation.

84. Review by Lisa Zeidner, "Private Parts," *New York Times Book Review*, 13 September 1998.

85. Andahazi, *The Anatomist*, 4.

86. Andahazi, *The Anatomist*, 3.

87. Despite her understandable skepticism regarding the literary language ("trumped-up, breathless language") and the frequent historical reductiveness, Lisa Zeidner's rejection of Andahazi's reconstruction of the history of anatomy ("So Andahazi does not waste much time on credibility") seems too absolute if one considers what Katherine Park has had to say about the subject.

88. Jürgen Schlaeger, "Der Diskurs der Exploration und die Reise nach innen," in *Weltbildwandel: Selbstdeutung und Fremderfahrung im Epochenübergang vom Spätmittelalter zur Frühen Neuzeit*, ed. Hans-Jürgen Bachorski and Werner Röcke, Literatur-Imagination-Realität 10 (Trier: Wissenschaftlicher Verlag Trier, 1995), 135–45, at 135.

Duden looking at anatomical drawings of the unborn human being (between 1493 and 1799). The process goes through many stages in which the anatomy of the pregnant woman is redefined over and over again, from the stage of discovery (in the sense of the physical revelation of the embryo) to the stage of seeing the actual fact.[89] Strikingly, in her context of optical history, she too parallels the events on the coasts of Hispaniola and in Italian anatomy, since in both discoveries one sees attempts to grasp the unknown using known categories: "[J]ust as the caravels transported the classically educated eye across the oceans of the world, the dissecting knife opens up the inside of the body to the gaze."[90]

Paracelsus evidently shares his contemporaries' interest in the undiscovered, the invisible and mysterious, in the macrocosm and the microcosm. He participates in the discourses of discovery, anatomy, and magic, which lead outwards to America and to Nature and inwards into the human, particularly female, anatomy of the woman.[91] He views the latent as a divine challenge to humanity, to the human powers of cognition. Humankind is challenged to know God and know itself, says Paracelsus: "God does not wish anything to remain secret or hidden, but rather that everything should be made manifest that he has created in nature, that this same should be experienced."[92] With his intention to see as deeply as a human being can, he participates in early modern attempts to "toughen up" the eye by sending it out into obscure, mysterious, unfamiliar, and unknown fields.

However, Paracelsus's participation is subtle. His ideal of the eye sharpened for "heimlichkeiten," provides a counterbalance to his contemporaries' instruments of conquest and colonization—"Kulter und Karavelle" (dissecting knife and caravels), as Barbara Duden puts it—even if, as history in Paracelsus's own lifetime and beyond has shown, these instruments did conquer the world. To what school, then, does Paracelsus send the human eye, particularly the physician's eye? It must differ both from the haptic school of the Italian anatomists and the cartographic school of the conquistadors (charting the known in the unknown). Briefly, one can say that the physician has to develop a sort of X-ray eye if he wants to track down the "heimlichkeiten der natur." He should therefore not merely "die Augen sättigen" (feast his eyes), but should be filled with wonder and embark on

---

89. Cf. Duden, *Anatomie der guten Hoffnung*, chap 6 and 7; and idem, *"Zwischen 'wahrem Wissen' und Prophetie."*

90. Duden, *Anatomie der guten Hoffnung*, 141.

91. See Williams, *Defining Dominion*, 3–12 and 45–65; also Wehrheim-Peuker, "Die Konstruktion eines kolonialen Raumes," 163–78.

92. "[D]as got nicht will, das ichts heimlich oder verborgen bleib, sonder alles offenbar werde, was er in der natur geschaffen hat, das das selbig erfaren wird." Paracelsus, *Sämtliche Werke*, I, 12:123.

research.[93] Paracelsus supports the school for deciphering and reading the books of the micro- and macrocosm. Paracelsus conceives of what the eye must learn there in a manner completely removed from the contemporary idea of delight for the eye. The eye must learn to read the world—the large world, the smaller, and the smallest world—by penetrating into the spiritual and thus divine dimension of each one.

Paracelsus leaves it to each individual to develop this visual competence. Only the physician who has sharpened his *own* eye and in this way deepened his *own* level of experience does not have to rely on books and medical traditions (which are wrong anyway) and can instead depend on himself: "Let each person himself look to his own skill and experience…look to yourself; I experience things for myself, not for you. If I give you some of my experience, then you should experience it as much as I; in that way you will be my equal."[94] Here Paracelsus remolds his own personal motto, "Alterius non sit qui suus esse potest" (Let no man belong to another if he may belong to himself), as a professional creed. One could demonstrate with quotations from throughout his works that here Paracelsus is speaking with the particular vehemence of autobiographical experience. The idea of training the eye, including training his own eyes, can be shown to be a sort of *basso continuo* underlying both his professional ethics and his theology. Visual competence also proves (and this must be judged in light of Paracelsus's lifelong wanderings, which, to a great extent, required him to leave material things behind) to be a personal possession which cannot be lost: "Had I not put myself in the way of experience, if I had had to rely on the old authors, I would have been born completely blind in medicine, without eyes."[95] He who learns to see with his own eyes thus not only makes experience his own, but also—in light of Paracelsus's life motto—takes possession of himself.

Deciphering and reading in Paracelsus's sense contain an additional dimension which is quite new in relation to "schools of vision" of his time: the aspect of the inchoate, the nonfinite. Against this theological background, seeing becomes a process which is intended to be emotional and which constantly begins afresh, since the mysteries of God inscribed in the cosmos are always hidden from

---

93. Cf. the foreword to Paracelsus, *Die Bücher von den unsichtbaren Krankheiten / De Causis Morborum Invisibilium* (The books on the invisible illnesses), I, 9:251–58.

94. "[S]chau ein ietlicher auf sein kunst und erfarnheit selbst…schau du auf dich selbst; ich erfar mir, dir nichts. gib ich dir von meiner erfarnheit, so erfars als wol als ich, so stest du mir gleich." Paracelsus, *De Matrice*, I, 9:246. For such references to the individual experience and authority in early modern history of science cf. Eamon, *Science and the Secrets of Nature*.

95. "[H]et ich mich selbs in die erfarnheit nicht geben, von den alten [scribenten] wer ich stockblint geborn in der arznei, on augen," quoted in Pörksen, "Paracelsus und der Augenschein," 4.

human interpretation. This means that the human being is not in pursuit of a particular "something" which he can then reveal, discover, label, name, or dismiss in whatever way he chooses. He is not looking for any individual phenomenon, but rather he seeks to visualize the mystery of the Living itself, the "signs" of which are merely inscribed onto individual phenomena: "There is nothing so secret in the human being that it does not have an outward sign."[96]

In addition, and this is important for Paracelsus, in this way the human being undertakes a journey of knowledge and experience which will lead him to himself. This does not mean only the individual *nosce te ipsum* (know thyself), mentioned above, and the expansion of one's treasury of experience. Via the individual dimension, this sort of seeing leads to the gender-specific conditions of both sexes and to generic humanity: seeing is, for Paracelsus, the most elevated, the most truly human office.[97] He who follows the chain of wonders and tracks down their signatures is fulfilling his true vocation.[98] If he sees properly, then, the physician also serves humanity *sub specie aeternitatis*. His eye service is a sort of service to God, "in which lie the secrets which should be revealed by a doctor; in this God is praised."[99] With this end in view, Paracelsus's training of the eye—using the object of the sexes and particularly the special case of the woman or microcosma—is revealed also to provide instruction in serving God in worship.

Acknowledgments: All translations were provided by me in collaboration with the translator of this article, Dr. Maria C. Sherwood Smith. I would like to thank her very much for her commitment.

---

96. "[N]ichts is so heimlichs im menschen, das nit ein auswendig zeichen an im hat." Paracelsus, *Von den natürlichen Dingen*, in *Sämtliche Werke*, I, 2:59–386, at 86.

97. Cf. the introduction to Paracelsus, *Liber de Nymphis* and Alois Maria Haas, "Hohenheims dynamisches Sehen," in *Theophrastus Bombastus von Hohenheim genannt Paracelsus: Standpunkt und Würde: II. Dresdner Symposium* (Dresden: Deutsche Bombastus-Gesellschaft, 1999), 12–21; and idem, "Unsichtbares sichtbar machen." For exploring issues of the male and especially the female sex in his special and original way, Paracelsus should have been mentioned in the *Handbook of Medieval Sexuality*, ed. James A. Brundage and Vern L. Bullough, Garland Reference Library of the Humanities 1696 (London: Garland, 1996).

98. See n. 13.

99. "[D]arin dan ligent die heimlikeit die in einem arzt offenbar sollen sein, dardurch got gelobt wird." Paracelsus, *De Caducis*, I, 8:272.

Paracelsus, frontispiece from Paracelsus, *Etliche Tractaten* (Cologne: Birckmann, 1567), permission of Becker Medical Library, Washington University, St. Louis, Missouri.

Paracelsus on Baptism and the Acquiring of the Eternal Body

Dane Thor Daniel

In order to understand better the anthropology, natural philosophy, and medicine of Theophrastus Bombast von Hohenheim, called Paracelsus (1493/4–1541), it is important to turn to his theology, for there is an essential unity between the natural and religious elements of his *Weltanschauung*.[1] One cogent example regards his sacramental theology, which may be termed the *untötliche philosophei* (immortal philosophy).[2] Paracelsus elaborates the scriptural "new

---

1. Many scholars take for granted the separation between Paracelsus's theology and medicine/natural philosophy, a fact conspicuous in the bipartite division of the modern edition of Paracelsus's collected works. See Paracelsus, *Sämtliche Werke, I. Abteilung: Medizinische, naturwissenschaftliche und philosophische Schriften*, ed. Karl Sudhoff, 14 vols. (Munich and Berlin: R. Oldenbourg, 1922–33). The first volume of the second division is Paracelsus, *Sämtliche Werke, II. Abteilung: Die theologischen und religionsphilosophischen Schriften, Erster Band, Philosophia magna I*, ed. Wilhelm Matthiessen (Munich: Otto Wilhelm Barth, 1923). Under the direction of Kurt Goldammer, several more volumes of the second division were published between 1955 and 1986. The Zurich Paracelsus Project aims to edit the rest of Paracelsus's theological manuscripts, thus providing the long-awaited completion of his collected works. The Zurich Paracelsus Project, Institute and Museum for the History of Medicine, University of Zurich, aims to edit the rest of Paracelsus's theological manuscripts, thus providing the long-awaited completion of his collected works. See Urs Leo Gantenbein, The Zurich Paracelsus Project [www.paracelsus.unizh.ch]. Hereafter I will refer to works using the Roman numeral to denote the part of the collected edition of the complete works of Paracelsus followed by the volume and page number.

2. Scholars such as Kurt Goldammer, Ernst Wilhelm Kämmerer, Hartmut Rudolph, and Michael Bunners, mostly in their discussions of the Lord's Supper writings (*Abendmahlsschriften*), have focused attention on the integral role of Paracelsus's sacramental thought within the whole of his works: Kurt Goldammer, *Paracelsus: Natur und Offenbarung* (Hanover: Theodor Oppermann Verlag, 1953); idem, *Paracelsus in neuen Horizonten: Gesammelte Aufsätze* (Vienna: Verband der wissenschaftlichen Gesellschaften Österreichs, Verlag, 1986); Ernst Wilhelm Kämmerer, *Das Leib-Seele-Geist-Problem bei Paracelsus*

creature,"[3] holding that the Christian attains an eternal body through baptism and then nourishes it via the Eucharist. The present study includes an examination of Paracelsus's baptismal teaching as found in *De genealogia Christi* (1530s), the *Abendmahlsschriften* (eucharistic writings), *Vom tauf der Christen* (ca. 1534), *Libellus de baptismate Christiano* (ca. 1531–34), *Vom tauf der Christen: De baptismate* (n.d.), and his magnum opus, the *Astronomia magna* (1537/38).[4] Although these texts provide fascinating insights into Paracelsus's stances on a variety of theological issues (e.g., *sola scriptura, sola fideism, ex opere operato*, sacramental hylomorphism [matter-form theory applied to the sacraments]),[5] the focus of this paper is limited to the sections of these books bearing on Paracelsus's views concerning the means and consequences of receiving immortal corporeality.

The first section will summarize the "untötliche philosophei," which Paracelsus shaped during the early to mid-1530s in his *Abendmahlschriften* (Lord's Supper writings), and the role of baptism therein. The second section engages some of the characteristics of Paracelsus's picture of baptism, including the elements of the ceremony, the effects of baptism on the communicant, and other means of receiving the effects of baptism. Paracelsus finds baptism important not only for its soteriological value, but also on epistemological and medicinal grounds.

---

und einigen Autoren des 17. Jahrhunderts (Wiesbaden: Franz Steiner, 1971); Hartmut Rudolph, "Hohenheim's Anthropology in the Light of His Writings on the Eucharist,"in *Paracelsus: The Man and His Reputation, His Ideas and Their Transformation*, ed. Ole Peter Grell (Leiden: Brill, 1998), 187–206; and Michael Bunners, "Die Abendmahlsschriften und des medizinisch-naturphilosophische Werk des Paracelsus" (Berlin: Humboldt University, inaugural diss., 1961). Regarding the use of the term "untötliche Philosophei," Rudolph notes that Paracelsus himself uses the phrase "volumen von der untötlichen philosophei" to describe a group of writings in his *Prologus et initium voluminis limbi aeterni*. See Rudolph, "Hohenheim's Anthropology," 188; and Leiden Codex Vossianus Chymicus, fol. 24, f. 7r, Leiden University Library.

3. 2 Cor. 5:17: "Therefore if any man be in Christ, he is a new creature: old things are passed away; behold, all things are become new." The English translation of the Bible used in this study is the King James Version.

4. Paracelsus, *De genealogia Christi*, in *Sämtliche Werke*, II, 3:59–164; idem, *Vom tauf der Christen*, in *Sämtliche Werke*, II, 2:327–66 (also II, 1:317–59); idem, *Libellus de baptismate Christiano*, in *Sämtliche Werke*, II, 2:367–78; idem, *Vom tauf der Christen: De Baptismate*, in *Sämtliche Werke*, II, Supp.: *Religiöse und sozialphilosophische Schriften in Kurzfassungen*, ed. Kurt Goldammer, Norbert Kircher, and Karl-Heinz Weimann (Wiesbaden, Franz Steiner, 1973), 113–23; and idem, *Astronomia Magna oder die ganze Philosophia sagax der großen und kleinen Welt samt Beiwerk*, in *Sämtliche Werke*, I, 12:1–444.

5. Paracelsus's divergence from sacramental hylomorphism is briefly addressed by Dane Thor Daniel, "Paracelsus—die Sakramentenlehre und das Verhältnis von Religion und Naturwissenschaften in der wissenschaftlichen Revolution," in *Manuskripte, Thesen, Informationen; Herausgegeben von der Deutschen Bombastus-Gesellschaft* 16 (2000): 17–26.

The Immortal Philosophy

Prominent within Paracelsus's works of the 1530s, ranging from the dogmatic, polemic, and eucharistic writings to the *Astronomia magna*, is the explication of the immortal philosophy, the segment of Paracelsus's thought regarding the new creation. Paracelsus understands the first creation to be the physical creation by God the Father as related in Genesis; the new creation is that enacted by Christ, God the Son. The immortal philosophy offers a case study of Paracelsus' tendency to inform his biblical exegesis with naturalism while at the same time invoking Scripture to explain fundamental components of cosmogony and cosmology. Paracelsus imagines the new creation, which exists on earth and in heaven, in terms of its analogy with the physical world, drawing parallels between the temporary and immortal creation, the worldly and heavenly bread, the old and new soil (*limbus*), and the mortal and eternal body.[6] Such analogy is found in Scripture, but Paracelsus, the physician and natural philosopher, makes the eternal more concrete by emphasizing its natural dimensions. He talks about the generation, growth, and nourishment of the new body, and even discusses the eternal soil in which the eternal food grows and from which comes the material of the new body. The basis of Paracelsus's two-creation concept, as well as its application to the new body, is scriptural. He is influenced by, for example, 1 Cor. 15, John 6, and, notably, Gal. 6:15: "For in Christ Jesus neither circumcision availeth any thing, nor uncircumcision, but a new creature."[7]

Based on a study of the eucharistic writings, Hartmut Rudolph explains Paracelsus's belief that a person needs both body and soul, whether in heaven or

---

6. Paracelsus often employs the term *limbus*, especially when discussing the new creation. The word is perhaps related to the *limus terrae* as found in the Vulgate version of Gen. 2:7: "[F]ormavit igitur Dominus Deus hominem de limo terrae."

7. See 1 Cor. 15:44–47, "It is sown a natural body; it is raised a spiritual body. There is a natural body, and there is a spiritual body. And so it is written, The first man Adam was made a living soul; the last Adam was made a quickening spirit. Howbeit that was not first which is spiritual, but that which is natural; and afterward that which is spiritual. The first man is of the earth, earthy: the second man is the Lord from heaven." See also John 6:27, "Labour not for the meat which perisheth, but for that meat which endureth unto everlasting life, which the Son of man shall give unto you: for him hath God the Father sealed." Goldammer and Kämmerer also discuss Job 19:26. Note the Vulgate version of Job 19:26: "[E]t rursum circumdabor pelle mea et in carne mea videbo Deum." Charles D. Gunnoe, Jr., addresses the subject of Paracelsus's anthropology, including a discussion of Thomas Erastus's criticism of Paracelsus's approach to the resurrection body, in "Erastus and Paracelsianism: Theological Motifs in Thomas Erastus' Assault on Paracelsianism," in *Reading the Book of Nature: The Other Side of the Scientific Revolution*, ed. Allen G. Debus and Michael T. Walton (Kirksville, Mo.: Sixteenth Century Journal Publishers, 1998), 45–65. See also Kämmerer, *Das Leib-Seele-Geist-Problem*, 24; and Goldammer, *Paracelsus: Natur und Offenbarung*, 85. Paracelsus addresses Job 19:26 in the *Astronomia Magna*, I, 12:296–97; and *De genealogia Christi*, II, 3:89.

on earth. Humans, born of the spirit of God, were supposed to remain eternally in "the unity resulting from God's marrying body and soul."[8] When the Fall ruined this marriage, God fashioned a new marriage through the Son of God, who provides the soul with an immortal body which is born out of the heavenly limbus; baptism, which completes this process, is a necessary "precondition of immortal life."[9] Although the elemental body dies, because of the acquisition of a new body, the saved human still exists in spiritual and bodily form after death.

Perhaps the most vivid expression of Paracelsus's depiction of the new creature is found in *De genealogia Christi*, a work in which, as Goldammer writes, Paracelsus constantly builds connections to his notion of how humans take on a holy form via participation in the "divine-spiritual corporeality" of Christ.[10] Paracelsus stresses that in order to become "a new human of an eternal, immortal body,"[11] a special type of faith is required: only those believing in the second part of the Trinity, Christ, receive the eternal body. Accordingly, he categorizes two types of believers: the believer in God as a creator, but not a father, and the believer in God as not only a creator, but a father. Paracelsus writes that everyone believes in God as creator of the physical world, but that those limited to this belief, namely, those not believing in the Trinity, may only participate in the creation of God the Father; they are excluded from the creation of the Son and Holy Spirit. These people desire the carnal and believe in only what they are able to sense physically. The other type of believer, the Christian believer, recognizes God as a father—more specifically, as a father to whom a son was born through a virgin. Although the two types of believers do equally receive that created by the Father, i.e., "sun, earth, and water," those of the first type do not acquire the reward (*belohnung*) of the Son and Holy Spirit, and thus not the Son's kingdom.[12]

---

8. Rudolph, "Hohenheim's Anthropology," 193.

9. Ibid., 192–202.

10. Goldammer, "III. Einleitungen zu den Schriften dieses Bandes, 2. De genealogia Christi," in Sämtliche Werke, II, 3:xxx–xxxi. "Dabei werden aber dauernd Brücken geschlagen zur Gestaltwerdung der Heiligkeit im Menschen allgemein durch Teilhabe an der göttlich-geistigen Leiblichkeit Christi und zur Frage der "Gemeinschaft der Heiligen," d.h. der Christen...."

11. See the application of 1 Cor. 15:44–53 in Paracelsus, *De genealogia Christi*, II, 3:71: "So far, however, as it concerns the son, the second person of the Trinity, his gift is that he has produced a new creature, a new human of an eternal, immortal body." (Sovil aber und als ein son betrifft, das ist die ander person der trinitet, ist das sein gab, daß er ein neue creatur geschaffen hat, ein neuen menschen eins ewigen, untödlichen leibs.)

12. Paracelsus, *De genealogia Christi*, II, 3:70: "Thus, the first gift is that of God the Father. He made heaven and earth for the nourishment of the human. Thus, the entire world, believer and nonbeliever alike, believes in the creator of heaven and earth; however, there is a difference in that the believers recognize God the creator as a father who bore his Son of a virgin.... Now the belief [in the creator]

Only believers in the Son, believers whom Paracelsus calls the chosen (*erwöhlten*), procure holiness.[13]

Thus, Paracelsus draws a dichotomy between the Father's creation, which is the realm of things involving time, and that of the Son, the eternal. In the creation of the mortal world, the Father created Adam last, and in the new creation, Christ created the eternal body last. This new body is made of flesh and blood, suffers neither death nor illness.[14] One may receive the new flesh and enter the eternal only through the Son, not through the Father.[15] Bringing us to the sacrament of the Lord's Supper, Paracelsus evokes Christ's statement, "Blessed is he that will eat bread in the kingdom of heaven."[16] Thus, there are two types of bread, that of the first creation, and the bread in the kingdom of heaven which is the bread of the Lord's Supper.[17] One provides nourishment to the earthly body, the other to the heavenly body.[18] Slightly modifying the spelling of *limo* in the phrase *de limo terrae* of Gen. 2:7, and infusing the microcosm-macrocosm theory, Paracelsus notes that God formed Adam "out of the macrocosm, of the dust of the earth" ("aus der großen welt von dem limbo terrae").[19]

---

is such that God gives all of them the use of the earth's things,... and he lets the sun shine on them all, even if they do not possess a complete faith; they do indeed have faith in a creator god. The benefit that they receive is in accordance with the extent of their faith. They believe what they see; what they see, they have from the creator. They do not believe in the Son and thus do not receive that given by the Son nor that given by the Holy Spirit." (Also ist die erst gab von gott dem vatter. Der hat himmel und erden beschaffen dem menschen zu seiner nahrung. also glaubt die ganz welt in den schöpfer himels und der erden, glaubig und unglaubig alle gleich, iedoch mit der undterschaidt, daß die glaubigen gott den schöpfer erkennen als ein vater, aus dem er seinen son geboren hat von einer jungfrauen.... Nun ist der glauben sovil, daß gott allen denen gibt von der erden den nutz,... laßt ihnen allen die sonnen scheinen, ob sie gleich nit eines volkomben glaubens seindt; so seindt sie doch des glaubens in gott schöpfer. Als vil als sie glauben, empfahen sie belohnung. Sie glauben, das sie sehen; das sie sehen, das haben sie vom schöpfer. Sie glauben nit in son; darumb empfahen sie auch die belohnung nit vom son. Auch nit in heiligen geist.)

13. Ibid.: "Therewith they are not included among the number of chosen after death, for nobody becomes blessed through faith in the Father." (Damit seindt sie nach dem tod nit in der zahl der erwöhlten, dann durch den glauben in vatter wird keiner selig.)

14. Ibid., II, 3:72.

15. Ibid., II, 3:75.

16. Luke 14:15: "And when one of them that sat at meat with him heard these things, he said unto him, Blessed is he that shall eat bread in the kingdom of God."

17. Paracelsus, *De genealogia Christi*, II, 3:72.

18. Ibid., II, 3:80.

19. Ibid., II, 3:77. The editors add, "Vgl. Zu dieser Vorstellung schon den Liber de podagricis, De limbo; 1. Abt., I, 316f.; etwa im Sinne des späteren Begriffs der 'prima materia'. Zur theologischen Auswertung siehe De sensu et instrumentis; Bd. II, 86, bes. Anm. A. Vgl. Auch unten S. 131."

The eternal *limbus* leads to the topic of baptism. Paracelsus writes that the eternal *limbus* is that from which the new human of the *wiedergeburt* (rebirth) is made, from which his flesh and blood are taken.[20] All Christians are born anew, receiving eternal flesh and blood; they are different beings from those possessing only mortal flesh. He says Christians should not live according to the lasciviousness (*Wollust*) of the earthly flesh, but should live in concert with the flesh that gives life.[21] Here Paracelsus begins to clarify his position on baptismal efficacy, emphasizing Mark 16:16: "He who believes and is baptized will be saved."[22] Faith is the beginning of the baptism, and this faith, ignited by the Holy Spirit (*der erleuchter*), effects the new birth. "Through water and the Spirit the human is born again in the power of baptism."[23]

---

20. Ibid., II, 3:131. "The same incarnation is the eternal human, and he is the limbus, from which the new human of the rebirth is made, from which he takes his blood and flesh." (Dieselbig incarnation ist der ewig mensch, und er ist der limbus, aus dem der neu mensch der wiedergeburt gemacht wird, von dem er sein blüet und fleisch nembt.) See also footnotes t and u to this passage: "Zur paracelsischen Anschauung des ewigen, himmlischen Leibes vgl. u. a. die Abendmahlschriften, z. B. De coena Domini ad Clementem VII. oder Von der Wiedergeburt des Menschen." "Urmaterie, Urstoff, Urgrund im Vorstadium der Schöpfung, wohl eine frühere Fassung des späteren 'prima materia'-Begriffs, zu unterscheiden vom 'limus (terrae)' der Vulg. (Gen. 2, 7). Vgl. auch unten, S. 134; K. Goldammer, Bermerkungen zur Stuktur des Kosmos und der Materie bei P., FS Artelt (1971), S. 128f. u. ö."

21. Ibid., II, 3:130: "First note that we have a gospel and a clear text (*lautere geschrift*), which shows us that we must be born a second time, that is, not of human seed nor again from the womb of women, rather from above through the water and the holy spirit. Since the birth of Christians only lies in this, that we are accordingly born into another flesh and blood, that is, into another type of human, we are able to recognize that we should not live according to the lust of the flesh and do what the earthly flesh wants. For we do not live for the earthly human, but we live for the eternal, and we should act in accord with the eternal. For we are born into this same humanity through water and the Holy Spirit and, thus, we should live in accordance with that which made us live. As the earthly human lives according to the earthly, the new reborn human should thus live according to the eternal by which he lives." (So merken anfänglich, daß wir haben ein euangelium und ein lautere geschrift, die da uns ausweist, daß wir müessen zum andern mal geborn werden, nemblich nit vom menschensamen noch wiederumb von dem bauch der frauen komben, sonder von oben herab durch das wasser und den heiligen geist. Dieweil nun die geburt der christen allein in dem ligt, daß wir dermaßen geborn sollen werden in ein ander fleisch und blut, das ist in ein andere menschheit, mügen wir das wol erkennen, daß wir nit nach dem wollust des fleisch leben sollen und das volbringen, das das irdisch fleisch will. Dann wir leben nit dem irdischen menschen, sonder wir leben dem ewigen, und nach dem ewigen sollen wir uns halten. dann in demselbigen menschen seindt wir durch das wasser und heiligen geist geborn, und darumb sollen wir leben dem nach, von dem wir lebent. als der irdisch mensch lebt dem irdischen nach, also soll der neu wiedergeborn mensch dem ewigen nach, von dem er ist.) The new human, created from this eternal *limbus*, is born from above through water and the Holy Spirit; see John 3:5.

22. Mark 16:16: "He that believeth and is baptized shall be saved; but he that believeth not shall be damned."

23. Paracelsus, *De genealogia Christi*, II, 3:130. "Also wird durch das wasser und durch den geist der mensch wieder neu geboren in kraft des taufs."

The Baptismal Ceremony: The Means of Receiving Immortal Corporeality

How and when one receives the eternal body will help indicate who has the eternal body. Theoretically, such questions are of immense significance not only within Paracelsus's theological schema, but also in his approach to medicine and general mind-body problems, for participation in the sacraments effects a change in the very constitution of the human. Unfortunately, it is questionable whether Paracelsus actually believed the eternal body could be perceived. For example, in *Libellus de baptismate christiano*, which seems to convey a more spiritualist mood in that Paracelsus does not consistently hold to the necessity of the sacrament, Paracelsus notes, "[A]nd we do not know who is baptized or not."[24] Also, in the writings specifically treating baptism reviewed in this study, Paracelsus himself does not actually make the connection between baptismal efficacy and the receiving of the eternal body. Yet, because Paracelsus gives little attention to the process of receiving the new body in *De genealogia Christi* and the *Astronomia magna*—only noting that baptism is a prerequisite for the acquisition—it is important to turn to the specifically baptismal tracts to gain insight into pertinent questions regarding the reception of the eternal body, even if Paracelsus might not have thoroughly considered the relation between his thoughts on baptism and the immortal philosophy. Indeed, the baptismal tracts, in which Paracelsus usually stresses the necessity of baptism for salvation, contain clues regarding the range of people whom Paracelsus believes to possess the eternal body. The usual means of entering the new creation is clearly a baptismal ceremony. He writes in *Vom tauf der Christen: De baptismate* that in addition to the soteriological requirement of faith and love for God, one needs baptism, adding, "The person who does not take on the outer does not believe, and the one who does not

---

24. Paracelsus, *Libellus de baptismate Christiano*, in *Sämtliche Werke*, II, 3:376. "For the Holy Spirit baptizes the heart and not the body and we do not know who is baptized or not. He is the illuminator, the teacher, and the greatest gift." (Dann der heilig geist tauft die herzen und nit den leib, und wir wissen nit, was getauft ist oder was nit getauft ist. Er ist der erleuchter, der lehrer, die höchste gab.) This is perhaps the only possible answer for someone denouncing the doctrine that the sacraments confer grace *ex opere operato*, that is, if the sacramental grace does not necessarily accompany the proper performance of the sacrament, then seemingly nothing a person could do would offer proof that a sacrament has been efficacious. Paracelsus, echoing a common Protestant theme, seems at times to be attacking the Tridentine position that the right use of the elements in the sacrament confers grace unless the reception is blocked by mortal sin. See *Encyclopædia of Religion and Ethics*, s.v. "baptism (later Christian)." Regarding the perception of the eternal body, see also Kämmerer, *Das Leib-Seele-Geist-Problem*, 24: "Am jüngsten Tag wird dieser Leib aus Christus, der bis dahin unsichtbar ist, vom Adamsleib abfallen, wie eine reife Birne vom Baum."

believe, does not love."[25] Indeed, in order to understand the means of attaining immortal corporeality within Paracelsus's cosmos, it is necessary to review the elements of his baptismal ceremony.

The first requirement of the baptismal ceremony is the use of water; one is to use either immersion or affusion, depending on what is physically appropriate; immersion may be inappropriate, for example, for an infant or ill person.[26] Then the person administering the rites says to the person being baptized, "I baptize you in the name of the Father, Son, and Holy Spirit" (Matt. 28:19).[27] The power of the sacrament is in the words: "No one is made holy through water alone. One needs Christ's baptism.... Only the word of God can make a person holy and nothing else."[28] Paracelsus writes that Christ has set the word of God with the water, adding, "The three names are the word of God, giving baptism its strength and virtue; those who remain in the baptism and keep their faith are holy and with God."[29] Paracelsus's discussion is reminiscent of Augustine's and Luther's stress on the relation of water and word in the sacrament.[30] Paracelsus claims that the water and

---

25. Paracelsus, *Vom tauf der Christen: De Baptismate*, II, Supp.: 118: "Belief in and love of God does make one blessed, but not without baptism. It is beforehand in the children. The person who does not take on the outer does not believe. He who does not believe, does not love." (Scilicet propter filium Christum: glaub und lieb in gott macht selig, aber nit ohn den tauf. Der ist der anfang zuvor in den kindern. Wer nit das außer annembt, der glaubt nit. Wer nit glaubt, liebet nit.)

26. The water-word approach is biblical; one need only combine a few verses, e.g., Matt. 28:19, "Go ye therefore, and teach all nations, baptizing them in the name of the Father, and of the Son, and of the Holy Ghost"; John 3:5, "Jesus answered, Verily, verily, I say unto thee, Except a man be born of water and of the Spirit, he cannot enter into the kingdom of God"; and Acts 10:47, in which St. Peter says, "Can any man forbid water, that these should not be baptized?"

27. Paracelsus, *Vom tauf der Christen*, II, 2:366: "The custom of baptism consists in the dunking into water or sprinkling, depending on the condition of the child or person, and saying to the recipient: 'I baptize you in the name of the Father, Son, and Holy Spirit.' Thus one is properly baptized." (Der brauch des taufs ist allein in das wasser stoßen oder überschutten, nach anseheung des kinds oder person, und im selbigen sagen: "ich tauf dich inn namen des vatters, suns und heiligen geists." Do ist es gnugsam getauft.)

28. Ibid., II, 2:332–33. "For nobody is blessed through this baptism with mere water. Rather, one becomes blessed through the baptism by Christ.... Now, however, only the word of God makes one blessed and nothing else." (Dann durch denselbigen tauf des bloßen wassers ist niemandts selig worden. aber durch den tauf Christi ist ein ieglicher selig worden.... Nun aber uns macht allein das wort gottes selig und sonst nichts.) See also Goldammer's remark in footnote d.

29. Ibid., II, 2:333. "Darumb hat Christus das wort gottes hinzu geton und gesetzt zu dem wasser.... Dise drei namen sind nun das wort gottes. Dise namen geben dem gesalbten und geben dem tauf sein kraft und tugend, daß ein ieglicher, der im tauf bleibt und seiner glübt, selig ist und bei gott." This does raise the question of whether one can lose immortal corporeality by not remaining in one's baptism.

30. Note especially Luther's earlier period, when he wrote that the grace of baptism is not assuredly imparted by the water; instead, it comes from the word of God, which accompanies and is beside

words suffice, but adds that it is appropriate to include the Lord's Prayer and the creed [*Apostolicum*]. Anything more than that, he says, is not in God's word. As a pragmatic afterthought, however, he writes that the name of the person being baptized should also be mentioned.[31] Paracelsus felt that the baptismal ceremony needed only to follow the concise formula provided by the Bible. Apparently, Paracelsus allows the efficacy of baptism when these few simple directions are followed, regardless of denomination, the clear implication being that he does not limit the heavenly body to Catholics. In *Libellus de baptismate christiano*, he adamantly attacks the stress on works within the Roman ceremony: "[I]t is clumsy of you to think that those who have been baptized through your hands are holy, even if they do not know of the Holy Spirit. And the fact that those *not baptized by your hands* are supposed to be damned or in blind heaven is completely against the faith."[32] Paracelsus definitely thinks the elaborate rituals of the Church are

---

the water, and from the faith and trust in the word of God in the water. Concerning the Protestant emphasis on faith, Paracelsus is clearly influenced by Luther's exasperation with the Catholic emphasis on works as well as his stress on the sufficiency of biblical faith. However, he does not seem to follow all of Luther's language concerning baptism, e.g., Luther's statement in *The Large Catechism* that baptism remains true even without faith, "for God's ordinance and word cannot be made variable or be altered by men." See Martin Luther, *The Large Catechism*, trans. F. Bente and W.H.T. Dau, in *Triglot Concordia: The Symbolical Books of the Ev. Lutheran Church*, XIIIA, Part Fourth, Of Infant Baptism (St. Louis: Concordia, 1921), 565–773: "Therefore let it be decided that Baptism always remains true, retains its full essence, even though a single person should be baptized, and he, in addition, should not believe truly. For God's ordinance and Word cannot be made variable or be altered by men." See also Adolf von Harnack, *History of Dogma* (London, 1896–99), 7:217; and Goldammer's note d in Paracelsus, *Vom tauf der Christen*, II, 2:333: "Das entspricht der von Luther aufgenommenen Theorie Augustins über das Verhältnis von Wort und Element (Wasser) bei der Taufe."

31. Paracelsus, *Vom tauf der Christen*, II, 2:366. "Regarding prayer, do not have anything but the Lord's Prayer and faith. The reason for the Lord's Prayer is so that he remains with him; and the faith is also required so that he remains. Nothing more should be added. Everything else is superfluous (*zusatz*) that comes from humans without God's word. Concerning the name, the baptizer should ask, 'What is [the baptized] to be named?' And the name is given to him. He should thus speak, '[Name of baptized], I baptize you [etc.] as above.'" (Des betens halben nichts als ein pater noster und glauben. das pater noster darumb, daß er bei dem bleibt; und den glauben, daß er auch darbei bleib. Weiter nichts mehr. Ist alls zusatz, der von menschen hinzufallt, on gottes wort. Des namens halben ist, daß der teufer fragt: wie soll es heißen? Und es wird ihm genennt. So soll er sprechen: N., ich tauf dich (etc.) ut supra.)

32. Paracelsus, *Libellus de baptismate Christiano*, II, 3:376 (my emphasis). "Darumb ist es ungeschickt an euch, da ir die getauften von euern hänigen selig schätzt ohne wissen des heiligen geists; und welche von euer hand mit getauft seindt, die sollen verdambt oder im blinden himmel sein. Es ist ganz wider den glauben." Paracelsus may be speaking of limbo with his reference to "blind heaven." See footnote c: "D. h. wohl in einem unsichtbaren, nicht vorhandenen.—Oder zu konj.: kinder himmel (limbus infantium), oder limbus himmel? D. h. in der nach kirchl. Lehre den ungetauft gestorbenen Kindern (und den vorchristlichen Frommen) zustehenden 'Vorhölle' (limbus)."

unnecessary ceremonial impieties: this wariness is further apparent in his chiding of the use of balsams. He proclaims that baptism should simply make the person wet, and stresses that there should be no further ceremony because a Christian does not need it—Christ did not command it.[33]

Regarding Paracelsus's penchant for extending the range of the saved, it is important to note the case of infants, the handicapped, and the mentally ill. The necessity of infant baptism is conspicuous in all his baptismal works; as in the following passage: "Therefore, each person becomes blessed through Christ's baptism of Christ, all are saved. This also applies to young children: through the power of the word they become angels before God. The Devil is unable to hurt them whether they die or not."[34] In addition to children, Christ, because of his benevolence, saves such people as the deaf, fools, the mentally retarded, and the senseless. These are exempt from Paracelsus's requirement that in order to be saved one need be cognizant of what one believes. Ever the defender of the ill and disadvantaged, Paracelsus notes that otherwise only the people of faith with understanding of their faith would be saved, and that would be contrary to the honor of God. Actually, the mark (*zeichen*) worn by the Christian belongs even more to the feeble.[35] Paracelsus also brings the mentally ill (*maniaci, phrenetici, vesani*) and possessed into the fold of the saved.[36] Baptism and its protection of the soul are very valuable for Paracelsus, and he pleads with the "heretics" who oppose baptism to accept it.[37]

Paracelsus, clearly disturbed by the exclusivity of the Church's approach to the sacraments, broadens the range of people enjoying God's saving grace. For him, the realm of the new creation is very inclusive. Although Paracelsus sometimes

---

33. Paracelsus, *Vom tauf der Christen*, II, 2:332.

34. Paracelsus, *Vom tauf der Christen: De baptismate*, II, Supp:114. "Darumb durch Christi tauf iederman selig wird. also auch die jungen kinder: aus kraft des worts, dadurch seindt sie engel vor gott; sie sterben oder nit, und der teufel vermag nichts wider sie."

35. Ibid.: "The children, minors (*unmündigen*), feeble-minded, dumb, deaf and blind, fools, naive and senseless people are robbed of the knowledge and strength of such faith, and therefore damned. However, because Christ is kind and has not changed the course of nature, even for the sick and for the children, therefore they too will be saved according to the baptism: that is the sign that redemption is theirs rather more than less." (Die kinder, unmündigen, toren, stumben, gehörlosen und plinden, narren, einfaltige leut, sinnlosen seindt des wissens und kraft solchs glaubens beraubt, et ergo damnati. weil aber Christus güetig, und der natur den lauf nit genomben, als da sein die bemelten presthaften leut neben den kindern,—so gehöret ihn auch die erlösung an und folgends die tauf. das ist: das zeichen gehört ihnen das mehrer zu, also vil mehr das minder. Sonst weren allein die glaubigen des verstands erlöst. Das were wider gottes ehre.) See Matt. 5:3: "Blessed are the poor in spirit: for theirs is the kingdom of heaven."

36. Ibid., II, Supp:115.

37. Ibid.: "Darauf weichent ab, all ir ketzer, so ihnen den tauf nembent!"

vituperatively condemns false Christians and dismisses the efficacy of their baptisms, he nevertheless recognizes that baptism can function as a tool promoting Christian unity. Certain passages in his baptismal works suggest that all people, regardless of confession, become Christians through an efficacious baptism, that is, if the simple water-and-word rules are followed. Where this places Paracelsus among the warring factions within Christendom is not entirely clear, but Paracelsus signifies his predisposition toward inclusiveness and unity in the conclusion of *De genealogia Christi* when he writes, "However, I know very well that the Devil destests the idea that the Christian church be one and work together as one community, and that people may and should help each other."[38] A great hope of his is that Christians can and will help one another, and he fights to remove barriers.

Other Means of Receiving the Effects of Baptism

Despite his stressing the necessity of baptism, Paracelsus does not always adamantly hold to the sacrament. His lowering of the barrier of the necessity of the sacrament is another example of his inclusiveness with regard to salvation; many more people may be said to participate in the new creation. For example, he sometimes goes as far as such theologians as Tommaso de Vio Gaetani Cajetan (1469–1534), Gabriel Biel (ca. 1425–1495), and Jean de Gerson (1363–1429) in finding that baptism is not always necessary to save a child.[39] He notes that through parents' faith and baptism, the entire house becomes healthy, faithful, and baptized. This includes children: "Thus, such children become baptized and are holy through the word of Christ without water; through the parents' faith, the children are brought to Him."[40] In the case of unborn children who die in the womb, Paracelsus writes that a lack of belief among the parents may condemn the

---

38. Paracelsus, *De genealogia Christi*, I, 2:164. "So viel weiß ich aber wol, daß dem teufel gar zuwider ist, daß die christenliche kirch soll einig sein und in einer gemeinschaft handeln und daß ie einer dem andern helfen mag und soll."

39. A nice summary of precedents related to this idea is in *The Catholic Encyclopedia*. "It is true that some Catholic writers (as Cajetan, Durandus, Biel, Gerson, Toletus, Klee) have held that infants may be saved by an act of desire on the part of their parents, which is applied to them by some external sign, such as prayer or the invocation of the Holy Trinity; but Pius V, by expunging this opinion, as expressed by Cajetan, from that author's commentary on St. Thomas, manifested his judgment that such a theory was not agreeable to the Church's belief." *Catholic Encyclopedia*, s.v. "Baptism."

40. Paracelsus, *Vom tauf der Christen: De baptismate*, II, Supp.: 117. "Also seindt auch etliche kinder durchs wort Christi, ohns wasser getauft worden und selig, und durch den glauben der parentum, so die kinder zu ihm gebracht."

children to damnation.[41] Believing parents, however, can bring about a baptism through the Holy Spirit with a request, "petite et dabitur vobis."[42] Another illustration involves pagans. Paracelsus writes, "It can happen even today that a pagan is converted through the preached word or Holy Scripture, and that this occurs in a case in which no water is available and the person then dies in the meantime (i.e., before he can be properly baptized). Thus, he is blessed without any medium (*mittel*)."[43]

Effects of Baptism

The effects of baptism on the person participating in the sacrament occur at both the theological and anthropological levels. First, perhaps harkening back to Augustine's belief that baptism leaves a mark on Christians comparable to that given soldiers in the imperial service, Paracelsus says that baptism is a sign of a Christian—it is like a badge indicating one's military identity.[44] He adds that baptism is not only a sign (*zeichen*) of being a Christian but also a consecration

---

41. Ibid., II, Supp.: 117–18. "From such stories it is passed on (*erbet*) that the children should be baptized. We have no scripture regarding the baptism in the mother's womb, but it is in accordance with the faith that the fruit will arrive one day, according to the faith and desire of the mother and father. Baptism through the Holy Spirit stands upon this request: 'ask and it will be given to you' (*petite et dabitur vobis*).... Therefore, those adults who do not believe damn their children together with themselves." (Aus solchen geschichten erbet es, daß die kinder sollen getauft werden. von der tauf in mutterleib haben wir kein geschrift, und ist doch dem glauben gemeß, auf vatter und mutter glauben und begern, daß die frucht an tag kombe. in diesem begern "petite et dabitur vobis" stehet der tauf durch den heiligen geist. Dann gott beraubt sie nit an der seel das, darumb sie die natur bringt. Deshalben welliche ältern nit glauben, verdamben sich mitsampt iren kindern.)

42. Ibid., II, Supp.: 117, quoting from the Vulgate of Matt. 7:7: "Ask, and it shall be given you; seek, and ye shall find; knock, and it shall be opened unto you."

43. Ibid., II, Supp.: 116. "Also mag noch heut geschehen, daß ein heid durchs gepredigt wort oder heilig schrift zu Christo bekeret wird. Und im fall, daß nit wasser verhanden und er dazwischen stürben, so ist er ohn alle mittel selig. darumb ist solcher tauf Christi vorbehalten allen märtrern, so nit geporne christen sein."

44. Paracelsus, *Vom tauf der Christen*, II, 2:329–30. "Baptism is a mark of a Christian; as a color of a lord indicates the lord's servant, or a military standard indicates military identity, the party or the captain [to whom a soldier belongs]. It is also a sign in the same manner as a pass; a person who possesses it is allowed to travel.... It is also [like] the garment of a priest: if he has it, nobody is permitted to strike him." (Der tauf ist ein zeichen eins christen, wie ein farben eins herrn der herrn diener anzeigt, oder ein kriegszeichen sein kriegszeichen, die partei oder haubtman anzeigt. auch ist es ein zeichen zu gleicherweis einer bolet: welcher sie hat, den last man wandern.... auch ist es ein kleid eines priesters: der ihn hat, den darf niemants schlahen.) Regarding the connection between baptism and receiving a mark or sign, see St. Thomas's *Summa Th.* (III:63:2), Session VII of the Council of Trent (can. Ix.), and Goldammer, in *Von tauf der Christen*, 329, note b: "[Bolet =] 'Ausweis, Paß'...— Zu dieser ganzen 'Zeichen'-Theorie könnte P. durch Zwingli angeregt sein. Vgl. etwas dessen 'Von der Taufe, von der Wiedertaufe und der Kindertaufe' (1525), *Corpus Reformatorum* 91, 217f. 226f. usw."

(*weihe*) that hinders the Devil from doing anything; it is also a pass (*boleten*) that protects Christians so that they may travel through Hell and come out uninjured. A person in Hell without this consecration and pass would have to remain and burn. The person with the sign is written in the Book of Life (*das feltbuech Christi*).[45]

Paracelsus's transferral of baptism to his natural philosophy in the *Astronomia Magna* is exemplary of the importance of sacramental theology in his larger approach to nature. In the second part of the work, he parallels baptism with the virgin birth of Christ. The Holy Spirit brought about the incarnation in the Virgin, who necessarily was born not from Adam's seed but only from his flesh. Thus, Christ became human without the seed of Adam. To be resurrected and enter the Kingdom of God, a human must receive the same type of flesh as Christ, for the old flesh has no value and must be left behind. Thus, one must be newly born from a virgin and faith. However, with Christians, baptism takes the place of virgin birth. During baptism the Holy Spirit who appeared to Christ when he was baptized by John the Baptist brings about the incarnation. If not baptized and

---

45. Rev. 20:12; Paracelsus, *Vom tauf der Christen*, II, 2:332. "Therefore, baptism is not only a sign to blessedness, that we are Christians, but also a consecration so that the devil may do nothing to us; it is also a pass which protects us from the devil so that we may roam and go through Hell—free and unmolested—and come out undamaged. Without this consecration and pass all humans remain and burn. Furthermore, the sign is also such that the person who takes it on is written in the Book of Life, in which there is an account to be given on the day of judgment concerning the person's chivalry and misdeeds (*retterschaft und missetat*)." (Und ist nichts als ein nasmachen, blos, on alle ceremonien. dann es mueß sein, und ein christ muß es haben. darumb was ein christ haben mueß, dasselbig ist nit ceremonien. dann Christus hat uns ceremonien nit gebotten.... darumb ist der tauf zu der seligkeit nit allein ein zeichen, daß wir christen seindt, sonder ein weihe, daß uns der teufel nichts tun mag, und ein boleten die uns vor dem teufel bewart, daß wir frei, unbeleidigt durch die hellen wandern und gon mugen, und unbeschedigt daraus kommen. on welche weihe und boleten alle menschen verbrennen und bleiben. und ist also auch ein zeichen: wer darein kombt, der wird geschriben in das feltbuech Christi, do rechnung zu geben am jungsten tag umb sein retterschaft und missetat.)

46. Paracelsus, *Astronomia magna*, I, 12:309–10: "And in the same manner, as Christ alone was born of a virgin, who has become human and taken flesh in her without the human seed of Adam, and become human through the Holy Spirit, thus do we humans, who want to enter into eternal life, have to leave behind our mortal flesh and blood, and be born for a second time of a virgin and faith, made incarnate by the Holy Spirit: then we are capable [of living] in the Kingdom of God. For at some point (*dan einmal*) a human must be blood and flesh in eternity. Therefore there are two types of flesh: the first is of Adam, which has no value, and the second is of the Holy Spirit, which makes living flesh, for he provides an incarnation from Heaven above; thus, his incarnation comes through us and returns to heaven. Now know thus, that baptism exists in place of the virgin and through baptism we are made incarnate by the Holy Spirit, namely, by the Holy Spirit who appeared descending upon Christ as John the Baptist baptized him; [the Holy Spirit] will also be with us and make us flesh (*und uns incarnirn*) in the new birth, in which there is life and not death. And if we are not born into this birth, then we are children of death and not life. For in the flesh that we receive from the Spirit we will see Christ our

made incarnate by the Holy Spirit, one enters into damnation. Paracelsus adds, "Thus we are compelled to recommend baptism."[46] Thereafter, this new mysterious flesh and blood begin to grow through the nourishment of the Eucharist. Educated in the new body by heavenly wisdom, the Christian develops special abilities. Rudolph describes the change: "In his new, heavenly body man can exercise the capacity to do divine works (1/XII, 315). In accordance with the 'two schools on earth' (*'zwo schulen auf erden'*) a change of 'schoolmaster' takes place; more precisely, there is a gradual transition, a progression to a higher school. Instruction in the 'wisdom of heaven' (*'himlische weisheit'*) is no longer given in the school of nature but is given in the newborn body 'down from above' (*'von oben herab'*)."[47]

■ ■ ■

Several elements regarding the integration of Paracelsus's *untötliche philosophei* and *ganze philosophia* have already been touched on, including the manner in which he endows the eternal body with physiological aspects, his belief in the sanctification of the feeble and mentally ill via baptism, and the special qualities and protection possessed by those participating in the new creation. The ideas contained in the two not-so-disparate parts of Paracelsus's thought, the medicinal/

---

savior; we will not rise again and enter the Kingdom of Heaven in mortal flesh, rather in living flesh. However, what is not baptized and made incarnate by the Holy Spirit will be damned. Thus we are compelled to receive baptism; if we do not, then we are not of eternal flesh and blood." (Und zu gleicher weis, wie Christus alein von einer jungfrauen geboren ist, der one menlichen samen von Adam mensch ist worden und in ir incarnirt, und mensch worden durch den heiligen geist, also wir menschen, die da wöllen in das ewig leben, das tötlich fleisch und blut verlassen müssen, wider zum andern mal geboren werden von einer jungfrauen und glauben, incarnirt vom heiligen geist: als dan seind wir fehig in das reich gottes. dan einmal so muß der mensch blut und fleisch sein in ewikeit. Darumb sind zwei fleisch: von Adam, das ist nichts wert, von dem heiligen geist, das macht lebendig fleisch, dan er incarnirt von oben herab, darum komet sein incarnation durch uns wider gen himel. Also wissent nun, das der tauf anstat der jungfrauen da ist und im tauf werden wir incarnirt vom heiligen geist, nemlich von dem heiligen geist, der erschinen ist auf Christo, da in der teufer Johannes Baptista getauft hat, der wird auch dan sein bei uns und uns incarnirn in die neue geburt, in welcher ist das leben und nit der tot. Und so wir in diser geburt nit werden geboren, so sind wir kinder des tots und nit des lebens. Dan im selbigen fleisch, das wir vom geist empfahen, werden wir sehen Christum unsern erlöser und nit im tötlichen fleisch, und in dem lebendigen fleisch werden wir auferstehen und eingehen in das reich gottes. was aber nit getauft wird und vom heiligen geist incarnirt, das selbig gehet in die verdamnus, daraus wir dan gezwungen werden, den tauf zu empfangen, wo nicht, so seind wir nicht des ewigen fleisches und bluts.)

47. Rudolph, "Hohenheim's Anthropology," 204.

natural philosophical and the theological, cannot be fully understood except in relation to each other. Although the use of natural analogies is commonplace within Scripture and its explication, Paracelsus's nonconformist approach bears the special stamp of a physician and natural philosopher. It will be useful to recapitulate a few of the medicinal/natural philosophical aspects of Paracelsus's exegesis.[48]

With regard to the natural elements of Paracelsus's theology, Hartmut Rudolph has recently pointed to a few examples within the immortal philosophy. Summarizing Paracelsus's notion that the eating of the body and blood of Christ is a digestive and metabolic process by which the heavenly body grows and is nourished, Rudolph stresses Paracelsus's abnormal representation of the actions of Doubting Thomas:

> The fullness and complexity of the forms of [Paracelsus's] argumentation ...grows out of the tension which consists on the one hand in the spiritual origin of this heavenly birth and its nourishment (that is to say, its origin in the spirit of God) and, on the other hand, in the attempt nonetheless to elucidate the mysterious happening of the birth and the growth of the new creature from the *limbus aeternus* as a *bodily* process, indeed, to explain it as a *comprehensible* process by forming an analogy with natural reason, with natural-Adamic corporeality. Paracelsus sees himself quite expressly as a successor to the apostle Thomas, who, so to speak, puts the heavenly-clarified body of Christ which has appeared to him to the test (John 20:24ff.); "*ein greiflich prob*" ("a tactile test"), i.e. proving by sensory experience, is possible and necessary.[49]

In the baptismal writings and passages dealing with rebirth, Paracelsus continues to employ this analogical motif. Even the central concept contained in the title *De genealogia Christi* is an example of Paracelsus elucidating theology according to physical analogy.[50] One might expect a discussion tracing Jesus' paternal and

---

48. On the general topic of the relation between Paracelsus's religious and natural-philosophical thinking, see Kämmerer, *Das Leib-Seele-Geist-Problem*, 24: "Goldammer nennt es eine ganz eigentümliche Art, in der bei Paracelsus naturwissenschaftliche und theologische Anthropologie und Abendmahlslehre zusammengefuegt sind. Hier habe sein vitales, ganzheitliches Denken, seine Erkenntnis der Zerbrechlichkeit des Organischen und seine Sehnsucht nach einem höheren, besseren Leben in der elementaren, von der psycho-physischen Totalität ausgehenden christlichen Sakramentsvorstellung die rechte lösung gefunden. Einerseits sei dei Trennung von Naturwissenschftlichem und Religion deutlich, die Überbietung der natürlichen Verhältnisse durch die schlechthin andersartige Sphäre des Glaubens, andererseits ihre Berührung und der Übergang von dem einen Bereich in den anderen."

49. Rudolph, "Hohenheim's Anthropology," 193.

50. See Paracelsus, *De genealogia Christi*, II, 3:62, 129, 134.

maternal heritage, much like the genealogies of Matt. 1:1–17 and Luke 3:23–38. Instead, Paracelsus looks at the process of generation, drawing a parallel between earthly procreation and the new birth; one concerns the branch of Adam and the other the New Adam, Christ. Those participating in the new birth and thus descending from Christ include the patriarchs, prophets, holy apostles, martyrs, and all saints.[51]

Two more examples illustrate Paracelsus's inclination to read Scripture in a physically literal sense. First, while treating rebirth and the Eucharist together in *De genealogia Christi*, Paracelsus, extending the analogy between the visible and invisible worlds, even compares Christ to soil (*erde*):

> We are all from the earth, and the earth is that from which we are made. Now, does not the bread also come from the earth, and does not the wine also from of the earth? Yes. These things come from the same place as the human. It is not that the human is to eat the material earth in its substance; rather, the human eats the food (*speis*) that comes from it, and we imbibe the drink that comes from it.... Now [with regard to the second birth], as we are born from God—that is, from the body of Christ, from his bones and flesh—thus is he also the field (*acker*) which gives the fruit of its body; [this fruit] that comes forth is food (*die nahrung*). Now, it follows that we do not eat Christ in the person; contrarily, we consume the food and drink that comes from him for our nourishment in the eternal life.[52]

Another interesting passage is his view on baptism by fire. Given a variety of interpretative possibilities, Paracelsus takes Matt. 3:11 as literally as possible.[53] The biblical text does seem to support his stance. He writes that there are two baptisms

---

51. Ibid., II, 3:134.

52. Ibid., II, 3:81: "wir seindt alle aus der erden. und die erden ist die, aus der wir gemacht seindt. nun kombt nit auch aus der erden das brot, nit auch der wein? ja; so komben sie ie auch aus dem der die speis, die von ir gehet, die essen wir, und das trank, das von ir geet, das trinken wir. also ist das die göttliche ordnung. so mügent ir wol auf das ermessen, daß Christus der ist, aus dem wir neu geboren werden in die ander geburt. dann aus dem seindt wir, aus dem werden wir geboren. so wir nun aus gott geboren werden, das ist aus dem leib Christi, aus seinen beinen und fleischen, so ist er auch der acker, der die frucht gibet desselbigen leibs, so aus ihm kombt, das ist die nahrung. aus dem folgt nun, daß wir nit Christum essen in der person, sonder: vom ihm die speis und drank zu unser nahrung in das ewig leben. zugleich wie von der erden speis und trank, also da auch."

53. Matt. 3:11–12: "I indeed baptize you with water unto repentance: but he that cometh after me is mightier than I, whose shoes I am not worthy to bear: he shall baptize you with the Holy Ghost, and with fire: whose fan is in his hand, and he will throughly purge his floor, and gather his wheat into the garner; but he will burn up the chaff with unquenchable fire."

of the word of Christ: one with fire and the other with the Holy Spirit. The baptism with the Holy Spirit brings salvation while the baptism of fire damns one to Hell.[54] Certainly, in his general approach to baptism and the immortal philosophy, Paracelsus, although in many respects a Spiritualist,[55] shows much concern for the body and the empirical. The new body is a combination of spirit and body.

Paracelsus's approach to baptism is a fundamental part of his immortal philosophy, a religious philosophy that is clearly related to his anthropology and natural philosophy. The scriptural new creation gives Paracelsus the opportunity to elaborate a physical body, one endowed with physiological functions; this body, received sacramentally, is attached to the soul and can be carried with the saved into eternity. Despite sometimes condemning false Christians, Paracelsus is generally inclusive with regard to baptismal efficacy, and thus latitudinarian with regard to who participates in the eternal creation. To participate in the new creation, it is unnecessary to follow elaborate rituals, be baptized in the Roman church, or have an ordained priest administer the baptism, and there are a number of acceptable exceptions to baptism. In addition, for Paracelsus, baptism is an important quasi-physical factor differentiating Christians, whom he wanted to see united, from the devotees of other religions. Conclusively, the case of baptism is another example of the difficulty of assigning Paracelsus a particular religious designation. Paracelsus's concepts of baptism and the immortal body, theological concepts fundamental to his natural philosophy, are largely rooted in his own literal, if unusual, biblical exegesis; indeed, the theological context seems far more important than the "Hermetic," "Neoplatonic," or "Gnostic" influences that Walter Pagel emphasizes.[56] The extent to which Paracelsus's views on sacramental efficacy can be incorporated into his immortal and natural philosophies should be the subject of further studies, as should the evolution of his thought on baptism. Indeed, the theological event of baptism is vital to the development of special

---

54. Paracelsus, *Vom tauf der Christen: De baptismate*, II, Supp:116.

55. A prominent example of the spiritualist designation is that of George Huntston Williams, who initially numbers Paracelsus among the radical reformers, specifically as a German Spiritualist alongside Sebastian Franck and Andreas Bodenstein von Carlstadt. See George Huntston Williams, *The Radical Reformation*, 3d ed. (1992; repr. Kirksville, Mo.: Truman State University Press, 2000), 721–22. Later (1238), he terms Paracelsus a Catholic Spiritualist: "In the course of the history of radicalities in the Reformation era we have encountered several Catholic Spiritualists and Evangelicals, as diverse as Erasmus, Valdés, Johannes Campanus, Paracelsus, Guillaume Postel, Dirk Coornhert, and George Witzel, who, remaining within or returning to the Roman Church, were spiritually closer to Schwenckfeld and to such spiritualizing Anabaptists as Entfelder and Bünderlin."

56. Walter Pagel, *Paracelsus: An Introduction to Philosophical Medicine in the Era of the Renaissance*, 2d rev. ed. (Basel: Karger, 1982).

epistemological qualities and divine abilities, and it is also a very important medicine, one that is significant even in a person's life on earth.

Acknowledgments: I would like to thank the Deutscher Akademischer Austauschdienst (DAAD) and the Indiana University Graduate Student Exchange Program with the Free University–Berlin for funding my research projects in Berlin. I also wish to thank Dr. Hartmut Rudolph for his generous and enthusiastic help. In addition, I want to relate my sincerest appreciation to William R. Newman, Gerhild Scholz Williams, and Charles D. Gunnoe, Jr., for their patience and indispensable guidance. Many thanks as well to C. Andrew Weeks, Urs Leo Gantenbein, and the Free University–Berlin scholar who graciously worked through Paracelsus's *De genealogia Christi* with me, Gabriele Jancke.

Paracelsus and van Helmont on Imagination

Magnetism and Medicine before Mesmer

Heinz Schott

When studying different psychosomatic concepts in the history of medicine, we generally notice a certain duality of aspect in psychosomatic processes. On the one hand there are ideas, images, pictures, and illusions *within* a person's mind which may imprint themselves upon his own body or infect other bodies by communication, producing a physiological disorder or a psychological epidemic. This may be called the power of imagination (*Vorstellungs-* or *Einbildungskraft*). On the other hand, combined with imagination, there are natural powers, which were called "(nervous) fluid" (*fluidum*) in the early and "psychic energy" (*psychische Energie*) in the late nineteenth century, correlating one part of the human body with another, one individual organism with another, or a human being with the whole of nature—linking microcosm (man) and macrocosm (world). The most essential idea of natural philosophy, from the Stoics to the Romantics, claims that all bodies, including the human organism, are connected to each other by networks of magnetic influence. This may be called the power of magnetism. This concept includes the concept of sympathetic interaction: the transference of vital powers within the body or from one body to another.[1]

This concept of imagination or fantasy is central to the history of medicine, especially since the Renaissance.[2] It derives from antiquity and was widely discussed

---

1. On the concept of sympathy and its metaphoric use in the history of medicine see Heinz Schott, "Sympathie als Metapher in der Medizingeschichte," *Würzburger medizinhistorische Mitteilungen* 10 (1992): 107–27.

2. Heinz Schott, "Die 'Imagination' als historischer Schlüsselbegriff der neuzeitlichen Medizin und (Para)Psychologie," in *Psychologiegeschichte: Beziehungen zu Philosophie und Grenzgebieten*, ed.

by Greek philosophers, especially Aristotle. It was traditionally used to differentiate imagination from the effects of demons, as can be still observed in the learned literature of the early modern period.[3] Imagination or fantasy was of great importance to philosophical discourse in the Renaissance, especially in the work of Ficino.[4] His model of "fascinatio," exemplified in *De amore* by the rays sent through the eyes, bringing about love sickness as well as the evil eye (*böser Blick*), was fundamental to later discussions of imagination as a process of natural magic.[5] The universities and courts of the Renaissance expressed a general ambivalence regarding magic, depending on the assumed moral status of the person who practiced it.[6] The issue of "white" versus "black" magic became an important topic of learned discussion. In his work on natural magic, Giambattista Della Porta (1535–1615) included a special chapter on the question of how to become possessed by "fascination" and how to protect oneself from it.[7]

The concept of imagination implies potential proximity to the traditional idea of demonic possession. A pathogenic imago, like a demon or parasite, is incorporated into the mind by a sort of injection. It may develop dynamic powers within the mind and body of the "possessed" person, but it can also be transferred to other individuals or even to natural things in general.

The idea of magnetism derives from the cosmological concept of interaction. Occult powers within the natural world influence the human organism. Magic medicine is used to cure sick persons by so-called magnetic techniques. They are believed to strengthen the vital powers of the organism—in other words, to accumulate vitality. But there are also destructive powers of magnetism. They weaken vitality and take away life energy—like vampires.[8] The long tradition of magical practices and sympathetic cures documents the importance of magnetic influence

---

Jürgen Jahnke, Jochen Fahrenberg, Reiner Stegie, and Eberhard Bauer (Munich: Profil, 1998), 395–403; and Heinz Schott, "Imagination," in *Der sympathetische Arzt: Texte zur Medizin im 18. Jahrhundert* (Munich: C. H. Beck, 1998), 28–38.

3. Pierre Le Loyer, *IIII. Livres des Spectres ou Apparitions et Vision d'Esprit, Anges et Demons se monstrans sensiblement aux hommes* (Angers: Nepveu, 1586).

4. Eugenio Garin, "Phantasia e Imaginatione fra Marsilio Ficino e Pompnazzi," in *Phantasia-Imaginatio* V octavo Colloquio Internationale, ed. M. Fattori and M. Bianchi, Lessico Intellettuale Europeo 46 (Rome: Edizione dell' Ateneo, 1988), 3–20.

5. Marsilio Ficino, *Über die Liebe oder Platons Gastmahl*, ed. Paul Richard Blum, 3d ed. (Hamburg: Meiner, 1994), 320 et seq.

6. György E. Szyöny, "Tradition of Magic: From Faustus to Dee at European Universities and Courts," *Cauda Pavonis*, n.s. 10 (1991): 1–8.

7. Giovanni Battista Della Porta, *Natürliche Magia, das ist Ein ausführlicher und gründlicher Bericht von den Wunderwercken Natürlicher Dinge* (Magdeburg: Rausch, 1612), 262–80.

as a medical idea. It is remarkable that the fundamental idea of magnetism as a vector for imagination (*imaginatio* or *phantasia*) has been largely ignored by historians of medicine. What has been published tends to take a philosophical or psychological perspective on imagination, neglecting the physical or substantial counterpart, which was stressed by Paracelsus and van Helmont, the most important thinkers on magic and (al)chemical medicine in the early modern period.[9]

NATURAL MAGIC: PARACELSUS ON IMAGINATION

This essay will show that the Paracelsian concept of imagination is closely related to the concept of magnetism. Imagination and magnetism are linked in the work of Paracelsus (= Theophrastus von Hohenheim [1493/4–1541]), to whom "The whole of heaven is nothing other than *imaginatio* influencing man, producing plagues, colds, and other diseases."[10] The same is true for the microcosm, the individual human organism, for which Paracelsus uses the metaphor of an inner sun: "Well, what else is *imaginatio* than a sun within man, having such an effect in his globum [body], that is, thereon it shines?"[11] Indeed, there are, according to Walter Pagel, "concordances in detail between the lore of the Cabalah and the teaching of Paracelsus."[12]

What, then, is the basic model of imagination and magnetism for Paracelsus? First, the attractive power of the magnet symbolizes the power of imagination;

---

8. The term *Od-vampirism* (*Od-Vampirismus*) describes the negative (weakening) powers of persons in the middle of the nineteenth century according to the "Od" theory (*Odlehre*) of the German chemist Carl Reichenbach; see Karl Spiesberger, "Justinus Kerners 'Seherin von Prevorst' in Betrachtung esoterischer Tradition und im Lichte psychischer Forschung," in Erich Sopp and Karl Spiesberger, *Auf den Spuren der Seherin* (Sersheim: Osiris, 1953), 64.

9. See, e.g., the following encyclopedia articles: J.H. Trede, "Einbildung, Einbildungskraft"; H. Mainusch, "Imagination," in *Historisches Wörterbuch der Philosophie* (1976); and Silvio Vietta, "Phantasie, Einbildungskraft," in *Literaturlexikon* (1993). In the early eighteenth century, imagination was still linked with the idea of a (magnetic) power: see Johann Heinrich Zedler, *Grosses vollständiges Universal-Lexikon* (1734) and (1735); whereas in the second half of the nineteenth century this link disappeared: see Jacob Grimm and Wilhelm Grimm, *Deutsches Wörterbuch* (1862), 3:152–53.

10. "[A]ls der ganz himel ist nichts als imaginatio, derselbige wirket in den menschen, macht pesten, kaltwehe und anderst." Paracelsus, *Sämtliche Werke, I. Abteilung: Medizinische, naturwissenschaftliche und philosophische Schriften*, ed. Karl Sudhoff, 14 vols. (1922–33; repr. Hildesheim: Olms, 1971), I, 14:311. Hereafter I will refer to this edition using the Roman numeral to denote the part of the collected edition of the complete works of Paracelsus followed by the volume and page number. All translations from Paracelsus are mine.

11. "[N]un was ist imaginatio anderst, als ein sonn im menschen, die dermaßen wirket in sein globum, das ist, do hin sie scheint?" Ibid., I, 14:310.

12. Walter Pagel, *Paracelsus: An Introduction to Philosophical Medicine in the Era of the Renaissance*, 2d rev. ed. (Basel: Karger, 1982), 217.

Paracelsus characterizes both powers this way: "As the magnet can attract steel, there is also a magnet in the imagination, which also attracts. There is an *imaginatio* like a magnet, and an *impressio* like the sun and heaven, making a man by the power *vulcani*."[13] Paracelsus provides an example (*Exempel*), a parable, to explain the identity of imagination and magnetism. The magnet is a metaphor for the imagination: "Without hands and feet, the magnet attracts iron. Like the magnet attracting the visible, the corpora (bodies) are invisibly drawn to the imagination by itself. But it is not the corpus (body) that enters, but what the eyes see and is not palpable, i.e., form and color."[14] The attraction (incorporation) of an object (*ding*) by the imagination (a quasi magnet) is followed by an impression of this introjected object, like the impression the sun and heavens put on a man. "What climbs up into heaven is *imaginatio*, and what falls down is *impressio* born out of the imagination." This movement describes a reflex action, tying together microcosm *and* macrocosm. A macrocosmic reflex occurs, for instance, when the (evil) imagination of a human individual poisons the stars, which send the poison back to the earth, causing plagues and disorders.[15]

In his treatise *De causis morborum invisibilium* (On the invisible diseases), Paracelsus uses the term *imaginatio* (imagination) to explain the correlation between body and soul. "The imagination is a master by itself and has the art and all instruments and all it wants to produce, for example as a cellarman, painter, metalworker, weaver, etc.... What does imagination need? Nothing more than a globe on which it can work, that is, the screen on which it paints what it wants to paint." In this way, the imagination of a pregnant woman can impress itself directly on the body of the child in the womb: "The woman with her imagination is the workmaster and the child is the screen on which the work is perfected. The hand of the imagination is invisible, the instrument also, and both work together.... So the imagination does its work at that place, in the way the imagination has decided it."[16]

---

13. "Dan kan der magnes an sich zihen stahel, so ist auch ein magnet do in der imagination, wie ein magnet und ein impressio, wie die sonn und wie der himel, der ein menschen macht in der kraft vulcani." Paracelsus, *Sämtliche Werke*, I, 14:313.

14. "[D]er magnet zeucht an sich das eisen on hend und füß. Zu gleicher weis wie also der magnet das sichtig an sich zeucht, also werden auch die corpora unsichtig durch die imagination an sich gezogen. nicht das das corpus hinein gang, sonder das get hinein, das die augen sehen und nicht greiflich ist, also die form und die farbe." Ibid., I, 9:290.

15. "Und das herauf kompt in himel, ist imaginatio und wider herab felt, ist impressio, die geboren ist aus der imagination." Ibid., I, 14:314.

16. Ibid., I, 14:317. See Heinz Schott, "'Invisible diseases'—Imagination and Magnetism: Paracelsus and the Consequences," in *The Man and His Reputation, His Ideas and Their Transformation*, ed. Ole Peter Grell (Leiden: Brill, 1998), 309–21.

In this context, Paracelsus defines the power of imagination as "belief" (*Glaube*). Belief is "like a workman's instrument" which can be used for good as well as for evil purposes. Belief can produce any disease. Paracelsus compares it to a weapon. Disease will be produced when the weapon is active against its originator. Paracelsus uses the parable of the man with a gun, which can be compared to the reversion of affections (*Affektverkehrung*) in modern psychology: "We produce our diseases, so we become similar to a man who has got all his weapons and guns. But when he meets a manikin aiming at him with a ready gun the big man is anxious about the weapon and is frightened by it—the same happens to us.... When we become weak, the power of our belief hits us as a shot from a gun, and we have to tolerate and to suffer what we have thrown against us."[17]

Paracelsus calls self-destructive belief "despair" (*Verzweiflung*). It is the reversal of our belief which makes us weak and sick. The gun is directed against ourselves. The pathological imagination may even give rise to an epidemic, such as the plague. The most important cause of the plague, therefore, is that people in despair may "poison heaven, so some will suffer from plague, depending on their belief."[18] Imagination represents great danger if it is combined with despair and returns to its own origin. As pointed out above, this mechanism constitutes a sort of reflex action.

It is worth noting that Paracelsus links his theory of imagination and magnetism to social and political phenomena, explaining events of mass psychology, such as the attraction that a leader exerts on a crowd of people. In this context, Paracelsus again uses the magnet as a metaphor: "You find a man who knows to speak, so that all the world runs to him and listens. Know, then, that his mouth (*Maul*) is a magnet, powerfully attracting the people."[19]

In his treatise *Philosophia magna* Paracelsus discusses the effect of the "right belief" (*rechter Glaube*), which could promote a strengthening and healing imagination.[20] This concept was stressed by Paracelsians such as Oswald Croll: "The belief produces an imagination, whereas the imagination produces the stars (by marriage with the imagination).... To add belief to the remedies gives the spirit or mind to the medicine: But the mind is the knowledge of medicine: The medicine or remedy is sanity: Therefore it is consequent, that the medic or physician

---

17. Paracelsus, *Sämtliche Werke*, I, 9:280.

18. "[D]as sie den himel vergiften, das er etlichen pestilenz gibt, nach dem ir glaub ist." Ibid.

19. "[D]u findest ein man, der kan reden, das im alle welt zulauf, und hört im zu. nu wiß, das das maul ein magnet ist, zeucht an sich die leut in der kraft." Ibid., I, 9:363.

20. Ibid., I, 14:371.

originates from the belief...."[21] Croll's annotation recommends Paracelsus's *On the invisible diseases* as appropriate reading.

Magnetism and World Soul: Gilbert, Kepler, and the Consequences

In the late sixteenth and early seventeenth centuries, after Paracelsus, magnetism became a central topic of natural philosophy, experimental science, and medical theory. The influence of the prominent authors Gilbert, Kepler, and Kircher on science and medicine until the eighteenth century can hardly be overestimated.

William Gilbert (1544–1603) published his studies on magnetic phenomena under the title *De magnete, magnetisque corporibus, et de magno magnete tellure* in 1600, the same year he became court physician to Elizabeth I. He was the first to distinguish "electrics" from "magnetics," that is, the attraction caused by the amber effect from that caused by a lodestone. Rubbed amber or substances that behaved in a similar way (electrics) emitted effluvia to pull small particles inward. Magnetic materials like the lodestone (magnetics) shared their ability of attraction with the earth as a giant lodestone. Gilbert called the spherically ground lodestone a *terella*, since it did not depend on the emission of *effluvia* for attraction. Every magnet was surrounded by an invisible orb of virtue. Magnetics within this orb would be attracted to the magnetic body. But the specific difference from other sources of attraction was "magnetic coition," the mutual action of the attracting and the attracted body, the "coming together of two bodies harmoniously."[22] Magnetism turned out to be the world soul, and the magnetic force appeared to be a psychic force: it "is animate or imitates a soul; in many respects it surpasses the human soul, while that is united to an organic body."[23] This idea corresponds to the principle of natural magic (*magia naturalis*) and the concept of magnetism in medicine, especially in regard to the so-called magnetic-sympathetic cures in the early modern period. But Gilbert did not emphasize the idea of "imagination" or "fantasy";[24] his physical approach to magnetism was quite different from the Paracelsian conception.

---

21. Oswald Croll, *Chymisch Kleynod* (Frankfurt: Schönwetter, 1647), 86.
22. Suzanne Kelly, "William Gilbert," in *Dictionary of Scientific Biography* (1972), 5:396–401.
23. Heinrich Feldt, "Der Begriff der Kraft im Mesmerismus: Die Entwicklung des physikalischen Kraftbegriffes seit der Renaissance und sein Einfluß auf die Medizin des 18. Jahrhunderts" (medical thesis, Bonn, 1990), 20.
24. Neither did Franz Anton Mesmer, when he explained "animal magnetism" about two hundred years later! "Imagination" or *Einbildungskraft* had a pejorative meaning then: it described only a subjective, psychic impression without a physical, detectable effect.

Johannes Kepler (1571–1630) worked as a mathematician at the court of the German emperor Rudolf II in Prague from 1599. The intensive discussion at the court of alchemy, astronomy (astrology), hermetism, and the Kabbalah had a great impact on the Rosicrucian movement in the second decade of the seventeenth century. Kepler suggested replacing the word "soul" by "force," that is, the force of magnetism.[25] First he explained the circulating movement of the planets by "moving souls"; later, shifting his philosophy from animism to mechanism, he assumed the presence of a physical force.[26] When he spoke of "natural, magnetic forces," he did not rely on his own investigations but made use of Gilbert's magnetic philosophy, which postulated that the earth is a body and its nature corresponds to the soul of an animal. The force of the soul (of the planets, the world, or the human body) radiates in all directions with straight beams from the center of each body in all directions. The human soul originates in the heart and radiates to all points of the body. Kepler identified the beams of the soul with the *spiritus* of Galenic medicine and put his theory in the contemporary medical context.[27] When the beams from different sources come together, such as the astral beam with the beam of the human soul, a revelation might result.

The Jesuit scholar Athanasius Kircher (1602–1680) was also very interested in magnetism. As a professor of philosophy, mathematics, Hebrew, and Syriac in Würzburg, he published his first book, *Ars magnesia*, in 1631, on his own magnetic experiments. Later, he wrote a series of five books on magnetism. In the *Ars magna lucis et umbrae* (1646) he identified light with the "attracting magnets of all things": it is connected ultimately with the heavens and works like the lodestone.[28] In this perspective, Isaac Newton's concept of force, his theory of gravity, combines natural philosophy in the sense of Neoplatonism and Hermeticism with the mechanistic, corpuscular theory of matter. Newton's influence on eighteenth-century medicine and science was tremendous. Newton's ether theory was a basis for many speculations on the so-called imponderables, extremely subtle substances that were supposed to be the physical carriers of all forces. In medicine this meant the forces of electricity, mineral magnetism, the concept of nervous fluid, irritability, and, last but not least, the concept of magnetic fluid (*fluidum*) as the active power of animal magnetism (or mesmerism).[29]

25. Feldt, "Der Begriff der Kraft im Mesmerismus," 9–30. Feldt's treatise shows very well how important the concept of (magnetic) force was to establish a theory of a psychic force in regard to mesmerism.

26. Ibid., 13.

27. Ibid., 24–25.

28. Hans Kangro, "Athanasius Kircher," in *Dictionary of Scientific Biography* (1973), 7:374–78.

29. Cf. Feldt, "Der Begriff der Kraft im Mesmerismus," chaps. 2 and 3.

In contrast to the Parecelsian tradition, this scientific (physical) tradition of magnetism did not include the concept of imagination, in contrast to the Paracelsian tradition, where magnetism and imagination can hardly be separated. The following section will show that van Helmont's theory of imagination represents a sophisticated refinement of Paracelsian theory.

The Power of Imagination: van Helmont's Theory

Johann Baptist van Helmont (1579–1644), a scholar and a wealthy man, lived in Vilvoorde, near Brussels. In contrast to Paracelsus, who traveled restlessly all his life throughout Europe, van Helmont stayed at home and worked continuously in his chemical laboratory.[30] The most important figure in the Paracelsian movement, van Helmont was in fact the founder of the so-called chemical philosophy.[31] He was a critical follower of Paracelsus, principally sharing his alchemical approach and his religious attitude as a philosopher and doctor. However, van Helmont rejected the medical astrology of Paracelsus and his analogy of microcosm and macrocosm. Van Helmont concentrated on laboratory work and investigated, for the first time in the history of science, the "gases." He coined the term *gas*, which he probably derived from *chaos*, a term used by Paracelsus to indicate the vital stuff for animal and human beings as well as the elemental spirits (*Elementargeister*). Pagel has shown how van Helmont combined chemical experimentation with a sort of natural mysticism.[32] Like Paracelsus, he considered the magnet to be a powerful instrument for magnetic (so-called sympathetic) cures. When he published *De magnetica vulnerum... curatione* (On the magnetic cure of wounds) in 1621, he recommended, like many other medical writers of his time, the use of the so-called weapon salve (*Waffensalbe*). For his criticism of the Jesuits and his obvious support of Paracelsian science, he was sentenced—but not imprisoned—by the church authorities.

In *Tumulus pestis* (The tomb of the plague), van Helmont explains the origin of diseases. They are caused, he says, by specific seeds, which have their own life

---

30. Heinz Schott, "Paracelsismus und chemische Medizin: Johann Baptist van Helmont zwischen Naturmystik und Naturwissenschaft," in *Meilensteine der Medizin*, ed. Heinz Schott (Dortmund: Harenberg, 1996).

31. Allen G. Debus, *The Chemical Philosophy: Paracelsian Science and Medicine in the Sixteenth and Seventeenth Centuries* (New York: Science History Publishers, 1977).

32. Walter Pagel, "Johannes Baptista van Helmont als Naturmystiker," in *Epochen der Naturmystik: Hermetische Tradition im wissenschaftlichen Fortschritt*, ed. A. Faivre and R.C. Zimmermann (Berlin: E. Schmid, 1979), 169–211.

principle. This so-called ontological concept of disease resembles the theory of Paracelsus, who also spoke of the seed of a disease (*Krankheitssamen*). As Pagel stressed: "There is no area in which van Helmont's inspiration by and dependence upon Paracelsus is as evident as in his ontological theory of disease."[33] But in regard to the Paracelsian concept of imagination, van Helmont took a different approach: the seed of a disease is activated by a pathogenic image (*idea morbosa*) hidden within the seed itself. The life spirit (*archeus; Lebensgeist*) takes on this evil image if it is weakened by bad influences from the outer world.

Those evil images behave like parasites within the body. Imagination, therefore, means an infection that produces a more or less severe disorder. Imagination (*imaginatio*) is also called "fanciful animal" or "animal fancy" (*animal phantasticum; thierische Phantasie*), terms that stress its parasitic character. Chapter 11 of the *Tumulus pestis* deals with this topic. In the Latin edition[34] the title of the chapter is *Animal phantasticum*; in the German edition by Christian Knorr von Rosenroth it is *"Von der Thierischen Phantasie: Oder was die Einbildung vor Würckungen hat"* (On the animal fancy: Or what effects the imagination has).[35] Compared to Paracelsus, van Helmont develops a much more sophisticated system of how the imagination may influence the human body, mixing contemporary ideas with specific Paracelsian assumptions.

Like his contemporaries and Paracelsus, van Helmont supposed that the imagination of a pregnant woman can produce a birthmark on the fetus. When she is overcome by a certain passion (represented by an image) and touches a part of her body, the analogous part of the child's body will be imprinted with the image of her passion. But a new aspect appears when van Helmont states that both genders can suffer from hypochondriasis (*hypochondriaca*, or *Wahn-Witzigkeit*): "This happens in men as well as in women."[36] Men send their images up to the heart and brain, whereas women send them down to the womb, where they provoke passions and emotions.[37] A "slow sadness" coins an image (*ideam*), from which "splenetic melancholy" (*melancholia hypochondriaca*) in women and jaundice in men

---

33. Walter Pagel, *Joan Baptista van Helmont: Reformer of Science and Medicine* (Cambridge: Cambridge University Press, 1982), 149.

34. J.B. van Helmont, *Opuscula Inaudita* (1644; repr. Brussels: Culture et Civilisation, 1966).

35. Christian Knorr von Rosenroth, *Aufgang der Artzney-Kunst, das ist: Noch nie erhörte Grund-Lehren von der Natur...* (1683; repr. Munich: Kösel-Verlag, 1971), 591–95 [German translation of Latin and Dutch works of van Helmont]. Quotations of van Helmont in this paper are referenced to Knorr's German edition; they have been translated into English by the author.

36. The section is entitled "On the imagination in men"; cf. Knorr von Rosenroth, *Aufgang der Artzney-Kunst*, 592.

37. The fact that both genders can be affected is reminiscent of Sigmund Freud's assumption of "male hysteria" in quite another medico-historical context.

come.[38] The spleen, corresponding to the vital spirit (*lebendiger Geist*) of Saturn, is where the imagination usually creates images.

Remarkably, van Helmont also uses astrological ideas to show how celestial bodies influence the physiology of men. The spleen is the seat of Saturn, who may evoke passions. In the spleen (*als eine Mutter* [like a womb]), the imagination starts disturbing our life. The impression of the mother's image on the child's body does not depend on knowledge of the true nature of the image. An inner knowledge within the image is transferred: a "seedal" science (*sämliche Wissenschafft*) of, for example, a cherry. Van Helmont emphasizes that the horror of the plague produces an image (*idea*) of the plague and then the plague itself. The image itself contains the poison that produces the plague, although people do not know its true character. Walter Pagel summarizes van Helmont's ontological conception of disease:

> Like any other seed, the morbid semen begotten by the archeus soon achieves independence of its parent. Once externalised from the archeus it attacks and penetrates him from outside like poison. It behaves like a parasite that is hatched in and "obsesses" a part of the archeus. As such it distracts the archeus from his domestic duties and may thus destroy the organism that is administered by him.... It is, then, the archeus who first conjures up or "imagines" an idea or image. This rebounds upon him, enshrined in a morbid seed. In the latter the spiritual idea or image has assumed corporality, and brings forth an overpowering monster, that is, the disease.[39]

Once again, van Helmont characterized imagination as a parasitic process by the term *animal phantasticum* (*thierische Phantasie* in Knorr's translation), which means literally "fantastic animal."

In his tract on the magnetic cure of wounds (1621) van Helmont emphasizes the effectiveness of the "weapon salve," a then very popular assumption of natural magic based on "sympathy" or the "sympathetic" correspondences of all natural things. Matter and spirit, which are involved in these correspondences, cannot be separated: "the magnet is endowed with various senses and also with imagination, a certain *Naturall phansy*."[40] Later, in his *Ortus medicinae* (published posthumously in 1648), van Helmont once again deals with magnetic-sympathetic

---

38. Cf. Knorr von Rosenroth, *Aufgang der Artzney-Kunst*, 993/12.
39. Cf. Pagel, *Joan Baptista van Helmont*, 144.
40. Cf., ibid., 10.

remedies analogous to the weapon salve.[41] But he refused the assumption that the magnetic-sympathetic force would derive from the stars. "I derive it from a more obvious thing. Namely from the leading images (*ideis*), which are produced by love as a mother or by affection. Therefore it happens, that this sympathetic powder is more effective, when it is handled by this one than by somebody else.... Therefore I estimate the stars of mind in sympathetic remedies more than those in heaven."[42]

Van Helmont states that man also has a magnet, by which the plague poison is attracted from an infected person. The magnet thus pulls death into the body. But there is an antimagnet preventing the infection: precious stones (sapphire and transparent agate), which are rubbed at special "planet" locations of the body.[43] In the *Ortus medicinae*, van Helmont returns to the weapon salve in order to point out the intrinsic connection between imagination and magnetism. Thus, the "imagination of the blood" is put into the magnetic salve and is awakened by the power of the latter. Because of the healing power in the salve, the imagination of the blood wants to pull all foreign impressions out of the whole blood using a spiritual magnetic tension. The salve requires certain ingredients to become healing and magnetic "by nothing else but its fantasy."[44] Van Helmont takes this argument further: he declares that it is possible to produce a magnetic needle only by imagination (or intention) of the smith. His imagination imprints the magnetic power on the steel at the moment of the "birth" of the needle, when it is still glowing: "Therefore those seals are without any power, which are not imprinted by a magician, who has a strong imagination." It is not the stars of the macrocosm that influence the steel magnetically, but the "stars of the heaven of the microcosm," that is, the magician's imagination.[45]

In this way, van Helmont's elaborate theory of imagination as a "magnetic" or "sympathetic" power constructs a dynamic model of the disease process (and its cure) that includes physiology, pathology, psychopathology, and psychosomatic medicine, in terms of modern medicine.

■ ■ ■

---

41. Van Helmont refers to the tract on the "sympathetic powder" by Mohy, published in 1639; cf. Knorr von Rosenroth, *Aufgang der Artzney-Kunst*, 1002/2.

42. Ibid., 1001/4.

43. Ibid., 1018/17.

44. Ibid., 1040/166.

45. Ibid., 1041/170.

In the Paracelsian perspective of magic medicine, which may also be called chemical philosophy, imagination remained a major topic of discussion during later periods and anticipated modern trends. The dualism between soul and body introduced by René Descartes, the anatomical and physiological research on the nervous system and especially the brain (e.g., by Thomas Willis),[46] the new physical paradigms from Kepler to Newton, and the development of the physical and chemical analysis of the human body all influenced the concepts of imagination and magnetism. During the Enlightenment, the power of imagination and the power of magnetism seemed to be self-evident. But in general, they were not systematically combined or even identified until the period of Romanticism in medicine and science, about 1800. Franz Anton Mesmer described "animal magnetism" as the effect of a physical force, analogous to mineral magnetism, electricity, (invisible) fire, or light. On the other hand, his critics in the scientific community rejected his "fluid" theory and explained the phenomena of the magnetic manipulations by "pure" imagination without any transfer of magnetic forces whatsoever.[47]

Imagination was increasingly viewed as an idea fundamentally affecting psychosomatic interaction. The modern theory of mass psychology is based on the concept of transferable ideas. The image (*imago*) seems to be a *contagion*, like the germ of an infectious disease.[48] Obviously, much of this tradition has survived in modern cultural sciences, especially in some social theories of psychoanalysis and cultural anthropology (ethnology). Most important remains the concept of suggestion, proposed by Hippolyte Bernheim in the 1880s and adopted by Freud. The original imagery of Paracelsus and van Helmont is more fascinating than the modern explanations with their abstract scientific language. Paracelsus and van Helmont did not distinguish clearly between magnetism and imagination, between matter and mind, physiology and psychology, belief and knowledge, fantasy and

---

46. The reflex model by Descartes and the brain research by Willis are thoroughly analyzed and illustrated in Edwin Clarke and Kenneth Dewhurst, *Die Funktionen des Gehirns: Lokalisationstheorien von der Antike bis zur Gegenwart* (Munich: Moos, 1973), 69–74.

47. Cf. Heinz Schott, "Über den 'thierischen Magnetismus' und sein Legitimationsproblem: Zum 250. Geburtstag von F. A. Mesmer (1734–1815)," *Medizinhistorisches Journal* 21 (1986): 104–12.

48. The transfer of pathogenic images in terms of mass psychology was an important explanation for the "emotional plague" produced by crowds, e.g., in revolutions or similar mass movements. This was pointed out, e.g., by Gustave LeBon in the late nineteenth century and Wilhelm Reich in the early twentieth century; on the problem of (suggestive) psychic infection see Heinz Schott, "Die 'Suggestion' und ihre medizinhistorische Bedeutung," in *Bausteine zur Medizingeschichte: Heinrich Schipperges zum 65. Geburtstag*, ed. Eduard Seidler and Heinz Schott, Sudhoffs Archiv; vol. 24 (Stuttgart: Franz Steiner, 1984), 111–21.

reality. Nevertheless, this confusion (as it seems to us today) was a creative one and contributed to the development of medicine and science in early modern times.

Natural Magic
and Natural Wonders

Unholy Astrology
Did Pico Always View It That Way?

Sheila Rabin

In 1486, Giovanni Pico della Mirandola wrote his *900 Theses* and presented them in Rome for public debate. They reflected his wide-ranging education—law in Bologna, philosophy in Ferrara and Padua, and Platonism in Florence.

Pico announces at the beginning of the work that the theses consist of "Dialectical, Moral, Physical, Mathematical, Metaphysical, Theological, Magical, and Cabalistic opinions, including his own and those of the wise Chaldeans, Arabs, Hebrews, Greeks, Egyptians, and Latins."[1] This series of statements comes from sources as diverse as Plato and Aristotle, Greek Peripatetics and Neoplatonists, Avicenna and Averroes, Pythagoreans, Orpheus, Hermes Trismegistus, Zoroaster, and Kabbalah, as well as many others, reflecting an immense spectrum of thought and opinion. As Pico notes in his "Oration on the Dignity of Man," which he intended as an introduction to the *900 Theses*,

> I ... have so prepared myself that, pledged to the doctrines of no man, I have ranged through all the masters of philosophy, investigated all books, and come to know all schools.... And surely it is the part of a narrow mind to have confined itself within a single Porch or Academy. Nor can one rightly choose what suits one's self from all of them who has not first come to be familiar with them all. Consider, in addition,

---

1. Giovanni Pico della Mirandola, *900 Theses*, trans. in S. A. Farmer, *Syncretism in the West: Pico's 900 Theses (1486): The Evolution of Tradition Religious and Philosophical Systems* (Tempe, Ariz.: Medieval and Renaissance Studies and Texts, 1998), 211.

that there is in each school something distinctive that is not common to the others.[2]

What is unusual about the list of areas covered in the *900 Theses* is Jewish mysticism, the Kabbalah. While at Padua, Pico had studied with the Jewish Averroist Elia del Medigo, who may have noticed a mystical strain in the young scholar and introduced him to Kabbalah. It became a strong interest, and already in the *900 Theses* Pico had begun the development of a Christian interpretation of the Kabbalah.[3] He continued his studies of Kabbalah, and in 1487 he wrote *Heptaplus*, an interpretation of the seven days of creation as set forth in the first chapter of Genesis. In seven books, each with seven chapters, *Heptaplus* recounts the creation story as it pertains to each of four worlds, the elemental, celestial, angelic, and human. The work is modeled on kabbalistic exegesis. Though the subject matter and form are Jewish, the purpose is Christian. "Just as with Moses the seventh day is the Sabbath and a day of rest," he writes, "so we have taken care that every exposition of ours shall always in the seventh chapter be turned to Christ, who is the end of the law and is our Sabbath, our rest, and our felicity."[4]

In the Renaissance, Kabbalah, Neoplatonism, and magic all had an astrological component. It therefore seems surprising that at the end of his life, in 1494, Pico wrote a massive work against astrology, the *Disputations against Judicial Astrology*, in which, in twelve books, he faulted astrology on religious, philosophical, technical, and historical grounds.

The question arises whether Pico changed his mind about astrology. He ended the *900 Theses* with the declaration, "Just as true astrology teaches us to read in the book of God, so the Cabala teaches us to read in the book of the Law."[5] In light of the *Disputations against Judicial Astrology*, is "true astrology" taken to mean astronomy as a science and "false astrology" or "judicial astrology" to mean astrology as a superstition? Or should it be assumed that Pico refuted

---

2. Giovanni Pico della Mirandola, "Oration on the Dignity of Man," trans. Elizabeth Forbes, in Ernst Cassirer, Paul Oskar Kristeller, and John Herman Randall, Jr., *The Renaissance Philosophy of Man* (Chicago: University of Chicago Press, 1948), 242.

3. An excellent study of Pico's Christian Kabbalah in the *900 Theses* is Chaim Wirszubski, *Pico della Mirandola's Encounter with Jewish Mysticism* (Cambridge: Harvard University Press, 1989).

4. Translation by Douglas Carmichael from Giovanni Pico della Mirandola, *Heptaplus*, in *On the Dignity of Man*, trans. Charles Glenn Wallis, *On Being and the One*, trans. Paul J.W. Miller, and *Heptaplus*, trans. Douglas Carmichael (Indianapolis: Bobbs-Merrill, 1965), 84. Latin edition, G. Pico della Mirandola, *Heptaplus: De septiformi sex dierum geneseos enarratione*, in *De hominis dignitate, Heptaplus, De ente et Uno*, ed. Eugenio Garin (Florence: Vellecchi Editore, 1942), 202.

5. Pico, *900 Theses*, 553.

only "judicial" astrology in the *Disputations* and perhaps accepted another kind of astrology as distinct from astronomy that he considers beneficial or "true astrology"? Finally, can one suppose that when Pico wrote the *900 Theses* in 1486, he believed that astrology had identifiable good and bad forms and that the good form of astrology was true, whereas by 1494, when he wrote the *Disputations*, he had become convinced that all forms of astrology were false and therefore repudiated his earlier stand? There are no letters or other documents from Pico's hand that can give direct insight into his mind concerning this issue, so one must extrapolate from the texts, which in this case has led to contradictory answers.

D.P. Walker shows how Pico's views on celestial influence as expounded in book 3 of the *Disputations* are similar to those expressed by Marsilio Ficino in his 1489 treatise *On Obtaining Life from the Heavens*.[6] Both express a belief in a causal connection between human souls and the spirits that move the heavens. Walker notes that Pico differs from Ficino in denying specificity to that influence: Pico denies that a particular astral condition could affect individuals differently because of an astrological predisposition. Nevertheless, Walker maintains that in the *Disputations* Pico accepts a modified version of Ficino's astrological theories, particularly that astrology has good and bad forms and is good when it did not interfere with human free will or divine providence. Walker adduces evidence from Pico's earlier writings to bolster this conclusion.[7] Thus, Walker places limits on Pico's reputed rejection of astrology in his final work.

S.J. Tester similarly limits Pico's rejection of astrology. Using statements about the heavens in *Heptaplus* and Pico's statement in the Proem to the *Disputations* that he was not attacking all study of the heavenly bodies "but 'that which foretells things to come by the stars,'" Tester claims that the *Disputations* only refute judicial or divinatory astrology, under which he includes "natal charts, progressions, elections, 'interrogations'...all the uses of astrology to discover what is hidden." Natural astrology, or what Tester calls "physical astrology," which dealt with medical and meteorological conditions and natural magic and alchemy, is not discussed in the *Disputations*, according to Tester's argument.[8] But Pico does refer to issues involving natural astrology in the *Disputations*. For example, he states that

---

6. I use the translation of Pico's *De vita coelitus comparanda* by Carol V. Kaske and John R. Clark in Marsilio Ficino, *Three Books on Life* (Binghamton: Medieval and Renaissance Texts and Studies, 1989).

7. D.P. Walker, *Spiritual and Demonic Magic: From Ficino to Campanella* (1958; repr. Notre Dame: University of Notre Dame Press, 1975), 54–57.

8. S.J. Tester, *A History of Western Astrology* (Woodbridge: Boydell Press, 1987), 209. Tester's reference is to Pico's *Disputationes* 1:40.

"no one should believe that the periodicity of the humors [that is, the critical days] during a fever should be attributed to the stars."[9] The idea that the course of a fever is affected by the phases of the moon is typical of medical belief at that time and is an example of natural astrology. Perhaps, rather than leaving judicial astrology out of the discussion, Pico was broadening the definition of "judicial astrology."

Frances A. Yates goes further than Walker and Tester in limiting Pico's rejection of astrology. She draws on Walker's arguments in contending that "Pico is really defending the Ficinian 'astral magic.'... [T]he book against astrology is really a vindication of *Magia naturalis*."[10] Yates provides no textual evidence to support this claim, beyond Pico's references in book 1 to Ficino and Plotinus as opponents of astrology. Her use of the term *astral magic* of course implies the manipulation of the stars for human benefit. Nowhere in the *Disputations* does Pico state or even hint that people could manipulate the stars in any way for any ends whatsoever. As noted, Walker suggests that this is a point of difference between Pico and Ficino, claiming that Ficino maintains that the wise man could use the astral properties of various things to control his own fate.

This is an important point. Pico writes in the *Disputations* that "a general cause does not distinguish among its effects."[11] This is a denial of the specificity of astral influence, and it means that in a very real sense he made it impossible to practice or use astrology. When the effects of the stars on human affairs are universal, the study of those effects no longer teaches us anything about ourselves as individuals. Attempting prognostications also becomes useless because in Pico's scheme the stars could indicate only the universality of human fortunes. This is absurd: if measurable action of the stars has no measurable effect on the individual human being, astrology becomes meaningless.

In addition, Pico asserts that the bodies of the stars themselves have only physical effects. "This is the chief function of the sky," he maintains, "the activity of circular motion, necessary not only in order to carry light and heat down to us and by turns more or less imparted to the earth with wonderful aptness, but also by motion to render us more capable of heat, which continuously flows from the moving bodies."[12] The heavenly bodies have physical effects; they provide light,

---

9. In bk. 3, chap. 17: Giovanni Pico della Mirandola, *Disputationes adversus astrologiam divinatricem*, ed. Eugenio Garin, 2 vols. (Florence: Vallecchi Editore, 1946, 1952), 1:348. All English translations from the Latin text in this paper are by the author.

10. Frances A. Yates, *Giordano Bruno and the Hermetic Tradition* (1964; repr., Chicago: University of Chicago Press, 1979), 115.

11. Pico, *Disputationes*, 1:188.

12. Ibid., 1:196.

heat, and motion. This conflicts with astrological thought (including Ficino's), which identified planets and stars with human personality traits through which in part they influence human affairs.

Lastly, the belief in "celestial spirits" or "intelligences" that move the heavens and affect the earth was not necessarily concomitant with a belief in astrology, although it certainly reinforced such a belief. Learned opinion generally accepted that the heavens were moved by "intelligences," although uneasiness with the concept had begun to surface.[13] Nicole Oresme, whose treatises against astrology were an important source for Pico,[14] supports celestial intelligences as motive powers.[15] These intelligences could be angels; indeed, Pico, like Oresme, identifies celestial spirits as angels, who intercede on behalf of humanity and cause chance events.[16] This would explain the vivifying and ennobling effect of the "celestial spirits" on the human soul: angels draw people closer to God. This idea is not necessarily astrological; it is definitely Christian. Nevertheless, the idea of a material heaven producing universal physical effects stood in uneasy juxtaposition to the belief in superior intelligences that united with the human soul and vivified it. Pico may have lacked the imagination and certainly lacked the training to enable him to break out of the established mode of scientific thought on this question, but as the *Disputations* progresses, the intelligences disappear from the discussion and the material universe seems to prevail. This concept is fundamental to his later discussions of such issues as the division of the heavens and cosmological analogies. Thus, the reader is left with the sense of having moved closer to a materialistic view of the universe, even though the author expresses his belief in an animistic universe.

Despite mutual assertions of their fundamental agreement, Pico made a major break from Ficino in his critique of astrology. Even those assertions seem tentative.

---

13. For discussions on the medieval and Renaissance contibutions to the opposition to the animistic view of the universe see Harry A. Wolfson, "The Problem of the Souls of the Spheres, from the Byzantine Commentaries on Aristotle through the Arabs and St. Thomas to Kepler," *Dumbarton Oaks Papers* 16 (1962): 65–94; Richard C. Dales, "Medieval Deanimation of the Heavens," *Journal of the History of Ideas* 41 (1980): 531–50; and E. J. Dijksterhuis, *The Mechanization of the World Picture: Pythagoras to Newton*, trans. C. Dikshoorn (1961; repr., Princeton: Princeton University Press, 1986), 238–39. Only Dijksterhuis mentions any possible contribution by Pico to this development.

14. An edition and translation of the French *Livre de divinacions* and an edition of the Latin *Tractatus contra judicarios astronomos* appear in G. W. Coopland, *Nicole Oresme and the Astrologers: A Study of His Livre de Divinacions* (Liverpool: University of Liverpool Press, 1952).

15. Oresme discusses celestial intelligences in bk. 2, chap. 2 of *Le livre du ciel et du monde*. See Nicole Oresme, *Le livre du ciel et du monde*, ed. Albert D. Menut and Alexander J. Denomy, trans. Albert D. Menut (Madison: University of Wisconsin Press, 1968), 284–97.

16. In bk. 4, chap. 4: Pico, *Disputationes*, 1:442, 444.

Pico refers to Ficino once in the *Disputations* as "our Marsilio" among a list of contemporary opponents of astrology, but he never elaborates on Ficino's thought as he does on others'. There may be a note of disingenuousness in Ficino's declaration that he agrees with Pico's treatise; Angelo Poliziano, Pico's friend and an opponent of astrology, received Ficino's declaration skeptically.[17] Ficino waffled on the issue of astrology, and appeared to reject it during the period when Pico was writing the *Disputations*.[18] Pico's and Ficino's close friendship may also have led them to emphasize their agreement over the issue of astrology's validity. And Ficino may have been loath to criticize the "last testament" of his dead friend.

On the other hand, Eugenio Garin suggests that Pico may always have rejected astrology and that the *Disputations* should not be considered surprising in light of his earlier works. Garin, noting that the only mention of the word "astrology" in the *900 Theses* is the indirect reference noted above, that "true astrology teaches us to read in the book of God,"[19] claims that this reference could in fact be taken to mean astronomy. He reinforces this claim with references to Pico's condemnation of astrology in other works.[20] Indeed, in 1487 Pico wrote in *Heptaplus* a passage reminiscent of the *Disputations*: "How are the stars placed in the firmament?" he asked:

> As its more noble parts, as the Peripatetics think, or like the animals in their spheres…, as Eusebius the Mede and Diodorus would have it? This point would require conversation with the astrologers, who, from Moses' statement that God placed the stars for signs, draw support for their science of divining by the stars and of foreknowing future events. This science not only has been sharply criticized by Christians like Basil, who rightly called it a busy deceit, and by Apollinarius, Cyril, and Diodorus, but also was spat upon by the good Peripatetics. Aristotle despised it and, what is more, according to Theodoretus, it was repudiated by Pythagoras and Plato and all the Stoics.[21]

---

17. Paul Oskar Kristeller, *The Philosophy of Marsilio Ficino*, trans. Virginia Conant (1943; repr., Gloucester, Mass.: Peter Smith, 1964), 310–11.

18. On this issue see Carol Kaske, "Ficino's Shifting Attitude towards Astrology" in *De vita coelitus comparanda*, the Letter to Poliziano, and the *Apologia* to the Cardinals," in *Marsilio Ficino e il ritorno di Platone: Studi e documenti*, ed. Gian Carlo Garfagnini (Florence: Leo S. Olschki, 1986), 371–81. I thank Professor Kaske for sending me a copy of her article.

19. See Pico, *900 Theses*, 553.

20. Eugenio Garin, introduction to Pico, *Disputationes*, 1:7–8.

21. Fifth exposition, chap. 4: Pico, *Heptaplus*, 132; Latin ed., 196, 198.

Subsequently, in *Astrology in the Renaissance*, Garin backs away somewhat from such an extreme position regarding Pico's early rejection of astrology, but he still maintains that Pico distrusted astrology throughout his career. He claims that Pico "was always hostile to the theory of conjunctions, far from 'naturalistic interpretations' of prophetic phenomena, convinced of the conflict between human liberty and astral determinism, persuaded of the difficulties in establishing causal relationships between general principles and individual events, loath to admit that that which is not the cause could be the sign."[22]

This shows that Pico did, indeed, criticize astrology before he wrote the *Disputations*. However, Pico may have agreed with much of Ficino's *On Obtaining Life from the Heavens* when it appeared in 1489.[23] Though Pico condemns astrology in *Heptaplus*, he also writes, in discussing the formation of the heavens:

> If we seek the elements in the sky, we consider as earth the moon, the lowest and most ignoble of all the stars, just as earth is lowest of all the elements, and very similar to the earth in the [space] opacity of its substance and in its blemishes. Then for water we take Mercury, a shifting star that changes its form, and therefore called by Lucan the lord of the title; for air, Venus, giving life by its tempered warmth; and for fire, the Sun, for very obvious reasons. Then, in inverse order, Mars for fire; Jupiter, related in nature to Venus, for air; for water, Saturn, aged by pernicious cold; it remains for us to call the eighth and unwandering sphere earth, as the very order of the computation demands.[24]

Pico continues, in the next chapter, to expound on the effects of this celestial order:

> Jupiter is hot, Mars is hot, and the sun is hot, but the heat of Mars is angry and violent, that of Jupiter beneficent, and in the sun we see both the angry violence of Mars and the beneficent quality of Jupiter, that is, a certain tempered and intermediate nature blended of these. Jupiter is propitious, Mars of ill omen, the sun partly good and partly bad, good in its radiation, bad in conjunction. Aries is the house of Mars, Cancer the dignity of Jupiter: the sun, reaching its greatest height in Cancer and its greatest power in Aries, makes clear its kinship with both planets.[25]

---

22. Eugenio Garin, *Astrology in the Renaissance: The Zodiac of Life*, trans. Carolyn Jackson and June Allen, rev. trans. Clare Robertson and Eugenio Garin (London: Routledge & Kegan Paul, 1984), 80.

23. I agree with Walker on this point, see Walker, *Spiritual and Demonic Magic*, 56.

24. Second exposition, chap. 3: Pico, *Heptaplus*, 97–98; Latin ed., 228.

25. Ibid., 100–1; Latin ed., 234, 236.

This quotation shows acceptance of the theory of conjunctions, belief in "causal relationships between general principles and individual events," and an admission that "that which is not the cause could be the sign," contrary to Garin's assertion.

Pico not only condemns astrology in the *Disputations*, but refutes the very sort of reasoning about the correspondences between the planets and the houses that he accepted in the second exposition of *Heptaplus*. In the *Disputations* he ridicules astrologers for making such correspondences and particularly attacks the influential Persian astrologer Abu Mashar. Because there were seven planets and twelve houses, Abu Mashar assigned a double house to five planets and gave Saturn the first house, life, and the eighth, death. Pico exclaims, "These things wrote our Abumashar, who, unless he were drunk when he wrote them, was definitely crazy!"[26] Pico chastises Ptolemy in the *Disputations*, alleging another example of "badly philosophizing about the wandering stars," because in the *Tetrabiblos* Ptolemy attributes the elemental qualities to the planets;[27] yet in *Heptaplus* Pico related each sphere to one of the elements. In the *Disputations* Pico denies that the zodiacal sign of the sun had anything to do with its power, noting that the fact that it "produces more heat for us in Leo is not due to the nature of the sign it is passing through, but to the point and position, because from that spot it strikes the earth more directly and is closer to our region";[28] yet in *Heptaplus* he assumed the sun's strength has an integral relationship to the sign. In the *Disputations* he denies that conjunctions could have any effect, claiming that if the planets had similar traits, their effect in conjunction would be redundant, but if they had opposite traits, they would cancel each other out in conjunction;[29] in *Heptaplus* he acknowledged that a conjunction could alter the effect of a planet. Ultimately, Pico suggests in the *Disputations*, "since an astrologer looks at signs that are not signs, and thinks about causes that are not causes, he is, therefore, deceived."[30]

Pico's reasoning in *Heptaplus* would have been consistent with Ficino's *On Obtaining Life from the Heavens*; it opposes the kind of reasoning about the nature of the heavenly bodies that produced the *Disputations*. What is seen in *Heptaplus* may be "true astrology," the belief that the planets had particular traits and the ascendancy of a planet would result in the ascendancy of that trait. It was useful in medicine and for forecasting the weather. True astrology attempted not to contradict divine providence or human free will. "False astrology," as Pico

---

26. In bk. 10, chap. 7: Pico, *Disputationes*, 2:396.
27. Ibid., 2:426, 428.
28. Ibid., 2:42. This argument is repeated in ibid., 2:252, 254.
29. Ibid., 1:546.
30. Ibid., 1:358.

stated in *Heptaplus*, was the "science of divining by the stars and of foreknowing future events."[31] It involved divination and contradicted both divine providence and free will and therefore had to be eradicated. Ficino also rejects this "false astrology."[32] But Pico's *Disputations* is more than a tract against false astrology. It denounces, albeit imperfectly, all astrology rather than refuting a "false" type and approving a "true" type. All astrology emerges as false and as inimical to divine providence and human free will. Nevertheless, the fact that Pico criticized a whole category of astrology in *Heptaplus* should make one wary of claiming that he did a volte-face on astrology, as Wayne Shumaker claims.[33] In the fifteenth century, astrology was not an undifferentiated whole. Pico was a critic of divinatory astrology in *Heptaplus*; his definition of divinatory astrology had changed by the time he wrote the *Disputations*.

Pico's rejection of all astrology does not mean that his *Disputations* is unsullied by superstition and "unscientific" arguments; its "science" can be as specious as the superstition it attempts to refute. What modern reader with an elementary knowledge of science and philosophy would not chuckle at Pico's "empirical" refutation of astrological statements concerning good and bad sailing days. Pico rejects the influence of the sky in favor of the unsupported assertions of sailors, and from their experiences alone proceeds to list the days sailors ought to remain in port.[34] As if weather were not a factor! Renaissance readers, however, whether they opposed or supported astrology, took the *Disputations* quite seriously. Contemporary astrologers believed that Pico's attack was effective and had to be refuted.[35] Giovanni Pontano and Luca Bellanti, both noted astrologers, attacked the *Disputations* as attempting to subvert the study of the stars altogether. Bellanti seemed sure that had Pico lived longer, he would have burned the *Disputations* rather than allow them to be published.[36] Much learned opinion assumed

---

31. Fifth exposition, chap. 4: Pico, *Heptaplus*, 132; Latin ed., 196, 198.

32. Melissa Bullard shows that Ficino had accepted a more divinatory form of astrology earlier but had modified his beliefs by the time he wrote *On Obtaining Life from the Heavens*. Bullard suggests that he developed a more psychological approach in astrology. See Melissa Meriam Bullard, "The Inward Zodiac: A Development in Ficino's Thought on Astrology," *Renaissance Quarterly* 43 (1990): 687–708.

33. Wayne Shumaker, *The Occult Sciences in the Renaissance: A Study in Intellectual Patterns* (Berkeley: University of California Press, 1972), 17.

34. Pico, *Disputationes*, 1:282, 284.

35. Thorndike was correct when he stated that the immediate reaction was primarily negative. See Lynn Thorndike, *A History of Magic and Experimental Science*, 8 vols. (New York: Columbia University Press, 1934), 4: 540–43.

36. Shumaker, *Occult Sciences*, 27.

the scientific validity of astrology, and Pico was a highly esteemed author. How could such a learned individual turn against this important science? A legend arose concerning why Pico wrote the *Disputations*, and two centuries after its appearance it was repeated by the astrologer Antonio Bonatti: Bellanti himself had predicted Pico's premature death, and in uncontrolled fury over this prediction Pico spewed out his attack, hoping to deny the validity of that prediction.[37] The legend had a long life indeed: as late as 1971, Erhard Oeser, in his biography of Kepler, compares Bellanti's putative Pico horoscope to Kepler's Wallenstein horoscope.[38]

Perhaps Pico's growing friendship with the Dominican Savonarola was important in bringing him to a total repudiation of astrology, as Franz Boll and his collaborators suggest.[39] The friar attacked all astrology as utterly incompatible with Christianity. Perhaps he pushed Pico toward an all-out attack on all forms of astrology. As Paul Oskar Kristeller rightly notes about the *Disputations*, "The basic impulse of his attack was religious."[40] On the other hand, Ernst Cassirer rejects the influence of Savonarola as decisive and alludes to Pico's "ethical humanity" and freedom of spirit that pervades his earlier works as well as the *Disputations*.[41] It is true that Pico had always been aware of tensions between astrology and Christianity, particularly on the issue of free will. After all, his passage in *Heptaplus* claims that astrology is not supported by good Christians like Saint Basil. But Cassirer's argument does not account for a change in Pico's attitude toward natural astrology. Moreover, Kaske has shown that Ficino's greatest rejection of astrology took place during the period when he supported the friar.[42] Savonarola could just as likely have persuaded Pico.

But if Savonarola convinced Pico that the only solution to this problem was total repudiation of astrology, it is less likely that Savonarola provided the specific arguments Pico used. Indeed, Savonarola's only major work on astrology is a 1495 Italian summary of Pico's work, and he concentrated on the religious aspects of

---

37. Ibid., 17. Kaske, "Ficino's Shifting Attitude towards Astrology," 375, suggests that Ficino's waffling on astrology may have been influenced by his "ominous" horoscope.

38. Erhard Oeser, *Kepler: Die Entstehung der modernen Wissenschaft* (Zurich: Musterschmidt Göttingen, 1971), 92.

39. Franz Boll, Carl Bezold, and Wilhelm Gundel, *Sternglaube und Sterndeutung: Die Geschichte und das Wesen der Astrologie*, 5th ed. (Stuttgart: Teubner, 1966), 41.

40. Paul Oskar Kristeller, *Eight Philosophers of the Italian Renaissance* (Stanford: Stanford University Press, 1964), 68.

41. Ernst Cassirer, *The Individual and the Cosmos in Renaissance Philosophy*, trans. Mario Domandi (1963; repr. Philadelphia: University of Pennsylvania Press, 1972), 115.

42. Kaske, "Ficino's Shifting Attitude towards Astrology," 375.

the attack.[43] Pico emphasizes reason and experience in addition, and he could draw on a cornucopia of sources to help him fashion those arguments. But it is in those arguments that Pico makes a real contribution. His repudiation of astrology provides the context, but his arguments on the physical effects of the stars, the need to prove, not just assert, the significance of patterns, and the lack of validity of the cosmological analogies not only repudiate astrology but foster rational scientific discourse on causality.

Contra such detractors as William Craven, who finds the *Disputations* "more often cited than read,"[44] the *Disputations* had an important effect on natural philosophy. Robert Westman shows that Pico's deriding the inability of astrologers to establish the true order of the planets helped spur Copernicus's reform.[45] Johannes Kepler, despite his continued support for some astrology, discusses the *Disputations* extensively and sympathetically in two of his works: *On the New Star* and *The Harmony of the World*.[46] Pico may have been impelled to attack astrology in the service of religion, but in furthering the cause of religion, he furthered the cause of natural philosophy as well.

Pico's *Disputations against Judicial Astrology* represents a break from his earlier attitude toward astrology, a change that he certainly made after 1487, when *Heptaplus* appeared, and possibly after 1489, when Ficino's *On Obtaining Life from the Heavens* appeared. Pico had previously expressed doubts about the subject and appears to have rejected judicial astrology, which was involved with divination. But he had accepted natural astrology, what he termed "true astrology" in the *900 Theses*. In this kind of astrology, special qualities were attributed to the individual planets and other heavenly bodies, and the ascendancy of a body could

---

43. Girolamo Savonarola, "Trattato contra li astrologi" in *Scritti filosofici*, ed. Giancarlo Garfagnini and Eugenio Garin (Rome: Angelo Belardetti Editore, 1982), 1:273–393. See also discussions in Garin, *Astrology in the Renaissance*, 84; and Cesare Vasoli, "Le débat sur l'astrologie à Florence dans la seconde moitié du Xve siècle: Ficin, Pic de la Mirandole, Savonarole," in *Divination et controverse religieuse en France au XVIe siècle* (Paris: L'École Normale Supérieure de Jeunes Filles, 1987), 19–27.

44. William G. Craven, *Giovanni Pico della Mirandola: Symbol of His Age: Modern Interpretations of a Renaissance Philosopher* (Geneva: Droz, 1981), 154.

45. In bk. 4, chap. 10. Robert S. Westman, "Copernicus and the Prognosticators: The Bologna Period, 1496-1500," *Universitas* no. 5 (December 1993): 1–5. I thank Professor Westman for providing me with a copy of this article.

46. Johannes Kepler, "*De Stella nova*" in *Gesammelte Werke*, ed. Max Caspar (Munich: C.H. Beck, 1938), 1:181–94; and "*Harmonice Mundi*" in *Gesammelte Werke*, ed. Max Caspar (Munich: C.H. Beck, 1940), 6:264–86. The latter has been translated into English as *The Harmony of the World*, trans. E.J. Aiton, A.M. Duncan, and J.V. Field (Philadelphia: American Philosophical Society, 1997), 358–85. For a discussion of Pico's effect on Kepler, see my article "Kepler's Attitude toward Pico and the Anti-astrology Polemic," *Renaissance Quarterly* 50 (1997): 750–70.

influence the earth and its inhabitants with its particular qualities. These influences could be useful in medicine or in predicting the weather or in dealing with other general occurrences. In the *Disputations* Pico no longer distinguishes between natural astrology and judicial astrology. All astrology has become judicial astrology.

WINE AND OBSCENITIES
Astrology's Degradation in the Five Books of Rabelais

Dené Scoggins

In Rabelais's *Pantagruel*, Gargantua advises his son, "Go on and learn the rest, also the rules of astronomy. But leave divinatory astrology and Lully's art alone, I beg of you, for they are frauds and vanities."[1] Critics such as Mikhail Bakhtin and Lucien Febvre have concluded from Rabelais's treatment of astrology in the five books and his satirical almanacs that he rejects it as a science.[2] Indeed, Rabelais's *Pantagrueline Prognostication pour l'an 1533* ridicules the almanacs of the day, which predicted overthrow of the states, famine, and disease and related them to the appearance of comets or the conjunction of planets. Also, in *Le tiers livre*, Rabelais pokes fun at divinatory arts through Pantagruel's and Panurge's repeated attempts to know whether Panurge's wife would make him a cuckold. Many critics would agree with Wyndham Lewis's conclusion that Rabelais "despised" astrology.[3] However, the very fact that Rabelais was a medical doctor qualifies this

---

1. François Rabelais, *The Histories of Gargantua and Pantagruel*, trans. J.M. Cohen (London: Penguin, 1955), 195. "[D]e astronomie saiche-en tous les canons; laisse-moy l'astrologie divinatrice et l'art de Lullius, comme abuz et vanitez." François Rabelais, *Œuvres Complètes*, ed. Guy Demerson (Paris: Éditions du Seuil, 1973), 247.

2. Febvre asserts that Rabelais "repeated over and over…that to attribute to the stars, without any reverence for the divine Majesty, a power that belongs to Him alone, and to place the liberty of men in thrall to the stars, were acts of impiety." *The Problem of Unbelief in the Sixteenth Century: The Religion of Rabelais*, trans. Beatrice Gottlieb (Cambridge: Harvard University Press, 1982), 272. Similarly, Bakhtin states that Rabelais took neither astrology nor natural magic "seriously." Mikhail Bakhtin, *Rabelais and His World*, trans. Helene Iswolsky (Cambridge: MIT Press, 1965), 366.

3. D.B. Wyndham Lewis, *Doctor Rabelais* (New York: Sheed & Ward, 1957), 202.

view.[4] During the Renaissance, doctors used astrology and specifically almanacs to determine the best time to let blood or perform other medical procedures. Furthermore, Rabelais wrote at least one serious almanac, "Almanach pour l'an 1541," which gives specific advice about the best times of the year to do specific medical procedures. Though only four of his almanacs survive (1533, 1535, 1541, and 1544), most critics agree that Rabelais probably published one every year. Also, as V.L. Saulnier has pointed out, the *Prognostication Perpetuelle* (1566–85 or 1586), which lists Rabelais as an expert astrologer, indicates the level of expertise that Rabelais's contemporaries attributed to him:

> It was quite an honor to have been placed among the ranks of Pythagoras and Ptolemy in a widely read book, above all during a time…when prognostications had such a wide distribution among readers and were an important part of daily life. The attention deservedly given to *Pantagruel* makes it easy to forget: Rabelais was not only, for the people of the sixteenth century, a great writer. He was a doctor, a diplomat and a poet, but also a great "astrologer," ranked here among the supreme masters of the art.[5]

Thus, Rabelais is not simply rejecting astrology when he ridicules it. He is criticizing its abuses as an expert in the field.

Most of all, Rabelais criticizes astrology for its prideful self-sufficiency. Reading signs of the future encourages people to place their trust in the astrologers' predictions rather than in God. Rabelais's own serious almanac gives practical advice about the stars' influence on the weather and the human body but makes no attempt to predict future events. Yet despite his hostility towards astrologers who promise to reveal the secrets of the future, Rabelais recognizes the interpretive

---

4. Regarding the use of astrology by sixteenth-century doctors, see Allan Chapman, "Astrological Medicine," in *Health, Medicine and Morality in the Sixteenth Century*, ed. Charles Webster (Cambridge: Cambridge University Press, 1979), 275–300. Furthermore, Tamsyn Barton points out that from the ancient world until the late Renaissance, astrology competed on an equal footing with other sciences such as medicine, now accepted as "legitimate." Tamsyn Barton, *Power and Knowledge: Astrology, Physiognomics, and Medicine under the Roman Empire* (Ann Arbor: University of Michigan Press, 1994), 12.

5. Translation mine. "[C]e n'est pas rien, d'avoir été placé entre Pythagore et Ptolémée, en un livret populaire. Surtout à une époque…où les pronostications ont tant de lecture, appartienne au domaine quotidien. L'attention justement portée au Pantagruel risque parfois de le faire oublier: Rabelais n'est pas seulement, pour le public du XVIe siècle, un grand conteur. C'était un médecin, et un diplomate, et un poète. Et un grand 'astrologue': le voici classé au rang des maîtres suprêmes, dans les sciences du prognostic." V.L. Saulnier, "François Rabelais, Patron des Pronostiqueurs (Une Pronostication Retrouvée)," *Bibliothèque d'Humanisme et Renaissance* 16 (1954): 124–38, at 130.

potential in astrological signs. He accepts that signs can have deeper meanings and that God chooses to manifest himself through them. Thus, Panurge's quest becomes a quest for the correct way to seek out and read signs.

In the last three books, Rabelais ridicules those diviners who promise knowledge beyond the material realm. Trying to escape their physicality, they become entrenched in the obscene realities of the body. His laughter directed at the astrologer and Renaissance magus unmasks the hubris of those who believe they can become Godlike by shedding their materiality.[6] Rabelais thus participates in the intellectual debates of sixteenth-century France by ridiculing the Neoplatonist Hermetic magus's effort to read signs beyond the materiality of the body.[7] Yet the ridiculed diviners also anticipate a new form of divination that allows for reading signs, but only through the body.[8] Panurge's search for knowledge is redirected to a knowledge of God that Panurge must drink in and absorb through his body. He must embrace his physical nature to know the secrets of God.[9]

In his satirical almanacs, Rabelais tries to dispel anxieties about the conjunctions, the eclipse, and the comets predicted to appear, encouraging his readers to put their trust in God rather than in astrologers. The "bad" signs of 1532 and 1535—comets, an eclipse, a retrograde Saturn, and conjunctions of the moon with Mars

---

6. In this search for knowledge of the divine, of the future, and of the cosmos, the distinctions between astrologer, diviner, mystic, and natural magician become blurred. As Frances A. Yates points out, "The cosmological framework which they take for granted is always astrological, even where this is not expressly stated." Frances A. Yates, *Giordano Bruno and the Hermetic Tradition* (Chicago: University of Chicago Press, 1964), 22.

7. I will qualify Richard M. Berrong's contention that Rabelais's last three books are trips into fantasy, efforts by Rabelais to escape the realities of a world of political and religious chaos. Richard M. Berrong, *Every Man for Himself: Social Order and Its Dissolution in Rabelais* (Saratoga, Calif.: Anma Libri, 1985), 94–95, 104–108.

8. See Robert Marichal, "L'attitude de Rabelais devant le néoplatonisme et l'italianisme," in *François Rabelais: Ouvrage publié pour le quatrième centenaire de sa mort 1553–1953* (Geneva: Droz, 1953), for a discussion of Rabelais's early enthusiasm for Hermeticism and Neoplatonism. Based on evidence from Rabelais's library in 1525, Marichal argues that he was an ardent Neoplatonist as a young man, but by 1546, with the third and fourth books, he rejected much of this philosophy. Marichal bases the second part of his argument on Rabelais's satire of Ficino. I would argue, however, that Rabelais's satirical laughter aimed at Neoplatonists and Hermeticism does not reject such philosophy but attempts a correction of abuses.

9. See Wayne A. Rebhorn, "The Burdens and Joys of Freedom: An Interpretation of the Five Books of Rabelais," *Études Rabelaisiennes* 9 (1971): 71–90, for a helpful discussion of Rabelais's attitude toward the body, multiple interpretations of signs, and the personal effort to search for communion with God in Rabelais's five books.

and Saturn—coincided with anxieties over the security of the state.[10] The Habsburg-Valois wars in Italy had turned disastrous for Francis I, who had been captured by the Spanish king and Holy Roman Emperor, Charles, who demanded a huge ransom. Francis had to concede Naples, Milan, and Burgundy to Charles in a humiliating peace settlement. The almanacs played on the public's fears. One such almanac, published in France for the year 1524, had predicted "floods, overthrows of the state, great changes in all areas of life."[11] As M.A. Screech points out, the antiastrology propaganda of Rabelais "was intended to reassure the citizens of France, who lived during troubling events and were seemingly at the mercy of menacing astrological signs."[12]

The message of the satirical almanacs such as his *Pantagrueline Prognostication pour l'an 1533* is clear: instead of trusting in these unreliable prognostications, people should look to God for support. God, not the planets and the stars, is the force that holds everything together.[13] Rabelais does not reject astrology—the heavenly bodies may still influence the earth—but readers must be aware that God controls any influence that the stars might have.

> I say, as far as I am concerned, that if the kings, princes, and Christian communities held the divine word of God in reverence, and according to it governed themselves and their subjects, we would experience a year of bodily health, peace of spirit, and growing wealth like none other in our lifetimes; and we would see the sky's face, the earth's raiment, and the people's own deportment more joyous, happy, pleasant, and good-natured than they have been in more than fifty years.[14]

---

10. M.A. Screech, introduction to *Pantagrueline prognostication pour l'an 1533 avec Les almanachs pour les ans 1533, 1535 et 1541: La grande et vraye prognostication nouvelle de 1544*, by François Rabelais, ed. M.A. Screech (Geneva: Droz, 1974), 7–38, at 16–17.

11. Quoted in ibid., 15: "inondations, bouleversements des Etats, mutations dans tous les domaines" (translation mine).

12. Screech, introduction to *Pantagrueline prognostication*, 19: "était destinée à rassurer ses compatriotes, qui vivaient à une époque troublée, soumise à des influences astrales menaçantes" (translation mine).

13. "Et ne aura Saturne, ne Mars, ne Jupiter, ne aultre planete, certes non pas les anges, ni les sainctz, ny les hommes, ny les diables, vertuz, efficace, puissance, ne influence aulcune, si Dieu de son bon plaisir ne leur donne." (And not Saturn, Mars, Jupiter, nor any other planet, certainly not angels or saints, nor men, devils, virtues, powers, nor any celestial being whatsoever will have any influence unless God in his good will allows it) (translation mine). Rabelais, *Histories of Gargantua and Pantagruel*, 7–8.

14. Ibid., 47: "Je dis, quant est de moy, que si les Roys, Princes et communitez Christianes ont en reverence la divine parole de Dieu, et selon icelle gouvernent soy et leurs sujets, nous ne veismes de nostre aage année plus salubre és corps, plus paisible és asmes, plus fertile en biens, que sera cette-cy:

Rabelais insists that a society that holds "en reverence la divine parole de Dieu" will be blessed by God, despite ominous signs in the heavens.

As the "Almanach pour l'an 1533" explains, some things will be hidden and people should look in awe at God's majesty rather than attempt to grasp the secrets of the divine: "But these are the secrets of the council of the eternal King, who, in all that exists and comes about, moderates according to his will and pleasure. Thus, we should stop speaking and adore these secrets in silence."[15] Rabelais raises the difficult question whether people know the future by reading signs. As the next section will argue, those who attempt to find deeper meaning in signs often become objects of ridicule in Rabelais's first three books.

■ ■ ■

Rabelais's skepticism about reading signs reflects the larger debate associated with the recovery of Hermetic texts.[16] For example, the Hermetic writer of the *Asclepius* explains that God cannot be known by any name since he encompasses infinity:

> [T]he whole of god's name also includes meaning and spirit and air and everything at once that is in them or through them or from them; no, I cannot hope to name the maker of all majesty, the father and master of everything, with a single name, even a name composed of many names; he is nameless or rather he is all-named since he is one and all, so that one must call all things by his name or call him by the names of everything.[17]

---

et voirons la face du ciel, la vesture de la terre et le maintien du peuple, joyeux, gay, plaisant et benin, plus que ne fut depuis cinquante ans en çà" (translation mine).

15. Ibid., 41: "Mais ce sont secrets du conseil estroit du Roy eternel, qui tout ce qui est, et qui se fait, modere à son franc arbitre et bon plaisir. Les quels vaut mieux taire et les adorer en silence" (translation mine).

16. See A.J. Festugière, *Le Dieu Inconnu et la Gnose*, vol. 4 of *La Révélation d'Hermès Trismégiste* (Paris: Lecoffre, 1954), for a discussion of the search for God within the Hermetic tradition. See also Yates, *Giordano Bruno*, 124–25. The Hermetic corpus, reputedly an ancient philosophy going back to Moses and prefiguring Plato's texts, actually was written in the second and third centuries by a collection of writers. The Pseudo-Dionysian *Celestial Hierarchies*, believed to be written by a contemporary of St. Paul, was actually influenced by the Hermetic gnosis and is dated sometime between the second and sixth century (cf. Yates, *Giordano Bruno*, 117–18).

17. *Hermetica: The Greek "Corpus Hermeticum" and the Latin "Asclepius" in a New English Translation, with Notes and Introduction*, trans. Brian P. Copenhaver (New York: Cambridge University Press, 1992), 20. The Hebrew kabbalistic tradition has a similar negative theology in which the highest name

The futile attempt to know God through a single linguistic signifier reveals the impossibility of reading signs of the divine. God cannot be limited to one name, since his infinity encompasses all possible existences and all names.

However, Hermetic texts assured their readers that humans can know God by knowing his signs. In the eleventh book of the *Corpus Hermeticum*, for example, God reputedly shows himself in all of creation: "And do you say, 'god is unseen'? Hold your tongue! Who is more visible than god? This is why he made all things: so that through them all you might look on him."[18] Philosophers such as Giovanni Pico della Mirandola argued that one can and should read the signs of God in the world. In his "Oration on the Dignity of Man," Pico boasts that his natural magic "brings forth into the open the miracles concealed in the recesses of the world, in the depths of nature, and in the storehouses and mysteries of God."[19] The theologian Nicolas of Cusa had also advocated reading signs of God in the cosmos.[20] However, Cusa qualifies this position by explaining that we can only hope to approximate truth:

> The relationship of our intellect to the truth is like that of a polygon to a circle; the resemblance to the circle grows with the multiplication of the angles of the polygon; but apart from its being reduced to identity with the circle, no multiplication, even if it were infinite, of its angles will make the polygon equal the circle.[21]

According to Cusa, God essentially remains hidden; therefore, in his words, "all we know of the truth is that the absolute truth, such as it is, is beyond our reach."[22]

The first two books of Rabelais seem to support Nicolas of Cusa's skepticism about knowing the truths of God. The Thaumaste debate in *Pantagruel*, for example, ridicules those who seek hidden secrets in signs. While Thaumaste, the

---

of God is Nothing or No Name (Yates, *Giordano Bruno*, 125). The Pseudo-Dionysian texts also posited a negative theology since we can only describe God in terms of what he is not (Yates, *Giordano Bruno*, 124).

18. *Corpus Hermeticum*, in *Hermetica*, 1–66, at bk. XII, 22.

19. Pico della Mirandola, "Oration on the Dignity of Man," in *The Renaissance Philosophy of Man*, ed. Ernst Cassirer, Paul Oskar Kristeller, and John Herman Randall, Jr. (Chicago: University of Chicago Press, 1948), 223–386, at 249.

20. See Ernst Cassirer, *The Individual and the Cosmos in Renaissance Philosophy*, trans. Mario Domandi (New York: Barnes & Noble, 1963), 53. Also see the first two chapters of this work for a helpful overview of Nicolas of Cusa's philosophy.

21. Nicolas of Cusa, *Of Learned Ignorance*, trans. Germain Heron (London: Routledge & Kegan Paul, 1954), 11.

22. Ibid., 12.

serious sign reader, tries to interpret Panurge's gestures as though they had some philosophical or magical significance, Panurge degrades the art of deciphering signs by responding with the most obscene gestures he can think of:

> Panurge raised his thrice mighty codpiece into the air with his left hand, and with his right drew from it a piece of white ox rib and two pieces of wood of the same shape, one of black ebony, the other of red brasil-wood, which he placed symmetrically between the fingers of that hand. These he struck together, making the sort of noise that the lepers of Brittany make with their clappers—but it sounded better and more harmonious. At the same time, with his tongue contracted in his mouth, he hummed joyously, all the while looking at the Englishman.[23]

What is interesting about this sign is the audience's reaction to it. The narrator explains that the theologians, university doctors, and surgeons take a literal approach, believing that Panurge is signaling Thaumaste's leprosy. In contrast, the counselors, lawyers, and canon-lawyers think that Panurge "thought that in doing so he wished also to imply that some kind of human felicity lay in the leprous state, as Our Lord once affirmed."[24] While the audience is discussing the appropriate interpretation, Panurge displays his codpiece in front of a distinguished audience of scholars.

As this contest continues, Thaumaste begins to show signs of anxiety over his ability to "argue" with Panurge: he grows pale, trembles, and sweats heavily, resembling a man possessed. When Panurge makes a ring with his left hand, puts his right forefinger in that ring, and makes gnashing sounds with his teeth (an obviously obscene gesture, which Thaumaste is unable to decipher), Thaumaste becomes alarmed. At this point, the physicality of Panurge's responses has stripped away the high art of Thaumaste's rhetoric: Thaumaste "let a great baker's fart—for the bran followed it—pissed very strong vinegar, and stank like all the devils."[25]

---

23. Rabelais, *Histories of Gargantua and Pantagruel*, 235, and idem., *Œuvres Complètes*, 295: "Panurge…tyra en l'air sa très-mégiste braguette de la gausch, et de la dextre en tira un transon de couste bovine blanche et deux pièces de boys de form pareille, l'une de ébène noir, l'aultre de brésil incarnat, et les mist entre les doigtz d'ycelle en bonne symmétrie, et, les chocquant ensemble, faisoyt son tel que font les ladres en Bretaigne avecques leurs clicquettes, mieulx toutesfoys résonnant et plus harmonieux, et de la langue, contracte dedans la bouche, fredonnoyt joyeusement, tousjours reguardant l'Angloys."

24. Rabelais, *Histories of Gargantua and Pantagruel*, 235, and idem., *Œuvres Complètes*, 295: They thought that he "vouloyt conclurre quelque espèce de félicité humaine consister en estat de ladrye, comme jadys maintenoyt le Seigneur."

25. Rabelais, *Histories of Gargantua and Pantagruel*, 236, and idem., *Œuvres Complètes*, 296:

Next, he descends to the level of Panurge's absurdity: "Thaumaste began to puff up both his cheeks like a bagpiper, and blew as if he were blowing up a pig's bladder."[26] Panurge ends this contest by making several more obscene gestures.

After all of this absurdity and obscenity, Thaumaste is convinced that Panurge is one of the greatest philosophers of his age. He praises Pantagruel, Panurge's teacher, quoting the gospel of Luke: "Et ecce plus quam Salomon hic"(And behold, one greater than Solomon is here).[27] It was Pantagruel's reputation that convinced Thaumaste "to confer with him on certain insoluble problems of magic, alchemy, the Kabbalah, geomancy, and astrology, also of philosophy, all of which had been long in my mind."[28] However, there remains an even more significant reason for Thaumaste's conviction that Pantagruel is a greater philosopher than Solomon: Thaumaste's statement alludes to a passage in the Gospel of Luke when Jesus rebukes the people for seeking proof of his authority.[29] Rabelais's readers would have recognized this allusion as a message condemning Jesus' opponents for demanding more signs. Similarly, Thaumaste does not interpret Panurge's clearly obscene signs as insults; instead, he looks for their deeper meaning. Thaumaste seeks hidden knowledge where there is none.[30]

The prophecy at the end of *Gargantua* further illustrates the problem of assigning correct meanings to signs. Gargantua interprets the prophecy as an apocalyptic

---

Thaumaste "fist un gros pet de boulangier, car le bran vint après, et pissa vinaigre bien fort, et puoit comme tous les diables."

26. Rabelais, *Histories of Gargantua and Pantagruel*, 236, and idem., *Œuvres Complètes*, 297: "Thaumaste commença enfler les deux joues, comme un cornemuseur, et souffloit comme se il enfloit une vessie de porc."

27. Rabelais, *Œuvres Complètes*, 298.

28. Rabelais, *Histories of Gargantua and Pantagruel*, 238, and idem., *Œuvres Complètes*, 298: It convinces Thaumaste to "conférer avecques luy des problèmes insolubles, tant de magie, alchymie, de caballe, de géomantie, de astrologie, que de philosophie, lesquelz je avoys en mon esprit."

29. "This generation is an evil generation; it asks for a sign, but no sign will be given to it except the sign of Jonah. For just as Jonah became a sign to the people of Ninevah, so the Son of Man will be to this generation. The queen of the South will rise at the judgment with the people of this generation and condemn them, because she came from the ends of the earth to listen to the wisdom of Solomon, and see, something greater than Solomon is here!" Luke 11:29–31 (NRSV).

30. However, there is a second strong allusion in this passage. By seeking a deeper meaning in Thaumaste's comment instead of taking it at face value, the reader is pulled into the same process of reading for a deeper, hidden meaning. This is the genius of Rabelais: as we laugh at Thaumaste's ridiculous struggles to interpret Panurge's gestures, we also try to decipher the "deeper" meaning of this passage. We would like to identify ourselves with Panurge and be in on the joke, but our immediate impulse to search out the "marrow" within this episode would align us with Thaumaste, the spectators, and those gullible readers that Rabelais seems to toy with in "The Author's Prologue" to *Gargantua: The Histories of Gargantua and Pantagruel*, 38.

vision of the persecution and final victory of those steadfast in the Faith; however, Friar John chides Gargantua for searching for such an obscure meaning:

> You can read all the allegorical and serious meanings into it that you like, and dream on about it, you and all the world, as much as ever you will. For my part, I don't think there is any other sense concealed in it than the description of a game of tennis wrapped up in strange language.[31]

Friar John brings up a point made by the Thaumaste episode: because people will always read signs differently, interpretations can never be trusted.

In *Le tiers livre*, Panurge finds that his various consultations with diviners illustrate the apparent impossibility of agreeing on any one interpretation. Panurge and Pantagruel go from one form of divination to another, trying to get a more direct answer about Panurge's marriage. The problem lies not in messages of the diviners, but in Panurge's and Pantagruel's different interpretations. Panurge will continue seeking diviners until he gets the answer he wants. Even so, the messages and signs of the diviners in the *Le tiers livre* are fairly consistent. Pantagruel interprets all of them to mean that Panurge would be cuckolded, beaten, and robbed by his future wife. However, Panurge will read them as he wishes.[32]

The problem of reading signs in the *Le tiers livre* seems to support Nicolas of Cusa's assertion that "all we know of the truth is that the absolute truth, such as it is, is beyond our reach."[33] However, laughter at the expense of diviners and others reading signs does not eliminate the possibility of knowing the divine through signs. As this search touches the physicalities of the body in Rabelais, it must be recognized that these are degraded attempts to know the divine that anticipate the kind of revelation Panurge ultimately finds in *Le cinquième livre*.[34] Rabelais

---

31. Rabelais, *Histories of Gargantua and Pantagruel*, 163, and idem., *Œuvres Complètes*, 206–7: "Donnez-y allégories et intelligences tant graves que vouldrez, et y ravassez, vous et tout le monde, ainsy que vouldrez. De ma part, je n'y pense aultre sens enclous q'une description du jeu de paulme soubz obscures parolles."

32. See Catherine Randall for a discussion of Panurge's motivations for the quest: "Panurge se construit un tableau de l'avenir pour pouvoir se donner l'illusion d'agir sur lui; c'est un acte de magie imitative qui lui permet de regagner un peu de son pouvoir (et peut-être aussi de sa puissance) sur le monde culbuté." (Panurge constructs a tableau of the future in order to delude himself that he can direct that future: it is an act that mimics magic, thus allowing him to regain some of his influence (and perhaps even his power) over a world upside down) (translation mine). Catherine Randall, "Le cocuage hypothétique de Panurge: Le monde à l'envers dans *Le tiers livre*," *Constructions* (1986): 77–86, at 81.

33. Cusa, *Of Learned Ignorance*, 12.

34. The "authenticity" of *Le cinquième livre* has been questioned by many critics. However, I am convinced by the argument of G. Mallary Masters, *Rabelaisian Dialectic and Platonic-Hermetic Tradition* (Albany: State University of New York Press, 1969), that *Le cinquième livre* should be accepted as

enters the debate initiated by the Hermetic texts of the Italian Neoplatonists: whether to reject the material realm as an imperfect state that must be sloughed off to attain divine insight, or to explore the material realm as a creation infused with signs of a divine creator. With the imagery of drinking and of the grotesque body in his five books, Rabelais relocates Panurge's search for signs of the future in the physical realities of the body. He thus reinforces the Renaissance paradigm of the body as unexplored territory—vast in its potential to open up knowledge to those who would explore and colonize it. Rabelais's ridiculing humor is aimed at those who, in their quest for divine insight, blindly ignore the revelatory potential of the body's materiality. As seen in *Le cinquième livre*, an interpretation of signs through the body will rejuvenate the body and reveal the secrets of God.[35]

■ ■ ■

As we have seen, the Hermetic writings that enthralled Quattrocento Neoplatonists do not present a consistent body of beliefs. A. J. Festugière, in his *La Révélation d'Hermès Trismégiste*, describes two opposing views, "une tendance optimiste et une tendance pessimiste," toward the material.[36] The optimistic view of the world posits a cosmos ordered by a creator such that "the view of the world directs one naturally to the knowledge and adoration of the God/demiurge of the world."[37] God encompasses all, giving it energy and substance. Giovanni Pico

---

the logical ending to the adventures of Pantagruel and Gargantua based on its thematic unity with the other books.

35. See Mikhail Bakhtin, *Rabelais and His World*, trans. Helene Iswolsky (Cambridge: MIT Press, 1965), 21, for his explanation of the potential for renewal in carnivalesque humor. This laughter can rejuvenate through its contact with the body. Although I do not accept the entirety of Bakhtin's argument, his paradigm of the carnivalesque grotesque will be useful later since it puts in relief the two opposing assumptions about materiality that are operating within the Hermetic tradition. See Paul Allen Miller, "The Otherness of History in Rabelais's Carnival and Juvenal's Satires, or Why Bakhtin Got It Right the First Time," in *Carnivalizing Difference: Bakhtin and the Other*, ed. Peter I. Barta, Paul Allen Miller, Charles Platter, and David Shepherd (New York: Harwood Academic Publishers, 1998).

36. Festugière, *Le Dieu cosmique* in *La Révélation d'Hermès Trismégiste* (Paris: Lecoffre, 1949), 2:xi. Festugière labels the following treatises of the *Corpus Hermeticum* as optimistic: II, V, VI, VIII, IX–XII, XIV, XVI, Asclepius. Those he labels pessimistic include I, IV VII, XIII. See ibid., 2:ix n. 1.

37. Ibid: "la vue du monde conduit naturellement à la connaissance et à l'adoration d'un Dieu démiurge du monde" (translation mine). For example, the eleventh treatise of the *Hermetica* emphasizes the presence of God in a material cosmos: "Is he in matter, then, father?" "If matter is apart from god, my son, what sort of place would you allot to it? If it is not energized, do you suppose it is anything but a heap? But who energizes it if it is energized? We have said that the energies are parts of god. By whom, then, are all living things made alive? By whom are immortals made immortal? Things subject to change—by whom are they changed? Whether you say matter or body or essence, know that these

della Mirandola emphasizes this optimistic tendency in his praise of natural magic that "brings forth into the open the miracles concealed in the recesses of the world, in the depths of nature, and in the storehouses and mysteries of God."[38] Once the world is viewed as God's creation, the material realm becomes a cosmos that God fills with signs of his presence.

In contrast, the pessimistic tendency of the Hermetic writings assumes that God will reveal himself only if the seeker casts off the material, rejecting the body and its sensual pleasures: "Unless you first hate your body, my child, you cannot love yourself."[39] God requires that one "[l]eave the senses of the body idle, and the birth of divinity will begin."[40] Pico della Mirandola seems to have incorporated the Hermetic contraction into his own philosophy, since his vision of the natural philosopher also emphasizes a rejection of the body.[41] Those who wish to be initiated into the secrets of the universe and of God must reject the material, trading the physical pleasures of the body for the enlightenment of God beyond the stars.[42]

The Hermetic cosmos consists of a succession of spheres occupied by demons and angels: "Do you see how many bodies we must pass through, my child, how many troops of demons, (cosmic) connections and stellar circuits in order to hasten toward the one and only?"[43] Spiritual man will ascend beyond the sublunar realm to the spheres beyond, knowing that the body will be abandoned as the magus gains insight into the divine. Likewise, Pantagruel's description of the diviner in *Le tiers livre* emphasizes the pessimistic attitude of most diviners who believe that they must abandon their bodies to know the divine:

> [S]o in order to be wise in their eyes—wise, I mean, in knowing, and foreknowing by divine inspiration, and fit to receive the gift of divination—a

---

also are energies of god and that materiality is the energy of matter, corporeality the energy of bodies and essentiality the energy of essence. And this is god, the all." *Corpus Hermeticum*, bk. XII, 48.

38. Pico, "Oration on the Dignity of Man," 249.

39. *Corpus Hermeticum*, bk. VI, 6.

40. Ibid., bk. XIII, 7.

41. "If you see a philosopher determining all things by means of right reason, him you shall reverence: he is a heavenly being and not of this earth. If you see a pure contemplator, one unaware of the body and confined to the inner reaches of the mind, he is neither an earthly nor a heavenly being; he is a more reverend divinity vested with human flesh." Pico, "Oration on the Dignity of Man," 226.

42. See Jonathan Sawday, *The Body Emblazoned: Dissection and the Human Body in Renaissance Culture* (London: Routledge, 1995), 16–22, for his discussion of the body-soul tension, a helpful overview of Platonic dualism as it helped construct ideologies of the body during the Renaissance.

43. *Corpus Hermeticum*, bk. IV, 17. See also *Corpus Hermeticum*, bk. I, 5–6, which describes the ascension through the spheres, allowing one to leave the sins of the flesh at each sphere—change, malice, desire, ambition, audacity, appetite for riches, deceit—and finally be joined with God.

man must forget himself, rise above himself, rid his senses of all earthly affection, purge his spirit of all human solicitude, and view everything with unconcern.[44]

According to Pantagruel, those who search for meaning beyond the physical realm, wishing to discover the mysteries of the cosmos, feel they must transcend their physical bodies, cutting all ties to earthly affection.

The body and its pleasures separate one from revelation because, according to the Hermetic creation myth, God had created man to be spiritual, giving him dominion over the celestial sphere. The Hermetic version of the Fall places female Nature at center stage. Her seduction of and subsequent union with man imprisoned his spiritual nature in a physical body, thus creating the duality of spirit and flesh. The pessimistic strand of Hermeticism is dualistic, assuming that the body separates physical man from a spiritual God. As the writer of the first treatise explains, "the one who loved the body that came from the error of desire goes on in darkness, errant, suffering sensibly the effects of death."[45] Thus, the goal of the diviner is to repudiate the physicality of the body, the original cause of man's fall from grace, and regain the divinity for which he is destined.

The Hermetic tradition assumes man's original divinity. The magus can regain his place in the supercelestial realm, beyond the stars and even the hierarchies of angels. In light of Rabelais's criticism of astrology in his satirical almanacs, one can anticipate his rejection of this dualistic tendency with its concomitant denigration of the flesh. Just as astrology encourages false confidence in the power to control the future by ignoring human dependence on God, the Hermetic ascent toward the divine reveals an audacious form of hubris. As will be seen, in Panurge's consultation with Her Trippa, Rabelais confounds all aspirations to transcend the sexual body. Her Trippa's language, thoughts, and art are imbued with the obscene. Similarly, the sibyl's grotesque body brings her art into contact with the powers of death and rejuvenation. Thus, Rabelais redirects Panurge's search for knowledge of the future to a recognition of the physical body.

---

44. Rabelais, *Histories of Gargantua and Pantagruel*, 391, and idem., *Œuvres Complètes*, 505: "faut-il, pour davant icelles saige estre, je dis sage et praesage par aspiration divine et apte à recepvoir bénéfice de divination, se oublier soy-mesmes, issir hors de soy-mesmes, vuider ses sens de toute terrienne affection, purger son esprit de toute humaine sollicitude et mettre tout en nonchaloir."

45. *Corpus Hermeticum*, bk. I, 4. The pessimistic strand of the Hermetic tradition, with its dualistic understanding of the Fall as a descent into the material body, has close connections with Manichaean dualism. Both Manichaeism and the Hermetic texts originate from the second century A.D., and their popularity should be understood within the context of Neoplatonic dualisms.

Panurge's consultation with Her Trippa, the astrologer who cannot see that his own wife has cuckolded him, illustrates the futility of Panurge's efforts to transcend the body and obtain access to divine mysteries. As Her Trippa was focusing on celestial mysteries, his wife was having sex with other men. His search beyond the stars had blinded him to the matters on earth. When Panurge forces him to acknowledge his wife's infidelity, his transcendental art descends to the level of copulation by the end of the chapter.

This forced recognition of the physical begins Her Trippa's downward spiral to the obscene body, which infects his own language. After Panurge makes the sign of the cuckold's horns and points his fingers at Her Trippa, Her Trippa tries to maintain his dignity as a diviner who has transcended the physicality of the world by announcing his expertise in all the arts of divination. He asks Panurge whether he would like to know the answer "by pyromancy, by aeromancy...by hydromancy, or by lecanomancy."[46] However, Her Trippa's Panurgian language betrays the fact that his art cannot transcend the reality of his wife's sexual exploits: "In a basin full of water I'll show you your future wife being rogered by two rustics."[47] He will show Panurge's wife having sex (a more obscene term would be a better translation for "brimballant") with country boys. Panurge answers in kind: "When you poke your nose up my arse...don't forget to take off your spectacles."[48] Undaunted, Her Trippa continues with his catalogue of divinatory arts, but his language is degraded to the level of crude sex: he will reveal Panurge's future wife "brisgoutant" (being poked), and he will show Panurge the image of his wife and "ses taboureurs" (her belly-drummers).[49] Panurge continues the obscene exchange by wishing Her Trippa's body to be sodomized and covered with the filth of old trousers.

Her Trippa's high art has been reduced to physicality; his own language descends from the expression of celestial mysteries to mouthing obscene insults at the man who will not let him forget his wife's infidelity. Similarly, Her Trippa, the astrologer, slides into very suspect arts of divination, prompting Panurge to call him "le coqu, cornu, marrane, sorcier au diable, enchanteur de l'Antichrist!" (cuckold, heretic, sorcerer of the devil, wizard of the Antichrist!)[50] Astrology has become

---

46. Rabelais, *Histories of Gargantua and Pantagruel*, 357.

47. Ibid., 358; and idem., *Œuvres Complètes*, 463: "Dedans un bassin plein d'eaue je te monstreray ta femme future brimballant avecques deux rustres."

48. Rabelais, *Histories of Gargantua and Pantagruel*, 358, and idem., *Œuvres Complètes*, 463: "Quand tu mettras ton nez en mon cul,...sois recors de deschausser tes lunettes."

49. Ibid.

50. Ibid., 465 (translation mine).

an art associated with the physical and the satanic. By the end of the chapter, Her Trippa has little in common with Neoplatonists like Ficino and Giovanni Pico.[51]

Similarly, Panurge's consultation with the sibyl brings Panurge's search for meaning into contact with the body. Florence Weinberg has tried to reclaim the sibyl as one of the few positive female figures in the five books by diminishing the importance of her grotesque body: "She is the only female shown using her mental, intuitive, spiritual, and inspired powers, even though her portrayal is so grotesque as to make her appear purely farcical."[52] Weinberg misses the point of the episode by dismissing the grotesque portrayal of the sybil. Her physicality is not a side issue or a distraction from her role as sibyl: her body and her connection with the earth will prove to be a parodic image that anticipates the body- and earth-centered search for knowledge in *Le cinquième livre*.

The sibyl's grotesque body and her poor surroundings prevent an ascent beyond the material realm. Her body and its physical needs focus attention on themselves. Her surroundings and diet are deficient, identifying her with the peasant class. Panurge, Pantagruel, and Epistemon find her "straw-thatched cottage, which was badly built, badly furnished, and filled with smoke."[53] She is cooking soup out of "some green cabbage, a rind of yellow bacon, and an old marrow bone."[54] Also, her body is sickly and malnourished: "The old woman was grim to look at, ill-dressed, ill-nourished, toothless, blearly-eyed, hunchbacked, snotty, and feeble."[55] Panurge's choice of gifts reveals that her physicality has adjusted his own expectations of her. The traditional gift, the golden bough, is no longer appropriate.[56] Instead, he brings her six smoked ox tongues, a pot full of

---

51. See Pico, "Oration on the Dignity of Man," 26, for his distinction between these two kinds of magic. Also, throughout *Giordano Bruno*, Yates discusses the fine line drawn between demonic magic and natural philosophy.

52. Florence W. Weinberg, "Written on the Leaves: Rabelais and the Sibylline Tradition," *Renaissance Quarterly* (1990): 709–30, at 709.

53. Rabelais, *Histories of Gargantua and Pantagruel*, 333, and idem., *Œuvres Complètes*, 430: "la case chaumine, mal bastie, mal meublée, toute enfumèe."

54. Rabelais, *Histories of Gargantua and Pantagruel*, 333, and idem., *Œuvres Complètes*, 431: "choux verds avecques une couane de lard jausne et un vieil savorados."

55. Rabelais, *Histories of Gargantua and Pantagruel*, 333, and idem., *Œuvres Complètes*, 431: "La vieille estoit mal en poinct, mal vestue, mal nourrie, édentée, chassieuse, courbassée, roupieuse, languoureuse."

56. As Weinberg, "Written on the Leaves," 719, points out, the golden bough was not the payment for the sibyl in book 6 of the *Aeneid*. Instead, it was used as "an offering to Proserpine and as a means to enter and again leave Hades." Weinberg further speculates that this mix-up in the use of the golden bough implies that the sibyl is as disgustingly ugly as Proserpine and that her cottage is "as desolate as an outpost of Hades."

dumplings, a bottle of wine, a golden ring, and a purse of gold coins.[57] Note that these gifts (like food and wine) emphasize the sybil's connection to the body.

The sibyl's body is a body in flux whose boundaries are constantly broken down by the process of eating and drinking. For example, during the ceremony to divine Panurge's future, "she took a long swig from the bottle."[58] Also, her movements reveal her close connection with the earth and its fertility: she is a spinner, a figure associated with the woman creator;[59] she has a mortar for millet, a grain associated with "fertility and health in marriage"; and she removes one of her shoes "to put herself in closer contact with the Earth."[60] The sibyl makes no attempt to transcend the body, and its imposing presence focuses her divination on physical realities.

After throwing the leaves of prophecy to the wind, she brings Panurge's own search for divinatory guidance to the obscenity of the body: "on the doorstep she hitched up her gown, petticoat, and smock to her armpits, and showed them her arse."[61] She lifts up her skirts and shows her anus, nullifying Panurge's grand ideas about finding truth in divination. To Panurge, the "trou de la Sibylle" (sibyl's cave) is a dark cavern with the connotations of the grave; however, the sibyl's cavern-like "cul" (ass), a symbol of putrefaction, decay, and death, also has associations with the womb.[62] As Jeffrey Masten has argued, Renaissance notions of the "fundament" as the grave are complicated by other etymological connotations, which lend it foundational or originary potential. [63] Anticipating Bakhtin's paradigm of the grotesque body, in which the forces of death and birth are both inherent in the image of the anus, the sybil's "cul," which Panurge sees only as a deep cavern of deathlike emptiness, resonates with images of eating and of the earth's fertility. This hole of decay is juxtaposed with the vagina and its forces of renewal.

With the sibyl's and Her Trippa's laughable attempts at divination, Rabelais repudiates the pessimistic tendency of Hermetic writers. Those who, like Her

---

57. Rabelais, *Histories of Gargantua and Pantagruel*, 334. Weinberg points out the sexual connotations of the "verge," the bough, and the ring. Cf. Weinberg, "Written on the Leaves," 719.

58. Rabelais, *Histories of Gargantua and Pantagruel*, 334.

59. Cf. Weinberg, "Written on the Leaves," 727.

60. Ibid., 721.

61. Rabelais, *Histories of Gargantua and Pantagruel*, 335, and idem., *Œuvres Complètes*, 433: "sus le perron de la porte se recoursa, robbe, cotte et chemise jusques aux escelles, et leurs monstroit son cul." When the sybil lifts her skirts, Panurge exclaims "look, there's the sybil's cavern." Rabelais, *Histories of Gargantua and Pantagruel*, 335.

62. Rabelais, *Œuvres Complètes*, 433.

63. "Is the Fundament a Grave?" in *The Body in Parts: Fantasies of Corporeality in Early Modern Europe*, ed. David Hillman and Carla Mazzio (New York: Routledge, 1997), 128–45, at 134.

Trippa, aspire to a level of divinity beyond the material realm are quickly forced to recognize their participation in the body's obscene physicality.

■ ■ ■

As a medical doctor, Rabelais viewed much of the world in relation to the body. In *Le tiers livre*, for example, Panurge likens the cosmic hierarchy of the universe to "our microcosm, our little world that is, which is man, with all his limbs lending, borrowing, and owing—that is to say, in his natural state."[64] The elaborate system of cosmic influences is reflected by the "retz mervelleux" (miraculous network) of the body—the hands, eyes, appetite, stomach, and veins work together to generate the blood, while the kidneys, spleen, bile duct, and heart work to purify it.[65] As Mettra explains, "for a man like Rabelais, who followed the evolution of the medicine of his time with passion, the astrological analogy constituted another helpful insight into the organization of the human body."[66] For Rabelais, the microcosm, the body, mirrors the macrocosm, the universe. Moreover, the microcosm/macrocosm paradigm initiated a new discourse on the body, one that constructed the anatomist as explorer of unmapped territory. As Jonathan Sawday has argued, between 1540 and 1640 the Vesalian body is an unexplored world, which resembles, imitates, and reflects the cosmic hierarchies.[67] As this "geographic body" holds the potential for colonization and domination for those willing to explore its anatomy, for Rabelais it promises knowledge of the divine, hidden within the coils and vast recesses of the body's marvelous network.[68]

The cosmic model for the body's intricate processes had a very practical application: Rabelais depended on astrology to know when to use certain medicines and perform procedures such as bloodletting. His "Almanach pour l'an 1541" reiterates the microcosmic view by representing the intimate links between the forces of the cosmos (the stars, moon, planets, etc.) and the body. The almanac is a system for organizing the layers of correspondences in the cosmos. In this particular

---

64. Rabelais, *Histories of Gargantua and Pantagruel*, 300, and idem., *Œuvres Complètes*, 387: "nostre microcosme, *id est* petit monde, c'est l'homme, en tous ses membres prestans, empruntans, doibvans, c'est-à-dire en son naturel."

65. Cf. Rabelais, *Œuvres Complètes*, 387–89.

66. "[P]our un homme comme Rabelais, qui avait suivi avec passion l'évolution de la médecine de son temps, la réflexion astrologique constituait un regard supplémentaire sur l'organisation générale du corps." Claude Mettra, *Rabelais Secret* (Paris: Grasset, 1973), 227 (translation mine).

67. See Sawday, *Body Emblazoned*, 23–24.

68. See Sawday's discussion of the "geographic body." He contrasts this paradigm with the "mechanical body" of later Cartesian thought. Cf. Sawday, *Body Emblazoned*, 22–32.

one, all the celestial signs, measurements, and dates are finally brought to bear on the human body, the focus of influence and the center of Rabelais's vision of the universe. The left column lists the days of the week and the corresponding religious holidays and feasts; the middle column gives the signs of the zodiac and the degree of the moon for that day; and the right column provides weather predictions, phases of the moon, and medical advice about which procedures one should perform on particular days, based on astrological influences. As Alain-Michel Boyer points out, in the right column "exterior nature," the weather and the phases of the moon, and "human nature...mix and intermix with one another; their elements or their attributes, in close interaction, cross over, as if these two natures were the expression of one another, two versions or two sides of the same reality."[69] Rabelais's serious almanac thus reinforces what Panurge assumes to be the astral influence on the body: that "other little world, which is man," will reflect the harmony or discord of the universe. The body is the universal hub where influences from the cosmos focus their energies and then are reflected back out into the universe.[70]

Moreover, Rabelais asserts that there exists an intimate connection between the physical and spiritual workings within the body. In *Le tiers livre*, Panurge argues that within "the miraculous network" of the body "the animal spirits" are produced, "which endow us with imagination, reason, judgment, resolution, deliberation, ratiocination, and memory."[71] Reasoning and imagination depend

---

69. "Nature extérieur et nature humaine...se mêlent et s'entre-mêlent; leurs éléments ou leurs attributs, en étroite interaction, s'entrecroisent, comme si ces deux natures étaient l'expression l'une de l'autre, les deux versions ou les deux versants d'une même réalité." Alain-Michel Boyer, "Les Architectures du Lisible: Quelques Réflexions autour de l'Almanach pour 1541 de Rabelais," *Textes et Langages* 14 (1987): 1–27, at 5 (translation mine). Moreover, the system of notation in the right column is a mixture of alphabetical letters and words, pictorial signs of the zodiac, and symbols for various medical prescriptions. It was common for the almanacs of this time to allow "une double possibilité de déchiffrage," Geneviève Bollème, *Les Almanachs Populaires aux XVIIe et XVIIIe Siècles* (Paris: Mouton, 1969), 13; an almanac like Rabelais's could reach a literate and illiterate readership: "En effet, cet almanach pour 1541 est peut-être l'un des lieux où l'on peut le mieux observer un passage ou un glissement de la culture orale à la culture alphabétique, dans la mesure où les deux ordres, juxtaposés, coïncident ou concordent un moment"; Boyer, "Architectures du Lisible," 2–3. The third column creates a space where different systems of meaning—the oral and alphabetic systems of signs—are in contact.

70. As Michel Foucault explains, in the Ptolemaic cosmos, the microcosm of the body "est toujours la moitié possible d'un atlas universel.... Il est le grand foyer des proportions—le centre où les rapports viennent s'appuyer et d'où ils sont réfléchis à nouveau." Michel Foucault, *Les mots et les choses: Une archéologie des sciences humaines* (Mayenne, France: Gallimard, 1966), 37–38.

71. Rabelais, *Histories of Gargantua and Pantagruel*, 301, and idem., *Œuvres Complètes*, 389: "le retz mervelleux...les espritz animaulx...moyennans lesquelz elle [l'âme] imagine, discourt, juge, résoust, délibère, ratiocine et remémore."

on the physical processes of the body. Pantagruel explains that the writings of the fasting hermits are therefore dry and uninspired because they have ignored their physical needs.[72] Inspiration comes through the nourishment of this marvelous network of the body, not through an ascetic denial of its physicality.

Laughter at the expense of diviners like Her Trippa does not eliminate the possibility of knowing the divine through signs in the cosmos. Rather, these absurd attempts to know the divine anticipate the kind of revelation made possible through a drunken engagement with the body. Even Panurge's reaction to Her Trippa suggests, in a ridiculous sort of way, that the sexual and obscene realities of the body can reveal deeper mysteries than those in distant stars. His violent reaction to the cuckold Her Trippa, moreover, reveals Panurge's own ambivalence toward his future as he continues to reject the diviners' prophecies that would make him a cuckold. Rabelais thus rejects the naive optimism of philosophers like Pico, but he also explores the potential for discovering divine secrets, even through ridiculously ambiguous signs of the body's physicality.

In opposition to Hermetic transcendence that would abandon the body, Pantagrueline drunkenness becomes synonymous with reading signs hidden in the mysteries of the physical body and the material creation. When the sybil takes "a long swig from the bottle" in preparation for her prophecy, her needy, even obscene body ironically anticipates the divinatory insight made possible through a drunken participation in the body.[73] Likewise, Panurge's comment after he trades obscenities with Her Trippa opens the door to a new kind of divination which gains its power through the body: "Dictez amen, et allons boyre" (Say amen, and let's drink).[74] Panurge's impulse to drink, to take nourishing wine and refresh his body, reflects the possibility for rejuvenation, both physical and spiritual.

Giovanni Pico della Mirandola also employs the metaphor of Bacchic drunkenness to describe the initiation into the mysteries of the world: "Bacchus, the leader of the Muses, by showing in his mysteries, that is, in the visible signs of nature, the invisible things of God to us who study philosophy, will intoxicate us with the fulness of God's house."[75] In Pico's optimistic reading of the imagery of wine, the philosopher, drunk "with the fulness of God's house," reads these mysteries of the divine in the material world. However, drinking wine is also synonymous with

---

72. "[D]ifficile chose estre bons et serains rester les espritz, estant le corps en inanition, veu que les philosophes et médicins afferment les espritz animaulx sourdre, naistre et practicquer par le sang artérial, purifié et affiné à perfection dedans le retz admirable." Rabelais, *Œuvres Complètes*, 416.

73. Rabelais, *Histories of Gargantua and Pantagruel*, 334.

74. Ibid., 465 (translation mine).

75. Pico, "Oration on the Dignity of Man," 234.

indulging the physical needs of the body, which the pessimistic strand of the Hermetic tradition criticizes as a descent into the lower depths of the material.[76] Wine leaves the drinker imprisoned within the physical pleasures of the body, preventing his ascent to spiritual revelation.

In contrast to the pessimistic tendency of some Hermetic texts, Rabelais will use the metaphor of wine as inspiration for the human participation in the material creation of God.[77] Thus, Pantagrueline drunkenness becomes synonymous with reading signs hidden in the mysteries of the body. In the Prologue to *Le cinquième livre*, Rabelais describes the acquisition of knowledge with metaphors of food and wine. In the style of Erasmus's *Praise of Folly*, he compares the books of high seriousness ("the flower of the bean...a pile of books in time of Lent. These used to appear as flowery, flourishing, and florid as so many butterflies, but were really tiresome, boring, perilous, prickly, and dark") to the "beans in the pod. These jolly and fruitful books of Pantagruelism."[78] The books of Pantagruelism are joyful and fertile, able to nourish those who consume them. For this reason, "all the world in fact is given over to the study of these books."[79] Rabelais uses the metaphor of wine and nourishment to make a similar comparison between ancient books of wisdom and his own writing. He has drunk of the books of Parnassus and found that they produced nothing of worth for the vulgar tongue:[80]

> I see that after hanging for a while around Apollo's school on Mount Parnassus and taking frequent swigs at the Caballine spring, all among the jolly Muses, they contribute nothing to the eternal fabric of our vulgar

---

76. "Where are you heading in your drunkenness, you people? Have you swallowed the doctrine of ignorance undiluted, vomiting it up already because you cannot hold it? Stop and sober yourselves up!" (*Corpus Hermeticum*, bk. VII, 24).

77. Masters, *Rabelaisian Dialectic*, as well as Florence Weinberg, *The Wine and the Will* (Albany: State University of New York Press, 1969), identify the metaphor of drinking with Dionysian and Hermetic efforts to effect a spiritual union with the mysteries of God.

78. Rabelais, *Histories of Gargantua and Pantagruel*, 602, and idem., *Œuvres Complètes*, 788: He compares "la fleur des febves...un tas de livres qui sembloient florides, florulens, floris comme beaux papillons, mais au vray estoient ennuyeux, fascheux, dangereux, espineux et ténébreux" to the "febves en gousse...ces joyeux et fructueux livres de pantagruélisme."

79. Rabelais, *Histories of Gargantua and Pantagruel*, 602, and idem., *Œuvres Complètes*, 788: "à l'estude desquels tout le monde s'est adonné."

80. For the debate between the ancients and moderns in France, see Joachim Du Bellay, "La défense et illustration de la langue française," in *Les regrets: Les antiquités de Rome*, ed. S. de Sacy (Bussière à Saint-Amand: Gallimard, 1990), 197–267. In this influential treatise first published in 1549, Du Bellay argues that the French language is inherently just as noble as the ancient languages and appeals to his contemporaries to enrich their national literature by writing in their vulgar tongue.

tongue but Parian marble, alabaster, porphyry, and firm royal cement. They treat of nothing but heroic exploits, great themes, arduous, weighty, and difficult subjects, and all this in a crimson satin style. In their writings they produce nothing but divine nectar, rare, light, and sparkling wines, delicate and delicious juice of the muscat grape.[81]

The wine of Parnassus cannot sustain or nourish; it has nothing to offer but cold marble and alabaster, the dead smoothness of abandoned temples. It treats the idealized exploits of heroes and "arduous, grave, and difficult matters." However, Rabelais's books should be consumed, gulped down, and absorbed into the body:[82]

Therefore, my dear boozers, I advise you in good time to lay up a fair store of them when the opportunity offers, and as soon as you find them in the booksellers' shops. When the chance comes you must not only shell them, but gulp them down as an opiate cordial and absorb them into your systems.[83]

The inscription on the door to the temple reads, "en vin vérité."[84] Yet drinking the wine of the fountain does not yield the kind of truth Panurge had sought earlier. The wine that each of them drinks will not give definitive answers, since each drinker seeks his own meaning.[85] Divination is locked in the subjectivity of the body as each person becomes his or her own interpreter.[86] Reminiscent of a Protestant introspection that relies on God's spirit to provide insight into the

---

81. Rabelais, *Histories of Gargantua and Pantagruel*, 604, and idem., *Œuvres Complètes*, 789–90: "Et voy que, par long temps avoir en mont Parnase versé à l'escole d'Apollo et du fons Cabalin beu à plein godet entre les joyeuses Muses, à l'éternelle fabrique de nostre vulgaire ils ne portent que marbre Parien, Alebastre, Porphire, et bon ciment Royal; ils ne traittent que gestes héroïques, de grandes choses, de matières ardues, graves et difficiles, et le tout en rhétorique armoisine, cramoisine; par leurs écrits ils ne produisent que nectar divin, vin précieux, friand, riant, de goût muscat, délicat, délicieux."

82. Rabelais's readers would have recognized allusions to the Old Testament Book of Ezekiel, where God instructs Ezekiel, the prophet to Judah, to eat the scroll before he preaches to the people. Ezek. 2:8–3:3 (NRSV).

83. Rabelais, *Histories of Gargantua and Pantagruel*, 605. "Pourtant, beuveurs, je vous advise en heure oportune, faictes d'iceux bonne provision soudain que les trouverez par les officines des libraires; et non seulement les égoussez, mais dévorez, comme opiatte cordialle et les incorporez en vous mesmes." Idem., *Œuvres Complètes*, 791.

84. Rabelais, *Œuvres Complètes*, 887.

85. Ibid., 904.

86. See Jonathan Sawday's discussion of the body's interiority, especially chap. 1, "The Autoptic Vision" and "The Rhetoric of Self-Dissection," *Body Emblazoned*, 110–129. He argues that viewing the interior of the body violates a taboo but also promises to open up an intimate knowledge of oneself.

Scriptures, each individual is called to drink in the sacred texts and find truth within. As Bacbuc explains, "à Dieu rien soit impossible" (to God, nothing is impossible).[87] God makes this revelation possible, allowing each drinker to find insight into the divine through the body's consumption of the wine.[88]

Panurge's search for knowledge of the future in books 3 and 4, motivated by the desire to control life rather than trust in God, is transformed by his experience at the temple. Panurge finally understands that another diviner cannot offer him sure knowledge of the future based on signs from the stars. Instead, he should drink in the signs of God, taste the presence of the divine, and swallow the glosses of veiled prophecies.[89] Panurge swallows the message of the oracle, for only through its bodily consumption can the gloss have meaning.[90] He wants to avoid the uncertainties of marriage by sure signs of the future from expert diviners. As Bacbuc explains, however, Panurge must look within himself to know the truth: "The Holy Bottle directs you to it. You must be your own interpreters in this matter."[91]

The potential for knowledge within the body, however, allows for a new kind of human domination, one that would map the secrets of the body and claim this new territory for the colonizing anatomist. Bacbuc's instruction to gulp down the wine of divine knowledge could suggest this potential for domination, a new kind of hubris made possible by the colonizing efforts of those who would stake out the secrets of the "geographic body." This potential for domination, however, seems to be contained by Panurge's surrender to physical vulnerability at the fountain.

When Panurge hears the prophecy of the oracle of the bottle, "Trink," he embraces the vulnerability and pleasures that a marital union promises. This new divination at the temple demands total participation in the material creation,

---

87. Rabelais, *Œuvres Complètes*, 904.

88. In *Le tiers livre*, the art of astrology is made stable to a certain extent by God. Pantagruel explains Bridoie's success in deciding court cases with dice by assuming that God influenced the results: "Conjecturallement, je référerois cestuy heur de jugement en l'aspect bénévole des cieulx et faveur des Intelligences motrices." Ibid., 530.

89. Ibid., 908.

90. I disagree with Sheila Rabin's analysis that Panurge reaches a state of enlightenment beyond the body. Though Panurge journeys toward a certain truth, the final destination, the oracle of the bottle, does not lift him to levels beyond the earth and the body, in a perfect union with the godhead beyond the limits of the physical. Instead, he is brought into the depths of the earth, and there he swallows the wine and interprets the prophecy of the oracle through his body. Sheila J. Rabin, "The Qabbalistic Spirit in Gargantua and Pantagruel," in *Voices in Translation: The Authority of "Olde Bookes" in Medieval Literature*, ed. Deborah M. Sinnreich-Levi and Allen Mandelbaum (New York: AMS, 1992), 183–90.

91. Rabelais, *Histories of Gargantua and Pantagruel*, 705, and idem., *Œuvres Complètes*, 909: "La dive Bouteille vous y envoye, soyez vous mesmes interprètes de vostre entreprinse."

rather than calculated decisions based on astrological readings of distant signs. As seen in *Le tiers livre*, when the body digests food and drink, it produces blood and its more refined derivative, semen, yielding not only spiritual insight but also the compulsion to indulge the flesh and procreate. Those who would deny "le debvoir de marriage" (the debt of marriage) will suffer the consequences in their own bodies: "Penalties are inflicted by Nature on those who refuse to pay, in the form of grievous vexation of the limbs and disturbance of the senses. But the reward assigned to the lender is pleasure, joy, and sensual delight."[92]

Panurge ends his quest as he experiences the mysteries of the wine and surrenders himself to his body's drunken participation in God's creation. Embracing this body—recognizing its vulnerability to death and its dependence on nourishment—opens the way to human communion with the divine.

Similarly, the temple's structure—a meeting of the bodies of the celestial spheres, the objects of astrological prognostication, and the earth—creates a vision of the cosmos that breaks down the separation between the earth and the realm of the stars, the human body and God. For example, the columns around the fountain of many wines are made of seven different stones, representing the seven planets.[93] The signs of the zodiac also cover the cupola above the fountain.[94] Moreover, the dimensions of the temple, fountain, and chapel of the oracle of the bottle are perfectly ordered to reflect the order of the universe, the perfect distance and association between celestial bodies.[95] The temple of the oracle re-creates the intimate connections between the body and the heavenly spheres, the earth and the heavens.

Inscribed in the depths of the earth, the structure of the zodiac—with its signs of astrological divination and planetary orbs—assures us that God will fill the material realm with his presence, even at the lowest levels of physicality:

---

92. Rabelais, *Histories of Gargantua and Pantagruel*, 301, and idem., *Œuvres Complètes*, 389: "Poine par nature est au refusant interminée, acre vexation parmy les membres, et furie parmy les sens; au prestant, loyer consigné, plaisir, alaigress et volupté."

93. Rabelais, *Histories of Gargantua and Pantagruel*, 900.

94. Ibid., 901.

95. The temple of the oracle could easily be seen as a talisman since its astrological arrangement channels good influences from the stars and blocks out bad ones. See *Hermetica*, II, 80–81, 89–91, for passages from the *Asclepius* that later Neoplatonists used to justify using these talismans. Also see Yates, *Giordano Bruno* for a discussion of Ficino's efforts to "alter, to escape from, his Saturnian horoscope by capturing, guiding towards himself, more fortunate astral influences" (60, 62–83). More recently, Jonathan Sawday has argued that the architectural conceit of the body as a temple that mirrored the structure of the cosmos was carried out in the anatomy theaters of the sixteenth century. For a discussion of this conceit in the anatomy theaters, see Sawday, *Body Emblazoned*, 66–78.

Go, my friends, under the protection of this intellectual sphere, the centre of which is at all points and the circumference at none, and which we call God; and when you come to your country bear testimony that great treasures and wonderful things are hidden beneath the earth.[96]

Because God's presence is like a sphere without a circumference, whose center is everywhere, the search for the divine, usually directed upward to the stars away from the earth, should go to the center of the earth for these mysteries.[97] As Bacbuc explains, "So much as you can see of the heavens, and which you call the *Phenomena*, so much as the earth reveals to you, so much as the sea and all the rivers contain, is not to be compared with what is concealed in the earth."[98] The earth is the center of all influences, and all points of contact in the universe finally meet in the depths of the earth.

It is here, in the caverns of the earth, that the tomb transforms into the womb and begins its cycle of death and rebirth. Thus, Rabelais corrects the tendency of astrologers and Hermetic philosophers to look beyond the earth to a transcendent understanding of the stars. Instead, he brings the diviner's gaze back to the body and the earth. It is only through the wine at the temple, suggestive of the wine of the Eucharist, that God reveals his secrets of divinity and truth: "[B]y wine one grows divine; there is no surer argument, no art of divination less fallacious.... For it has the power to fill the soul with all truth, all knowledge, and all philosophy...the truth lies hidden in wine."[99] Rabelais's new divination through the body celebrates the material cosmos, which God has reconciled to

---

96. Rabelais, *Histories of Gargantua and Pantagruel*, 709, and idem., *Œuvres Complètes*, 915: "Allez, amis, en protection de ceste sphère intellectuale de laquelle en tous lieux est le centre et n'a en lieu aucun circonférance, que nous appellons Dieu: et venus en vostre monde portez tesmoignage que sous terre sont les grands trésors et choses admirables."

97. This saying originated in the Pseudo-Hermetic, twelfth-century *Book of the XXIV Philosophers*, and echoes of this concept reappeared in the writings of Nicholas of Cusa and Ficino. Regarding this metaphor's origin and its reappearance in Ficino, see Yates, *Giordano Bruno*, 247 n. 2, n. 3. For a discussion of the analogous conception in Nicholas of Cusa, see Alexandre Koyré, *From the Closed World to the Infinite Universe* (Baltimore: Johns Hopkins University Press, 1957), 5–27, esp. 18. Masters also cites A. J. Krailsheimer, *Rabelais and the Franciscans* (Oxford: Clarendon Press, 1963), 98–99; Abel Lefranc, *Grands écrivains français de la renaissance* (Paris: Champion, 1914), 174–85; and Hiram Haydn, *The Counter-Renaissance* (New York: Harcourt-Brace, 1960), 335–36 (Masters, 11 n. 12).

98. Rabelais, *Histories of Gargantua and Pantagruel*, 710, and idem., *Œuvres Complètes*, 916: "Ce que du ciel vous apparoist, et appellez Phénomènes, ce que la terre vous exhibe, ce que la mer et autres fleuves contiennent, n'est comparable à ce qui est en terre caché."

99. Rabelais, *Histories of Gargantua and Pantagruel*, 705, and idem., *Œuvres Complètes*, 908–9: "[D]e vin divin on devient, et n'y a argument tant seur, n'y art de divination moins fallace;... Car pouvoir il a d'emplir l'âme de toute vérité, tout savoir et philosophie.... [V]in est vérité cachée."

himself and imbued with signs of his presence. To understand the secrets of God, one must digest the wine, allowing it to flow through a living, breathing, copulating, defecating, conceiving body.

■ ■ ■

In his satirical almanacs, Rabelais criticizes the self-assurance that astrological predictions promote, as people begin trusting in their own ability to take control of their future. He also ridicules those astrologers and diviners who believe they can access the secrets of the divine by reading ambiguous signs. As a medical doctor, Rabelais does believe that the stars influence humans, but he is careful not to overstep his bounds as others have, searching in vain for those secrets of God that will never be revealed to humans. It also follows that Rabelais would oppose pessimistic dualism as sometimes expressed by Pico della Mirandola, who advised readers to reject the body, with "the excrementary and filthy parts of the lower world," and ascend to "the higher forms, which are divine."[100] Instead, Rabelais aligns himself with the new discoverers of the body, those anatomists who compare its hidden interiority to cosmic signifiers of divine knowledge. Rabelais thus counters a Hermetic hubris that would aspire to gain access to God with a new kind of divination centered in the body. This new "astrology" must embrace the ambiguities and absurdities of human physicality, rather than insist on transcending it. For Rabelais, the search for signs of God must begin with joyous participation in the physical pleasures of the body. Thus, Panurge's search for knowledge of the future is transformed by the tangible presence of God at the fountain, a presence that can be grasped like a bottle, drunk, and digested.

---

100. Pico, "Oration on the Dignity of Man," 204, 205.

Robert Boyle, "The Sceptical Chymist," and Hebrew

Michael T. Walton

The large number of Hebrew devices on title pages and Hebrew words in texts show that sixteenth- and seventeenth-century natural philosophers were much taken with the Christian-Hebrew tradition. Although many "virtuosi" appear to have known little Hebrew and used it as mere dilettantes,[1] Robert Boyle not only seriously studied the holy tongue, but he also applied his Hebraic understanding to the biblical text, especially to Genesis, which he believed was literally true.[2] Boyle

---

1. For my thoughts on the use of Christian Hebraism among several natural philosophers, see Michael T. Walton, "Genesis and Chemistry in the Sixteenth Century," in *Reading the Book of Nature*, ed. Allen G. Debus and Michael T. Walton (Kirksville, Mo.: Sixteenth Century Journal Publishers, 1998), 1–14; Michael T. Walton and Phyllis J. Walton, "Being Up Front: The Frontispiece and the *Prisca* Tradition," *Cauda Pavonis* 17 (Spring–Fall 1998): 8–12; and idem., "The Geometrical Kabbalahs of John Dee and Johannes Kepler," in *Experiencing Nature*, ed. Paul H. Theerman and Karen Hunger Parshall (Dordrecht: Kluwer Academic, 1997), 43–60.

2. I originally wrote this paper, with the encouragement of Professor Harris Lenowitz of the University of Utah, in 1999, before two revolutionary Boyle studies appeared. *The Works of Robert Boyle*, ed. Michael Hunter and Edward B. Davis, 14 vols. (London: Pickering & Chatto, 1999–2000) adds new organization and data. Almost as important to understanding Boyle's complex intellectual development is Michael Hunter, *Robert Boyle, 1627–1691: Scrupulosity and Science* (Woodbridge: Boydell Press, 2000); this collection of essays joins Lawrence M. Principe, *The Aspiring Adept: Robert Boyle and His Alchemical Quest* (Princeton: Princeton University Press, 1998) in revealing Boyle's chemical interests and how they related to his religiosity. Chapter 2 of Hunter, *Robert Boyle*, relates directly to this essay. It deals with Boyle's "conversion" to being a naturalist. Hunter describes how Boyle's complex personality and intellectual development led him to become a priest in the temple of nature. The failure of Steven Shapin in *A Social History of Truth: Civility and Science in Seventeenth-Century England* (Chicago: University of Chicago Press, 1994) to understand Boyle's personality, especially in regard to religion, is also discussed by Michael Hunter. The present essay can now be seen as a detailed analysis of one facet of Boyle's complexity.

believed that Hebrew provided an accurate understanding of the creation, and this insight conditioned his interpretation of experiments. He found in his reading of Genesis support for his beliefs in corpuscularianism and in active principles, or *semina*, which shaped corpuscles into various forms.

■ ■ ■

It is useful to look at the goals for the study of Hebrew among the various sets and subsets of sixteenth- and seventeenth-century Christian Hebraists, English Christian Hebraists, and, especially, well-read laymen like Boyle. Christians did not view Hebrew and Jewish texts with the same eye as did the rabbis. For the most part, they did not focus on the Talmud and rabbinical commentaries. They wanted access to the Hebrew Bible and texts that would enhance biblical study from a Christian perspective. Some desired a knowledge that would help in the conversion of Jews. Others were enamored of mystical texts and their promise of insights into the *prisca doctrina*, the kabbalistic knowledge revealed at Sinai.[3]

---

3. The study of Hebrew among Christians was an ancient tradition. The quest for the *veritas hebraica* was a goal of Jerome which continued into the Middle Ages with the important studies of Nicolas of Lyra (1270–1349) and Paul of Burgos (d. 1435). This medieval tradition and how it moved into the fifteenth and sixteenth centuries is discussed by Heiko A. Oberman, "The Discovery of Hebrew and Discrimination against the Jews: The *Veritas Hebraica* as Double-Edged Sword in Renaissance and Reformation," in *Germania Illustrata*, ed. Andrew C. Fix and Susan C. Karant-Nunn (Kirksville, Mo.: Sixteenth Century Journal Publishers, 1992), 19–34. Many of the most prominent Christian Hebraists followed the biblical exegetical tradition dating from Jerome. In the sixteenth century, Sebastian Muenster, Conrad Pellican, Paul Fagius, and Sanctu Pagnini all worked on the Hebrew Bible and Jewish commentaries. Converts from Judaism, such as Immanuel Tremellius and Anthonius Margaritha, also worked with the Hebrew Bible, using it in debate, exegesis, and the converting of Jews.

Moshe Goshen-Gottstein discusses the Christian and Jewish study of the Bible text in the seventeenth century in "Foundations of Biblical Philology in the Seventeenth Century: Christian and Jewish Dimensions," in *Jewish Thought in the Seventeenth Century*, ed. Isadore Twersky and Bernard Septimus (Cambridge: Harvard University Press, 1987), 77–94.

Pico della Mirandola's study of Hebrew at the end of the fifteenth century focused on the mystical tradition. While Pico read little Hebrew, the translations he commissioned from the convert Flavius Mithridates were a gold mine of esoteric rabbinical literature. See the superb study of Chaim Wirzubski, *Pico della Mirandola's Encounter with Jewish Mysticism* (Cambridge: Harvard University Press, 1989). Pico influenced Johannes Reuchlin's Hebrew studies. The biblical exegetical tradition and the study of the Kabbalah were united in Guillaume Postel, perhaps the most erudite, and certainly the most idiosyncratic, Christian Hebraist of the sixteenth century; see Marion Leathers Kuntz, *Guillaume Postel, Prophet of the Restitution of All Things: His Life and Thought* (The Hague: Martinus Nijhoff, 1981). In the early seventeenth century the work of Johannes Buxtorf, the Elder and the Younger, along with that of Dionysius, Gerardus, and Isaac Vossius, was influential in Jewish ethnography and in producing Latin translations of texts such as Maimonides's *Guide for the Perplexed*. All of these continental Christian Hebraists touched those studying Hebrew in England.

In 1523, with the appointment of Robert Wakefield, at Cambridge, as the first paid professor of Hebrew in England, the study of Hebrew in England was placed on an academic footing. Wakefield's object was to use Hebrew to advance biblical studies. In his text *On the Three Languages* (London, 1524), he quoted Augustine:

> For understanding Scripture a command of Hebrew and Greek is necessary.... [M]y primary aim and purpose...is to expound Hebrew, especially those parts of it which, by the singular gift and very great providence of God, express the peculiar quality, idiom and genius of the sacred language.[4]

The importance of knowledge of Hebrew for properly understanding God's word took root not only in English universities but also in secondary schools. There the new learning was fostered by wealthy patrons. As G. Lloyd Jones notes at the end of his study, *The Discovery of Hebrew in Tudor England*,

> [S]chools did make provision for Greek. Hebrew was not so popular...it made little progress as a school subject before the early decades of the seventeenth century. Nevertheless, by the end of Elizabeth's reign, it had taken root in some of the leading schools.... The motive for its study was primarily religious. As Ubaldini observed in 1551, the rich made their children learn Hebrew because the Protestant "heresy" emphasized the importance not only of having the Bible in the vernacular, but also of being able to read it in the original.[5]

The secondary school and university limited themselves to teaching Hebrew as a means of reading the Bible and seldom created profound scholars. Nonetheless, some students developed an interest in Hebrew that led them to more intensive study outside of academe. As Matt Goldish notes in *Judaism in the Theology of Sir Isaac Newton*,

> Neither Jews nor university teaching, then, were driving forces in the peak of high-level English Hebraism. Most English Hebraists of the seventeenth century learned only the rudiments of the language in grammar school or university and studied further on their own or were complete autodidacts. Thus, for example, John Lightfoot began his advanced Hebrew studies at

---

4. Augustine, *On Christian Doctrine* 2.11, quoted in Robert Wakefield, *On the Three Languages*, ed. and trans. G. Lloyd Jones (London: Wynken de Worde, 1525; Binghamton, N.Y.: Medieval & Renaissance Texts & Studies, 1989), 64–66.

5. G. Lloyd Jones, *The Discovery of Hebrew in Tudor England: A Third Language* (Manchester: University of Manchester Press, 1993), 243–44.

the suggestion of his patron Sir Rowland Cotton (a student of Hugh Broughton), *after* his university education. [John] Spencer and [John] Selden seem to have learned Hebrew and Jewish studies essentially on their own as well. Only [Edward] Pococke learned Hebrew with Jews, and this was not in England, but in Aleppo and Constantinople.[6]

■ ■ ■

The Renaissance–early modern interest in Hebrew influenced an older tradition of natural philosophy. That tradition linked the text of Genesis to the chemical study of nature. George Ripley's *The Compound of Alchymie*, composed in the fifteenth century, sets the search for the philosopher's stone in the context of the creation.

> In the begynnyng when though madyst all of nought,
> A globose Mater and darke under confusyon,
> By thee Begynner mervelously was wrought,
> Conteynyng naturally all thyngs withoute dyvysyon,
> Of whych thou madyst in six Dayes dere dystynction;
> As Genesys apertly doth recorde
> Then Heavyn and Erth perfeytyd were wyth thy word.[7]

The sixteenth-century adept Solomon Trismosin (Trithemius), attributed the origin and continuation of all corporeal things to God. Growing metals and minerals followed processes "as first created at the Beginning by God, the Creator."[8] In his *Meteorology*, Paracelsus stated that "God began creation with the heavens, or chaos, the ethereal realm which is above the other elements."[9] The pseudo-Paracelsus *Three Books of Philosophy to the Athenians* is a commentary on the creation meant to correct the Athenian Aristotle.[10] It explains how God created everything from the Great Mystery by means of the chemical process of separation.

The "Mosaical" philosophy of the Paracelsians was systematized and placed in the context of mainline academic medicine by the Dane Peter Severinus. In his

---

6. Matt Goldish, *Judaism in the Theology of Sir Isaac Newton* (Dordrecht: Kluwer Academic, 1998), 21.

7. Elias Ashmole, *Theatrum Chemicum Britannicum*, introduction by Allen G. Debus (1652, repr., New York: Johnson Reprint, 1967), 122.

8. Solomon Trismosin, *Splendor Solis*, ed. J.K. (London: Kegan Paul, Trench, Trubner, 1921), 15.

9. Paracelsus, *Meteorologicae in Philosophiae magnae* (Cologne, 1567), 35 (my translation).

10. Paracelsus, *Philosophiae ad Atheniensis libri tres*, translated into English in *Philosophy Reformed*, trans. H. Pinnel (London, 1657).

discussion of the elements in *Idea medicinae philosophicae,* Severinus used biblical language:

> On these four [elements], by nature incorporeal, unorganized, void, the Creator established light and number, the seeds of all things. [This was done] completely incomprehensibly by power of his word and spirit which hovered over the water.[11]

The use of Hebrew to refine the chemists' understanding of Genesis and the creation process was virtually inevitable and appears in both Robert Fludd and Jean Baptiste van Helmont. At the end of the sixteenth and beginning of the seventeenth century, Fludd was using concepts based on Hebrew words and the Christian adaptation of the Jewish Kabbalah to explain natural phenomena. In his *Mosaicall Philosophy* (published posthumously in English, 1659), Fludd wrote of Moses, "whose Philosophy was originally delineated by the finger of God, forasmuch as the fiery characters thereof, were stamped out or engraven in the dark Hyle, by the eternall Wisdom, or divine Word."[12] The heavens and stars were made by the action of the "*Ruach Elohim* or Eternall Spirit" by the chemical actions of "Contraction and Rarefaction."[13]

Van Helmont also used Hebraic knowledge in explicating his detailed chemical philosophy. From his discussion of his original terms, *Gas* and *Blas,* it is clear that the text of Genesis was a factor in his formulation of chemical ideas. He believed that chemistry, indeed any science, could be known only by means of the "divine kiss" or, as the rabbis called it, "*Binsicam.*"[14] Helmont's discussion of *gas* and *blas* is a biblical commentary.

> Gas and Blas are new terms introduced by me which were unknown to the ancients [the German translation defines *Gas* as a subtle water vapor or spirit and *Blas* as a wind or movement from heavenly bodies]. Gas and Blas are among the first physical necessities; therefore, these new concepts must be amplified and expanded. First it must be shown how Gas is produced from water, that it is different from [the products produced by]

---

11. Peter Severinus, *Idea medicinae philosophicae* (Basel, 1571), 40–41 (my translation). See also, Jole Shackelford, "Paracelsianism in Denmark and Norway in the 16th and 17th Centuries" (Ph.D. diss., University of Wisconsin, 1989). This excellent source on the work of Severinus demonstrates that Erastus drew heavily on Severinus for his understanding of Paracelsian ideas.

12. Robert Fludd, *Mosaicall Philosophy* (London, 1659), bk. 3, p. 40.

13. Ibid., 67.

14. Walter Pagel, *Joan Baptista van Helmont: Reformer of Science and Medicine* (London: Cambridge University Press, 1982); Jean Baptiste van Helmont, *Ortus medicinae* (Amsterdam, 1652), 24.

heating water until vapor rises. To understand how this happens, one must learn the anatomy of water; therefore, I will repeat that the most glorious God in the beginning created the heavens and the earth and the abyss of waters. The abyss began at the vault of the heavens [the Dutch reads "at the far end of the dwelling place of the souls"] and ends above the globe of the earth. Nothing is written about the creation of air, which is only a body and was made into an element. Only after the six days of creation, and there was a place [for it] to fill, was there air. Thus "heaven" means air and the matter of the heavens is air, something not known before [I discovered it]. [The Dutch reads "and after the beginning, the air was with the water together in one body and known by one name, heaven."] The Eternal created the eternal firmament and separated the water which ought to remain below from the water which ought to remain above. Now this firmament was not a cataract or passive division of the waters, but by the power of its operation, the very principle of separation. Just as the sun is not a dividing wall between day and night but is rather itself the source of division, the heavens or air is the source of division between the waters.[15]

The use of Hebrew and Hebraic "knowledge" extended beyond chemistry to other areas of natural philosophy. John Dee saw that the numerical value of Hebrew and vulgar letters was connected to the harmony of the cosmos and could be used to study nature. The kabbalistic techniques of *gematria, notaricon,* and *tzyruph* could be applied to geometry, a higher and more perfect art than alphabetic manipulation.

> [T]he Hebrew cabbalist who, when he will see that (the three principal keys to his art, called) Gematria, Notariacon, and Tzyruph, are used outside the confines of the language called holy... will call this art [geometrical kabbalah] holy, too....[16]

This mathematical (neo-Pythagorean/neo-Platonic) approach bore fruit in the rigorously data-centered work of Johannes Kepler. Kepler termed his *Harmonices mundi* a geometrical kabbalah.

---

15. Jean Baptiste van Helmont, *Aufgang der Artzney-Kuenst* (Sultzbach, 1683), 109. The *Aufgang* is an interpretive translation of the *Ortus*. The passage in the *Ortus* begins on page 59. The translation is mine, based on both texts. The "Dutch" text referred to is in the German translation.
16. C. H. Josten, "A Translation of John Dee's *Monas hieroglyphica* (Antwerp, 1564) with an introduction and Annotations," *Ambix* 12 (1964): 84–221, at 132–33.

> I also play with symbols; I have begun a little work, a *geometrical kabbalah*, which deals with the ideas of the thing of nature in geometry. But when I play I do not forget I am playing. For nothing is proved with symbols, nothing hidden is shown in natural philosophy through geometrical symbols except that known before, unless it is shown by certain reasons to be not only symbolic but also a description of the connection between things and causes.[17]

In addition to Dee and Kepler, Hebrew and the Hebraic tradition influenced Robert Boyle's younger contemporaries, Isaac Newton and Gottfried Wilhelm Leibniz. Matt Goldish has examined Newton's study of religion in detail. He has shown Newton was deeply influenced by the structure of the Temple of Solomon with its central fire, as well as by other aspects of the Hebrew Scriptures. Newton, unlike Fludd, van Helmont, Dee, Kepler, and Leibniz, rejected the Kabbalah, regarding it as a late idea that partook of Gnostic heresy.[18]

The details of how Leibniz used and was influenced by Hebrew are only beginning to emerge. Using the work of Anne Becco, Allison P. Coudert has shown that Leibniz wrote the commentary on Genesis attributed to J.B. van Helmont's son Francis Mercury. She connects the Kabbalah to Leibniz's monadic philosophy. The Leibniz–van Helmont text discusses, among other things in the Genesis text, the "hatching" of seeds as a creative process (an idea used by J.B. van Helmont), rejecting creation *ex nihilo*.

> And so creation *ex nihilo* is not based on sacred Scripture but rather on a certain tradition and can be accepted in a reasonable way. But as commonly received, it is not without error. Indeed, it is true that chaos or atoms did not exist or any other matter coeternal with god from which the world was made. But, nevertheless, it is false that the world was strictly speaking made from nothing, as if from some material. It is an eternal truth that nothing can be made from nothing. Therefore it is more correct to agree with the author of the epistle to the Hebrews, cap. 11, v. 3, that visible things are made from invisible. That is, in the lofty realm of Aelohim himself the seeds of this corporeal world were hidden

---

17. Johannes Kepler, letter to Joachim Tanckius, May 12, 1608, in Johannes Kepler, *Gesammelte Werke*, ed. W.V. Dyck and Max Caspar (Munich: C.H. Beck 1938–), 16:158, translation and emphasis mine. For a discussion of Dee and Kepler and Kabbalah, see Michael T. Walton and Phyllis J. Walton, "The Geometrical Kabbalahs of John Dee and Johannes Kepler," in *Experiencing Nature*, ed. Paul H. Theerman and Karen Hunger Parshall (Dordrecht: Kluwer Academic, 1997), 43–59.

18. Goldish, *Judaism in Newton*, 85–107, 146.

in an ideal or spiritual fashion. At some time these seeds were finally produced and hatched.[19]

Newton and Leibniz, as younger contemporaries of Robert Boyle, not only illustrate the importance of Hebrew and the Hebraic tradition for natural philosophy, but they also demonstrate that the interplay of theology with natural philosophy, the subject of this essay, was a mainstream phenomenon.

■ ■ ■

Robert Boyle fits into both the English pattern of Hebrew study and the tradition of relating natural philosophy to Genesis. He attended Eton and was trained in Latin and introduced to Greek and Hebrew, but it was not until Bishop James Ussher encouraged him to study Greek and Hebrew that he undertook to master languages.[20] He did not attend university but had tutors and traveled. Like Lightfoot, Spencer, and Selden, he educated himself in the many areas that interested him. He met the prominent rabbi, Menasseh ben Israel, in Holland in 1648.[21] The extent of Boyle's knowledge of Hebrew and Jewish tradition, and their influence on his thought, can be seen in and inferred from his writings and what others said about him.

Although Boyle is viewed today, as he was in his own day, as principally a natural philosopher, it is clear that his intense Christian belief and theological ideas were of primary concern to him. He was a supporter, along with his literary agent, Henry Oldenburg, of a translation of the Bible into Turkish, a serious missionary

---

19. F.M. van Helmont, *Quaedam praemeditatae et consideratae cogitationes super Quator Priora Capita Libri Moysis Genesis nominati*, trans. in Allison P. Coudert, *Leibniz and the Kabbalah* (Dordrecht: Kluwer Academic, 1995), 85.

20. Jan Wojcik, *Robert Boyle and the Limits of Reason* (Cambridge: Cambridge University Press, 1997), 55–59.

21. Richard Popkin, "Some Aspects of Jewish-Christian Theological Interchange in Holland and England 1640– 1700," in *Jewish-Christian Relations in the Seventeenth Century Studies and Documents*, ed. J. van den Bert and Ernestine G. E. Van der Wall (Dordrecht: Kluwer Academic, 1988), 5–6. The role of Menasseh Ben Israel in Jewish-Christian dialogue and studies is addressed by Aaron L. Katcher, "Menasseh Ben Israel the Apologist and the Christian Study of Maimonides' *Mishneh Torah*," in *Jewish Thought in the Seventeenth Century*, 201–20. Boyle's literary agent and Latin translator, Henry Oldenburg, had met Menasseh Ben Israel when he was in England. Oldenburg wrote Menasseh a highly laudatory letter on July 25, 1657. The letter speaks of the prophecies of Jewish return to "Judea" and hopes for the illumination of the Jews. These statements from one so close to Boyle show an affinity with Boyle's position on the Jews. See *The Correspondence of Henry Oldenburg*, ed. and trans. A. Rupert Hall and Marie Boas Hall, 10 vols. (Madison: University of Wisconsin Press, 1965–75), 1:123–27.

project.[22] As with Newton and other Christian virtuosi, the study of nature was an aspect of his lifelong preoccupation with God, morality, and the creation. For Boyle, theology dealt with the nature of God and general principles, while science revealed *particular* aspects of the creation. He considered science a sacred calling and himself a priest in the temple that was the "commonwealth of nature."[23]

In *The Excellency of Theology or the Pre-eminence of the Study of Divinity above that of Natural Philosophy*, Boyle explicitly stated his belief that Scripture is superior to the study of nature. He argued that "the Book of Scripture discloses to us much more of the attributes of God, than the Book of Nature."[24] The Book of Nature reveals things as through a glass, darkly. The naturalist could know "things corporeal" in their "particulars."[25] Theology derived from Scripture gives comprehensive knowledge far beyond the particular.

Because the revelation of Scripture is superior to the particular knowledge gained from "reading" nature, it behooved both theologians and natural philosophers to understand the nature of their objects of study. For Boyle, as a natural philosopher, a precise understanding of the Book of Scripture was important for a correct study of the Book of Nature. Boyle conceded that physics was, perhaps, superior to theology in one way: physics had a "certainty and clarity," whereas theology was subject to controversy.[26]

Given the relationship between theology and natural philosophy in Boyle's view, a natural philosopher had first to understand Scripture. Hebrew was a matchless tool to that end. How Boyle understood and functioned in Hebrew can be glimpsed in his *Some Considerations touching the Style of the Holy Scriptures*.[27] Boyle regarded many of those who analyzed and criticized the Holy Scriptures as

---

22. Boyle to Oldenburg, ca. 2 October 1664, *Correspondence of Henry Oldenburg*, 2:245–46. In another letter, Oldenburg to Boyle, 16 January 1665/66, *Correspondence of Henry Oldenburg*, 3:18, Oldenburg discusses whether the text should be printed with vowels and whether Boyle had had the type cut.

23. Boyle's priestly calling is mentioned in his papers, Boyle Papers, Royal Society, vol. 3, fol. 128. Jacob Pries discusses this in "Boyle, Young Theodicean" (Ph.D. diss., Cornell University, 1969), 158. Steven Shapin builds on the idea in "The House of Experiment in 17th Century England," *Isis* 79 (September 1988): 383–84, as does Harold Fisch in "The Scientist as Priest: A Note on Robert Boyle's Natural Theology," *Isis* 44 (1953): 252–65.

24. *The Excellency of Theology or the Pre-eminence of the Study of Divinity above that of Natural Philosophy*, in *The Works of the Honorable Robert Boyle*, ed. Thomas Birch, 6 vols. (London: W. Johnson et al., 1772), 4:7.

25. Ibid., 4:10.

26. Ibid., 4:41.

27. Ibid., 2:247.

"want[ing] of skill in the original especially in the Hebrew."[28] Boyle himself was free of this debility because he understood the "keri" (how the text was read), the "cathib" (how the text was written), and the Masoretic markings, as well as the Ashkenazic and Sephardic textual traditions.

> I am not unacquainted with the Keri, and the Cethib, nor the *Tikkum Soph'rim* in the old testament: nor yet with the *Variae Lectiones* (especially those of the Eastern and Western Jews, as they are called) taken notice of by modern criticks in the Hebrew text of the old, as well as in the Greek of the new testament.[29]

Boyle spent pages discussing the difficulties in translating the Scriptures. Important problems arose because of lack of information about Hebrew word meanings and Jewish history.

> [F]or I consider in the second place, that not only we have lost divers of the significations of many of the Hebrew words and phrases, but that we have also lost the means of acquainting ourselves with a multitude of particulars relating to the topography, history, rites, opinions, fashions, customs, &c. of the antient Jews and neighbouring nations, without the knowledge of which we cannot, in the perusing of books of such antiquity, as those of the Old Testament, and written by and (principally) for Jews; we cannot, I say, but lose very much.[30]

Other difficulties came from the lack of congruence between Hebrew and English.

> [T]here being in the idiotisms of all languages peculiar graces, which (like those most subtil spirits, which exhale in pouring essences out of one vessel into another) are lost in most (especially if literal) translations; and the holy tongue being that, which God himself made choice of to dignify with his expressions, having divers whose penetrancy is as little transfusable into any other as the sun's dazzling brightness, or the water of a diamond can be undetractingly painted; and having divers words and phrases, whose pithiness and copiousness none in derived (or other) languages can match. Some of the Hebrew conjugations, as those called *Hiphil* and *Hitpael*, give significations to verbs, which the want of answerable conjugations in western languages makes us unable to fill or

---

28. Ibid., 2:257.
29. Ibid., 2:259.
30. Ibid.

> equal without paraphrases, which are very rarely so comprehensive as the original words.[31]

How Boyle translated Hebrew in order to solve theological problems and how competent he was can be seen in Sir Peter Pett's manuscript on Boyle.[32] After noting Boyle's pride in the practical application of his experimental knowledge, Pett makes several references to Boyle, Hebrew, and biblical interpretation. Of importance is Pett's report of the opinion of the noted orientalist Thomas Hyde on Boyle as a Hebraist.

> And for this I can referre to a letter I lately had from Dr. Hyde the Keeper of the Bodleian Library, and one whome fame speakes as more knowing in those Languages than any Christian in the world, in which letter the Doctor, having mentioned his having been acquainted with Mr. Boyle for about 30 years, thus goes on viz Mr. Boyle besides his skill in moderne Languages, was excellently versed in...the Latin tongue.... He was also well skilled in Greek and Hebrew, and which he hath told me that when the Chapters were reading in the Church, he alwaies had in his hand the original, wondering to heare our English translation so different from it.[33]

Boyle was associated not only with Hyde but also with other orientalists, like Edward Pococke, Robert Huntington, Samuel Clarke, Thomas Smith, and Thomas Greaves.[34] Indeed, discussions of Hebrew and other Near Eastern languages are mixed with scientific interests, like tides and chemistry in the Boyle-Oldenburg correspondence.[35] This picture of Boyle agrees with that of other English students

---

31. Ibid., 2:297.

32. The Sir Peter Pett notes, BL Add. MSS 4229, fols. 33–49, are printed in Michael Hunter, *Robert Boyle by Himself and His Friends* (London: William Pickering, 1994), 55 ff.

33. Ibid., 65.

34. *Correspondence of Henry Oldenburg*, 4:141.

35. Oldenburg to Boyle, 4 February 1667/68, *Correspondence of Henry Oldenburg*, 4:145:

> On Thursday last ye [Royal Philosophical] Society met not, because it was ye 30th of January. At our next meeting we are like to have Experiment to determine, whether there may be made a Body more ponderous than gold, by trying, whether any Quicksilver will penetrate into ye pores of Gold: And another, to shew ye passage from the Ear to ye Mouth, by cutting away ye Tympanum, and blowing Tobacco-fume through yt organ into ye mouth.
>
> I have now receaved Stenonis Musculi Descriptio Geometrica, and Alphonsus Borrelli De vi percussionis; of both wch, and some two or three new books more I intend, God willing, to take notice of in this months Transactions.
>
> Since I wrote this, I was visited by R. [Jacob] Abendana, who show'd me a letter from his

of Hebrew who wanted to understand the biblical text and whose Hebrew allowed them to see the flaws in the authorized translation. Pett gives several examples of how Boyle came to understand the Bible better through Hebrew. The first deals with Exod. 11:2, where the Jews are commanded in the English to "borrow" jewels, silver, gold, and clothing from the Egyptians. This "borrowing" or "despoiling" was the subject of Ralph Cudworth's Bachelor of Divinity thesis and of a discussion between Pett and Boyle.[36] Borrowing, not to return, seemed a moral problem even though all belongs to God, who can dispose at will. Boyle told Pett that the passage had disturbed him, but that the meaning was clarified when he weighed the words in the "original Hebrew," where the Israelites had not "borrowed," but "demanded."[37] Boyle did not look either to the rabbinical commentaries or to the larger context of the passage; rather, as a Christian, he focused

---

> Brother, now at Oxford, mentioning, that you had expressed to him yr wondering at my not taking notice of the Catalogue of Hebrew Books, brought over by yesd Abendana to be sold here. I was surprised to hear this, who, if I am not extreamly mistaken, mentiond in my very last yt busines to you, expressing, how I had endeavord to recommend it to Dr. Barlow here in London, and making it my request to you to assist in it, as much as yr conviency would permit. 'Tis true, before I was spoken to by ye Rabbi, I was silent of it in my letters to you, but yt was upon ye consideration of yr manifold other engagements, wch I thought would not suffer you to medle wth Books of yet nature, as for yr owne use.

See also, Wallis to Oldenburg or Hevelius, 9 December 1668, ibid., 5:235:

> I would have already sent the same thing with the notes of Mr. Hyde (the Sub-Librarian of the Bodleian) as then printed, if Sir Robert Moray had not begged to be allowed to take this upon himself, and I have no doubt he took care of it long ago. However we have elsewhere in our libraries, I hear, other tables of the same kind in Persian or Arabic and of these, it I convenietly can, I shall have transcripts made by men learned in those languages, and translated into Latin, so that there may be plenty of them also. And I am the more inclined to think this should be done because I find the Persian tables to be praised by the publisher of Tycho's observations [in the recently published *Historia coelestis*] and the fragments of them to be deemed worthy of publication; and so I guess that the entire tables will not be unwelcome. But indeed I can only undertake this, which must depend upon others (for I am not so skilled in those languages that I can take this upon myself alone), if I shall be able to command the leisure of such scholars, who are not numerous amongst us. Especially as Mr. Pocock (second to none in knowledge of the oriental tongues) has now been long weakened by illness; Mr. Hyde will be for some years occupied in completing a new catalogue of all the books in the Bodleian Library; and Mr. Bernard will be going abroad immediately to Holland, in order to collate the seven books of Apollonius of Perga's *Conics*, which we have here in Arabic, with some other codices of the same work so that he may in due course publish them in an amended version; and Mr. Clark will be for some time engaged in preparing a further volume of the Polyglot Bible and in revising the text of Abulfeda's Arabic and Persian Geography from various manuscripts which we have here.

36. Pett, quoted in Hunter, *Boyle by Himself,* 65.
37. Ibid., 65–66.

on the words themselves. His scholarship was not as deep as that of a Selden or a Lightfoot, but Hyde agreed with his interpretation.[38]

Just what Boyle knew about rabbinical commentaries and what he thought of that intellectual tradition is difficult to assess. He was in favor of allowing Jews back into England, but believed that it would present problems because of learned rabbis. In a letter to John Mallett, November 1651, Boyle stated that the readmitted Jews would "seduce many of those numerous Unprincipled Soules, who [were] never…solidly or settledly grounded in the truth."[39] Given the trepidation with which Jews approached making converts, Boyle's fear was both unfounded and indicative of a lack of knowledge concerning contemporary Jewry. Pett remarked on Boyle's solution to the perceived conversion problem:

> In discoursing with him about the Jews applying to Cromwel for their being tolerated here, I found he wished well to the same, provided that the Government would take care to give good Salarys to two or three men for their studying the Orientall Languages, and the text of the Old Testament most accurately, that so our Ministers might be the better enabled to confute their Rabbis, which few of them were then able to do. For Mr Boyle observed that the Jewish Rabbis on all occasions of discourse did continually fly to false acceptions of the Hebrew words, and particularly where it is said a Virgin shall conceive, they replyd that by a Virgin was meant a young woman. He said that the Rabbis lay much stresse on the perpetuall continuance and lasting of Moses his Law, whereas the words for ever (said Mr Boyl) do signify only a very long time. And he therupon directed me to some authors who writ ex professo of the various acceptions of the Hebrew words.[40]

In the same conversation, Boyle expressed an opinion about Jewish temple sacrifices that he may have drawn from Maimonides. Maimonides taught that sacrifices were instituted to wean the Jews from idolatry. In his *Guide for the Perplexed*, he distinguished the unleavened bread and salt used in the temple from the unsalted, leavened bread offered by pagans.[41] Whether Boyle had read Maimonides

---

38. Ibid., 66.

39. Letter to John Mallet, November 1651, L.L. MS Harley 7003, fol. 179, cited in J.R. Jacob, *Robert Boyle and the English Revolution* (New York: Burt Franklin, 1977), 97.

40. Pett, quoted in Hunter, *Boyle by Himself*, 66.

41. Maimonides, *Guide for the Perplexed (Moreh Nevuchim)*, bk. 3, chap. 56. While some scholars, such as Dionysius Vossius, studied and used the rabbinical tradition, others, such as Jean Moran, saw

in Hebrew (a translation, used by Jews, from the original Arabic) or in the 1629 Latin translation by Buxtorf is unknown. Pett wrote,

> [Boyle] further observed to me the condiscention of the divine benignity to the Jewes in suffering them to have Sacrifices, since they were so much addicted to them, on their finding all the Nations of the World to use them, and as indeed (said he) they naturally do, which appears in their useing them in the West Indies. He then further remarked that the Heathens in their Sacrifices used hony & no Salt, and that God required Salt to be used and no hony.[42]

Boyle clearly had some knowledge of rabbinical thought, but he seems to have felt that it was too corrupt and difficult a tradition on which to rely. Whatever his stance, Boyle was obviously well versed enough in Hebrew to read and interpret the Bible. In the tradition of some Christian Hebraists, Boyle believed that the Bible and Hebrew were not limited to the sphere of theology, but also could be useful in reading and interpreting the Book of Nature.

■ ■ ■

How the Bible and the study of Hebrew conditioned Boyle's natural philosophy is reflected in his attempt to develop a theory of matter that comported with experimentation. Boyle's theory of matter, or corpuscularianism, is a variant of atomism. He conceived that corpuscles combined to form the substances, such as sulfur and earth, which had traditionally been viewed as elements.

> To begin then with the first of these, I consider that if it be as true, as 'tis probable, that compounded bodies differ from one another but in the various textures resulting from the bigness, shape, motion, and contrivance of their small parts, it will not be irrational to conceive that one and the same parcel of the universall matter may by various alterations and contextures be brought to deserve the name, sometimes of a sulphureous, and sometimes of a terrene, or aqueous body.[43]

---

it as corrupt and misleading. Boyle seems to have viewed rabbinical works as suspect. See Goshen-Gottstein, "Foundations of Biblical Philology," for a discussion of both approaches to rabbinical thought.

42. Pett, quoted in Hunter, *Boyle by Himself*, 66.

43. Robert Boyle, *The Sceptical Chymist*, Everyman's Library, Science, no. 559 (London: Dent, 1911), 64. While this paper is concerned only with Boyle's thought up to and including *The Sceptical Chymist*, it does not appear to me that his corpuscularianism changed much in his later publications.

William R. Newman has shown that Boyle's corpuscles are descended from corpuscular alchemy as presented by "Geber" in *The Summa Perfectionis*.[44] In *The Excellency of Theology*, Boyle stated that the origin of the idea of corpuscles was the atomism of Moschus the Phoenician and the doctrines of Descartes.[45] Descartes, according to Boyle, showed that such particles allowed the rational study of the created world by showing it to be the product of mechanical processes.

> For that hypothesis, supposing the whole universe (the soul of man excepted) to be but a great automaton, or self-moving engine, wherein all things are performed by the bare motion (or rest), the size, the shape, and the situation, or texture of the parts of the universal matter it consists of; all the phenomena result from those few principles, single or combined.[46]

Boyle's corpuscularianism was not, however, the pure mechanism of Descartes. Boyle objected to pure mechanism, in part on theological grounds, because it divorced matter from spirit, body from soul. His objections to Descartes and classical atomists like Leucippus, Democrates, and Lucretius were similar to those of Joseph Glanville, who wrote:

> If the notion of a Spirit be absurd as is pretended, that of a GOD and SOUL distinct from matter, and immortal, are likewise absurdities. And then, that the World was jumbled into this elegant and orderly Fabrick by chance; and that our Soules are only parts of matter that came together we know not whence nor how, and shall again shortly be dissolv'd into those loose Atoms that compound them; That all our conceptions are but the thrusting of one part of matter against another; and the Idea's of our minds mere blind and casual motions. These, and a thousand more the grossest impossibilities and absurdities.[47]

---

44. William R. Newman, "Boyle's Debt to Corpuscular Alchemy," in *Robert Boyle Reconsidered*, ed. Michael Hunter (Cambridge: Cambridge University Press, 1991), 107; idem., *The Summae Perfectiones of Pseudo-Geber* (Leiden: Brill, 1991); and idem., *Gehennical Fire: The Lives of George Starkey, an American Alchemist in the Scientific Revolution* (Cambridge: Harvard University Press, 1997).

45. Boyle, *Works*, 4:48. On the religious genealogy of atomism, see Danton B. Sailor, "Moses and Atomism," *Journal of the History of Ideas* 25 (1964): 3–16.

46. Boyle, *Works*, 4:49.

47. Joseph Glanville, *Saducismus Triumphatus*, 2d ed. (London: Thomas Newcomb, 1682), 6. Theological similarities between Glanville and Boyle are touched on by Jan W. Wojcik, "The Theological Context of Boyle's *Things Above Reason*," in *Robert Boyle Reconsidered*, ed. Michael Hunter (Cambridge: Cambridge University Press, 1994), 138–51. Boyle's rejection of pure mechanism runs against the mechanist interpretation of his work by John Hedley Brooke, *Science and Religion* (Cambridge:

Theology demanded something more than corpuscles in motion, as did chemistry. Mere mechanical motion seemed inadequate to account for dynamic processes seen in nature. Boyle, like van Helmont and his predecessors, Paracelsus and Severinus, conceived of active forces as *semina* or seeds, which were the agents that acted to combine corpuscles.[48] Like Gassendi and Charlton, Boyle also mosaicized the origins of atomism.[49] In *The Sceptical Chymist*, he tied atomism and its originator, Moschus, to the ancient Hebrews. He further posited that the probable source of atoms for the creation were the primordial waters of Genesis.

> And the like opinion has been by some of the antients ascribed to the Phoenicians, from whom Thales himself is conceived to have borrowed it; as probably the Greeks did much of theologie, and, as I am apt to think, of their philosophy too; since the devising of the atomical hypothesis commonly ascribed to Leucippus and his disciple Democritus, is by learned men attributed to one Moschus a Phoenician. And possibly the opinion is yet antienter than so; for 'tis known that the Phoenicians borrowed most of their learning from the Hebrews. And among those that acknowledge the Books of Moses, many have been inclined to think water to have been the primitive and universal matter, by perusing the beginning of Genesis, where the waters seem to be mentioned as the material cause, not only of sublunary compound bodies, but of all those that make up the universe.[50]

Section II above shows that Boyle was not unusual in seeing Genesis as important to natural philosophy. Boyle, however, is more literal than most other chemists in his biblical belief and is hence very literal in his understanding of Genesis. For him, the details of the creation account are fact, not allegory.

Cambridge University Press, 1991), 134–35; and Herbert Breger, "The Paracelsians—Nature and Character," in *Paracelsus*, ed. Ole Peter Grell (Leiden: Brill, 1998), 105.

48. The importance of *semina* as active chemical forces in late-sixteenth- and early-seventeenth-century chemistry is discussed by Jole Shackelford in "Seeds with a Mechanical Purpose," in *Reading the Book of Nature*, ed. Debus and Walton, 15 ff. Christoph Meinel, "Early 17th-Century Atomism: Theory, Epistemology, and the Insufficiency of Experiment," *Isis* 79 (September 1988): 68–103, at 70, claims that Boyle's theory consisted only of "matter" and "motion." The following passages from *The Sceptical Chymist* show that view to be incomplete.

49. For my view of Gassendi and Charlton on atomism and the literature dealing with them, see Michael T. Walton, "Genesis and Chemistry in the Sixteenth Century," in *Reading the Book of Nature*, ed. Debus and Walton; and idem, "Boyle and Newton," *Ambix* 27, pt. 1 (March 1980): 11–18. Meinel, "Early 17th-Century Atomism," provides a detailed discussion of the development of atomism and refers to most the of major studies on the topic.

50. Boyle, *Sceptical Chymist*, 71.

> I see no just reason to embrace their opinion, that would so turn the two first chapters of Genesis, into an allegory, as to overthrow the literal and historical sense of them. and though I take the scripture to be mainly designed to teach us nobler and better truths, than those of philosophy; yet I am not forward to comdemn those, who think the beginning of Genesis, contains divers particulars, in reference to the origin of things, which though not unwarily, or alone to be used in physicks, may yet afford very considerable hints to an attentive and inquisitive peruser.[51]

Understanding the true meaning of Genesis required knowledge of Hebrew. As both a theologian who could read Hebrew and as a natural philosopher, Boyle was well equipped to peruse Genesis and use his knowledge to read the Book of Nature. Of course, Hebrew alone could not reveal the natural philosophical basis of Genesis. Nature and Genesis had to be read together. For Boyle, Genesis supported experimental data showing that *semina* worked on material corpuscles. Boyle saw *semina* in the primordial waters. After the creation, they continued to be found in common water, as chemical experiments proved.

Boyle saw *semina* in the waters of Genesis because the text uses the verb *merachephet* to describe the action of God's spirit *hovering* over the primordial waters. By looking at another place in the Bible where the verb is found (Deut. 32:11), where it is used to describe a female eagle fluttering or hovering over her nest, Boyle concluded that God hatched the world by hovering like a female bird over the seed-imbued waters. The Authorized Version's translation of *merachephet* as "moved" thus did not convey the proper meaning, either theologically or philosophically.

> [The] component parts did orderly, as it were, emerge out of that vast abysse by the operation of the Spirit of God, who is said to have been moving Himself, as hatching females do (as the original, *Merachephet*, is said to import, and it seems to signifie in one of the two other places, wherein alone I have met with it in the Hebrew Bible) upon the face of the waters; which being as may be supposed, divinely impregnated with the seeds of all things, were by that productive incubation qualified to produce them.[52]

---

51. Boyle, *Excellency of Theology*, in *Works*, 4:11.
52. Boyle, *Sceptical Chymist*, 71–72. Johann R. Glauber also believed God impregnated the waters by hovering over them. *Dess Teutschlands-Wohlfahrt*, pt. 3 (Amsterdam, 1659), 147.

While Boyle's corpuscularianism was conditioned and supported by Genesis, he recognized that theological and philological discussions were out of place in natural philosophical discourse.[53] While this is correct, it is undeniable that his theology and reading of Genesis were crucial to his justification of the seminal-corpuscular theory of matter. His understanding and use of Genesis were dependent, in large measure, on his knowledge of Hebrew. Boyle believed that his experiments in plant growth, inspired by Helmont's tree experiment, revealed seminal action as the organizer of corpuscles. These experiments further proved that all the "elements" of the chemists and Aristotelians could be derived from the action of *semina* in common water.

> But whether or no we conclude that all things were at first generated of water, I may deduce from what I have tried concerning the growth of vegetables, nourished with water, all that I now proposed to myself or need at present to prove, namely that salt, spirit, earth, and even oyl (though that be thought of all bodies the most opposite to water) may be produced out of water; and consequently that a chymical principle as well as a peripatetick element, may (in some cases) be generated anew, or obtained from such a parcel of matter as was not endowed with the form of such a principle or element before.[54]

Experimental data left no doubt that common water contained *semina* that formed corpuscles into both mineral and living matter.

> [T]hough as for the generation of living creatures, both vegetable and sensitive, it needs not seem incredible, since we find that our common water (which indeed is often impregnated with variety of seminal principles and rudiments) being long kept in a quiet place will putrifie and stink, and then perhaps too produce moss and little worms, or other insects, according to the nature of the seeds that were lurking in it.[55]

The text of Genesis also indicated to Boyle that the primordial waters were more fecund in their seminal and corpuscular makeup than postcreation, or

---

53. Walton, "Boyle and Newton"; and idem., "Genesis and Sixteenth Century Chemistry." Robert K. Merton, in "Science, Technology and Society in Seventeenth Century England," *Osiris* 4 (1938): 360–632, states that religion did not lead to specific scientific discoveries, but that it was supportive of science. Thomas F. Giery discusses Merton's ideas in "Distancing Science from Religion in Seventeenth Century England," *Isis* 79 (December 1988): 594–605.

54. Boyle, *Sceptical Chymist*, 76.

55. Ibid., 73.

common, water. Indeed, the creator implanted the entire creation into the abyss *in potentia*. In common water the creation is glimpsed only dimly.

> For I see no necessity to conceive that the water mentioned in the beginning of Genesis, as the universal matter, was simple and elementary water; since though we should suppose it to have been an agitated congeries or heap consisting of a great variety of seminal principles and rudiments, and of other corpuscles fit to be subdued and fashioned by them, it might yet be a body fluid like water, in case the corpuscles it was made up of, were by their creator made small enough, and put into such an actuall motion as might make them glide along one another.[56]

■ ■ ■

The natural philosopher, in Boyle's view, could see only traces of the creation in our present world. The chemist, however, with the insights of Genesis, could better understand physical processes seen in the laboratory, and hence better know both the creation and the creator. Chemistry was not an end in itself. Boyle believed that natural philosophy ultimately had a theological purpose. Boyle's seminal-corpuscular theory, however, was useful for chemists, not because of its conformity with Genesis, but because it provided a fruitful way to analyze natural phenomena. Although the biblical text and Hebrew conditioned Boyle's understanding of matter, the corpuscular theory needed no theological underpinning to explain natural phenomena. For example, Newton used Boyle's matter theory in his *Opticks* and in his speculations on comets in the *Principia*.[57] Beneath this practical theory of matter lay Boyle's reading of both the Book of Scripture and the Book of Nature, and his use of each to illuminate the other.

---

56. Ibid., 75.

57. In the *Principia*, Newton suggested that the tails of comets replaced the corpuscles used up by seminal action and respiration. See Walton, "Boyle and Newton."

Prognostications: Rest, from Paracelsus, *Prophecien und Weissagungen* (1549), fol. 19, permission of Becker Medical Library, Washington University, St. Louis, Missouri.

Johannes Praetorius

Early Modern Topography and the Giant Rübezahl

Gerhild Scholz Williams

Johannes Praetorius, *poeta laureatis* from Leipzig (1630–80), was a popular purveyor of stories about a myriad of topics, ranging from witch tracts and tales about natural and unnatural wonders, to comic, rather misogynist stories about women and their superstitious nature. The Indians of the New World are as much an object of his interest as ghosts, *Poltergeister*, water spirits, and moon people. The marvels his century was filled with and his lively curiosity kept him gainfully producing and selling books, and he benefited from Leipzig's position as a city prominent in the national and international book trade that also boasted a well-respected university and lively commerce.[1] His passion for observing, collecting, categorizing, understanding, and explaining marvels of all sorts prompted Praetorius to query travelers passing through town about anything new and unusual they might have encountered en route. Contemporaries reported that he was especially eager to hear tales about the Silesian giant Rübezahl, who was said to haunt the Riesengebirge, the mountain range separating Germany from Bohemia.[2]

Praetorius's diligent efforts at gathering information about Rübezahl yielded three volumes on the giant's life and times; the first edition was published in 1662

---

1. Helmut Waibler, *Johannes Praetorius (1630–1680): Ein Barockautor und seine Werke*, Archiv für die Geschichte des Buchwesens, 20 (Frankfurt am Main: Buchhändler-Vereinigung, 1979), 951–1152.

2. Carl Friedrich Flögel (1786) notes: "[Praetorius] hat zumahl einen seltsamen Collectaneen-Witz [*sic*].... Vom Rübenzahl einem schlesischen Gespenst hat er wunderliches Zeug zusammengetragen, womit ihn die schlesischen Kaufleute auf den Leipziger Meßen aus Spaß bewirtheten." See Helmut Waibler, *M. Johannes Praetorius, P. L. C.: Bio-bibliographische Studien zu einem Kompilator curieuser Materien im 17. Jahrhundert* (Frankfurt am Main: P. Lang, 1979), 67.

in Leipzig.[3] In the first two volumes Praetorius reviews at length the nature of Rübezahl, his genealogy, the history and topography of his domain, and a myriad of facts and tales loosely related to the giant and to other wonders. Along with much information about this preternatural being, he tells about an unusual people who live in the inaccessible valleys of the Riesengebirge and about those people living in the surrounding lands who venture across the mountains and thus, knowingly or unknowingly, invade the giant's territory.

Volume 3 of the *Daemonologia* was published several years after the first two; it is limited to a collection of tales about the giant—sometimes funny, sometimes scary, and mostly didactic. In the preface, Praetorius notes that he collected and published these tales in response to the immense popularity of the first two tracts, a popularity that persisted into the eighteenth and nineteenth centuries, when the Rübezahl stories were still alive as fairy tales.

This essay represents a first version of a larger study that analyzes Rübezahl as a specimen of the early modern marvelous and Praetorius as a popularizer of such marvels, which are, according to Lorraine Daston, "not so much violations of as exceptions to the natural order."[4] This essay will highlight the role of topography and demonology in the genesis of the giant Rübezahl as a *Gespenst* (demon, ghost). Where it is pertinent and possible, this essay will touch on the history and mythology of the Riesengebirge as it bears on understanding the giant's wondrous nature. It will become apparent just how Praetorius fits his giant into the topography, history, mythology, and demonology of a real space, the Riesengebirge, and a real time, the mid-seventeenth century, the period of the Thirty Years' War (1618–1648). In the process, Praetorius situates his giant at the intersection between wilderness and civilization, affirming Rübezahl as a phenomenon of early modern views of nature, science, and culture.[5] While a vigorous champion of the value of eyewitness reporting, Praetorius admits that in spite of much time spent in the mountains, even after dark, he has never seen the giant. He compensates for this handicap by gathering a huge number of testimonials vouching for the veracity of the marvel Rübezahl. Although the giant may be difficult to categorize, his very special nature, his rarity (Rarität), befits Silesia, whose inhabitants are said to be excellent poets, accomplished musicians, and

---

3. I am using the third edition: Johannes Praetorius, *Daemonologia Rvbezalii Silesii… zusammegezogen durch M. J. Praetorius*, 2 vols. (Leipzig: Verlegung J. B. Oehlers, 1668–73).

4. Lorraine Daston, "The Nature of Nature," *Configurations* 6, no. 2 (1998): 149–72, at 155.

5. Neil Evernden, *The Social Creation of Nature* (Baltimore: Johns Hopkins University Press, 1992), 41.

hardworking men—in short, strong, articulate, beautiful, and good.[6] Wonders tend to be useful in fostering many things, local pride not least among them.

Based on the authority of ancient and medieval science and cosmology, early modern natural history allows for the existence of preternatural beings such as Rübezahl.[7] Thus, Praetorius encounters little difficulty in constructing his gigantology against the background of early modern knowledge, based on a number of well-rehearsed theological, historical, scientific, and cultural facts. Together these facts make up the consciousness that, making use of George Lakoff's cognitive categories, have been called *experiential realism*, or *experientialism*.[8] Experiential realism is characterized by four assumptions: a commitment to the existence of the real world; a recognition that reality places constraints on concepts; a conception of truth that goes beyond the internal coherence of an idea; and a commitment to the existence of a stable knowledge of this world and of the universe.[9] Representing the totality of our beliefs, experiential realism determines the ways in which the mind forms categories and classifications that help to articulate, organize, name, and accept as true an individual's understanding of and interaction with society. In one of his more recent studies of cognition and experience, George Lakoff refers to this process as "experientially grounded mapping," whereby subjective experiences affect the development of reason and historically determined truth, which are influenced by the movement of historical contingency.[10] Experiential realism relies on "a large but not infinite number of *experiential repertoires*"[11] that rest on shared memory, previous experiences, and communal learning, making possible the integration and interpretation of new experiences into a given culture. The early modern experiences with the witch-hunts, with the discovery of a whole new continent and its strange inhabitants, and with religious dissidence point to the difficulties that all people had in trying to

---

6. "[F]ürtreffliche Poeten / vnd liebliche Musici...ein arbeitsames Volck...starck / beredt / schön / ...gut." Praetorius, *Daemonologia*, 1:6–7.

7. For the most comprehensive work on the topic of marvels, wonders, and the supernatural, see Lorraine Daston and Katharine Park, *Wonders and the Order of Nature, 1150–1750* (New York: Zone Books, 1998).

8. Gerhild Scholz Williams, *Defining Dominion: The Discourses of Magic and Witchcraft in Early Modern France and Germany* (Ann Arbor: University of Michigan Press, 1995), 18.

9. George Lakoff, *Women, Fire, and Dangerous Things: What Categories Reveal about the Mind* (Chicago: University of Chicago Press, 1987), xv.

10. George Lakoff and Mark Johnson, *Philosophy in the Flesh: The Embodied Mind and Its Challenges to Western Thought* (New York: Basic Books, 1999), 47; and Daston, "The Nature of Nature," 153.

11. David Herman, "Scripts, Sequences, and Stories: Elements of a Postclassical Narratology," *PMLA* 112 (1997): 1046–59, at 1047.

understand what was new in terms of what had been known and held to be true for ages. All the while, if social chaos was to be avoided, the existing state of knowledge had to remain both stable and open if it was to allow for the new to find its niche within the established.[12] During the early modern period, this openness was assured because of the firm conviction shared by many of the established authorities—scholars and clergy alike—that the preternatural, including monsters, strange births, violent weather, wondrous plants, and marvelous stones, had their rightful place in God's creation. Such wonders were not located outside of the natural or even in the realm of the unnatural; rather, they were part of the dominion of phenomena not yet fully understood.[13]

Praetorius was by no means the first to mention the strange creature Rübezahl, variously described as *Gespenst*, monk, satanic ghost, and shapeshifter. Almost a century earlier, Rübezahl appears in Georg Widman's reworking of the popular Faust tale *Warhafftige Historien... So D. Iohannes Fausts... hat getrieben* (The true tales about... Dr. Johannes Faustus) of 1599. Influenced by the story of Faust's pact with the Devil, first published in 1587 and instantly a great success, and his damnation and destruction, as well as by the widely circulating witch literature, Widman associates Rübezahl with the Devil, who, it was said, occasionally appears to people in the shape of a monk named Rübezal.[14] Widman notes that the giant often mocks travelers on their way across the rough mountains of the Riesengebirge; occasionally he even leads them dangerously out of their way. Widman counts Rübezahl among the satanic familiars that are said to be companions to witches and black magicians. Such creatures include the demon Mephistopheles, who kept company with Johannes Faustus, and Auerhan, who similarly assisted Faust's pupil Christoph Wagner.[15]

Working against the background of the topography, history, mythology, and demonology of the Riesengebirge, Praetorius begins by reviewing Rübezahl's

---

12. Steven Shapin, *A Social History of Truth: Civility and Science in Seventeenth Century England* (Chicago: University of Chicago Press, 1995), 21, 23.

13. Praetorius, *Daemonologia*, 1:82–83. Among the more unorthodox who firmly held to this view was Paracelsus (1493/4–1541). Knowledge to him was never forbidden, only veiled, not yet accessible to the human mind. Humankind needed to read carefully in the Book of Nature to understand all of nature's and thus God's arcana. See Williams, *Defining Dominion*, 51.

14. Georg Widman, *Erster Theil der Warhafftigen Historien von den grewlichen vnd abschewlichen Sünden und Lastern... So D. Johann Faustus Ein weitberuffener Schwartzkünstler... hat getrieben* (Hamburg: Herman Moller, 1599), 83.

15. In an adaptation of Widman's Faust text published in 1674, physician Johann Pfitzer compares Rübezahl, "ein abentheuerlichee Geist," with Christoph Wagner's Auerhan; in other words, he considers him a demon. Johann Pfitzer, *D. Johannis Fausti...* (Nürnberg, 1674), 88.

person and the habitat he shares, although reluctantly, with the people in and around the mountains.[16] Praetorius writes to "satisfy curious minds";[17] his methodology is "part poetry, part philology."[18] Establishing the authenticity of Rübezahl's life, Praetorius draws on the writings of many of his contemporaries and near-contemporaries. He points with obvious pride to himself as the first who took up the challenge of producing a comprehensive and factual history of the giant and the region he inhabits, even though the task was made very difficult by the dearth of reliable sources. An earlier writer who had promised a Rübezahl history had failed to deliver.[19] Moreover, according to the *Autor catalogi autumnalis* of 1658, a Latin *Daemonologia* had been announced but, to Praetorius's knowledge, never appeared, a fact that Praetorius notes with pointed derision.[20]

Handicapped by the lack of reliable and accessible sources, Praetorius makes use of the scant *Vorrath* (resources) and a selection of *Loci communes*, commonplaces that may, in part, have consisted of well-known Rübezahl tales. He found these materials in a library (possibly his own) whose contents were less comprehensive than he would have wished or needed. In addition, he draws on and freely quotes from popular geographic and topographic, historical, and demonological writings by some of the best-known and widely-read writers of the sixteenth and seventeenth centuries: Opitz, Schwenckfeldt, Schickfuss, Fechner, Zeiler, and Melanchthon, as well as Bodin, Weyer, and even Krämer of *Malleus maleficarum* fame.[21] As one who lived by the sale of his books, he employed many tales told

---

16. Claire Colebrook, *New Literary Histories: New Historicism and Contemporary Criticism* (New York: St. Martin in the Fields Press, 1997), 68, 73.

17. "[H]alb poetisch, halb philologisch,...die lüsternen Gemüter ziemlich stillen." Praetorius, Vorrede, in *Daemonologia*, 1.

18. "[P]hilology is the matrix out of which all else springs." Stephen G. Nichols, "Introduction: Philology in a Manuscript Culture," *Speculum* 65, no. 1 (1990): 1–10, at 1.

19. "[G]roß geschrey und wenig Wolle / sagt jener da er eine quickende Sau schur." Praetorius, "Die Vorspücknisse," in *Daemonologia*, 1.

20. "[N]on-Entia oder ungelegte Eyer," Narratio Theologico-Historica de spectro Rubezal, vulgo der Riebezahl /quod in Montanis Bohemiae, Selesiae & Moraviae iter fatiendibus saepiuscule apparet: Coloniae apud Autorem, Praetorius, "Die Vorspücknisse," in *Daemonologia*, 1.

21. Martin Opitz (1597–1639), *Schäffery von der Nimfen Hercinie* (Breslau: D. Müller, 1630); Caspar von Schwenckfeldt (d. 1606), *Thermarum Hirschbergensium* (1607); *Catalogum Silesiorum doctrina illustrium virorum etc.* and a Natural History of Silesia (1603); Jakob Schickfuß (1574–1636), *New vermehrte Schlesische Chronika und Landes Beschreibung* (1625); Johannes Fechner (1604–1686), *Deutsche Übersetzung der...lateinischen Gedichte...vom schlesischen Riesengebirge* (1675); Martin Zeiler (1589–1661), *Topographiae Germaniae* (1642); Philipp Melanchthon (1497–1560), *De predictionibus astronomicis* (1553); Jean Bodin (1529/30–1596), *De la demonomanie des sorciers* (1580); Johann Weyer (1515–1588), *De praestigiis daemonum* (1563); Heinrich Institutoris (Krämer) (1430–1505?), *Malleus maleficarum* (1487).

about the simple people—the artisans, students, herb gatherers, poor women, and old folks—who reportedly had encountered the *Gespenst* as they traversed the mountains.

Constructing his authority as historian of the preternatural, Praetorius employs three categories on which he bases his construction of truth, categories that would allow him to draw on topography, history, science, and mythology with equal ease. These are "shared ideas and modes of thought; shared forms of externalization; and forms of social distribution."[22]

Shared ideas and modes of thought include the belief in the idea of an activist yet inscrutable nature in which nonhuman spirits and human beings coexist and occupy the same space and from time to time communicate with each other. Such is the case with water spirits, who are said to resemble humans the most. Occasionally such water spirits are said to enter into intimate relationships with and even marry human beings.[23] Also frequently mentioned in contemporary discussions of preternatural wonders are mountain sprites, *Berggeister*, who, according to the mining expert and natural philosopher Georg Agricola, come in two varieties.[24] There are the *daemones malos* who mock and frighten miners, and *Mites*, a more benign category, well-meaning little grey creatures who never hurt anyone except if mocked. Praetorius groups his shape-shifting Rübezahl among the latter.[25]

Certain culturally shared forms of externalization are also needed to construct truths about the exploration of the marvelous and wondrous. These are the ways in which people express what is true in harmony with their understanding of their world.[26] Praetorius finds suitable forms of such constructions of truth in various kinds of learned sources: the Old and New Testaments, ancient and contemporary literature, chronicles, and myths. Furthermore, he relies on the experiences that dependable informants share with him. His *Daemonologia* abounds with stories these witnesses either tell him personally or convey to him through reliable sources.

The third ingredient needed for the effective construction of truth categories is located in the social distribution of the tellers and the tales, the "ways in which the

---

22. Ulf Hannerz, *Cultural Complexity: Studies in the Social Organization of Meaning* (New York: Columbia University Press, 1992), 7.

23. Melusine of Lusignan and the history of her family come to mind. On such intermarriage see Gerhild Scholz Williams on Melusine and Paracelsus in *Defining Dominion*, 23–45, 58–59.

24. Georg Agricola (1494–1555), *De animantibus subterraneis liber* (1549).

25. Praetorius, *Daemonologia*, 1:105–6.

26. Lakoff and Johnson, *Philosophy in the Flesh*, 107.

collective cultural inventory of meanings and meaningful external forms is spread over a population and its social relationships."[27] These forms of meaning are only in part based on religion or theology; increasingly, they depend on the experiential realism implicit in changing views of the body, of forms of social interaction, history, and social position of the witnesses who record these meanings.[28] Furthermore, they are significantly affected by level and manner of education, by age, and by gender.[29]

These three ingredients necessary for the construction of truth and validation of experience influence how Praetorius employs topography, history, mythology, and demonology and explores and explains the preternatural, of which giants are an important part. These creatures live in certain geographical areas and not in others; towns are less their natural habitat than mountains, caves, and other out-of-the-way places. These preferences bring them into (un)easy association with early modern demonology, a fact that accounts for Praetorius's insistent efforts to disassociate his giant from the satanic. Scientific support for the topography of gigantology is provided by records of giants' bones supposedly found, for example, in Italy.[30]

Although skeptics are never silenced for long, the sixteenth and seventeenth centuries produced an extensive chronicle literature discussing the giants' genesis and role in history and mythology. The great majority of these creatures appear as the evil Other, like Gog and Magog, the evil giants that Brutus captured and chained to the gates of the royal palace in London, David's adversary Goliath, or the giants of courtly romance and heroic verse. Giants form one of the genealogical links between early modern Northern European culture and ancient history.[31] From the early Middle Ages, when Geoffrey of Monmouth published his fanciful protonationalistic historical fiction, the *History of the Kings of Britain*, to Boccaccio's *Genealogy of the Gods*, Jean Lemaire de Belges's *Les illustrations de Gaule et*

27. Hannerz, *Cultural Complexity*, 8. Such early modern forms of externalization have been explored in detail by Michael Giesecke: *Der Buchdruck in der frühen Neuzeit: Eine Fallstudie über die Durchsetzung neuer Informations- und Kommunikationstechnologien* (Frankfurt am Main: Suhrkamp, 1991).

28. Lakoff and Johnson, *Philosophy in the Flesh*, 128.

29. Lakoff, *Women, Fire, and Dangerous Things*, 128.

30. About the staged excavation of gigantic bones at Viterbo, see Walter Stephens, *Giants in Those Days: Folklore, Ancient History, and Nationalism* (Lincoln: University of Nebraska Press, 1989), 103.

31. In some traditions, Gogmagog appears as a repulsive giant who killed many Trojans as they tried to settle in England. Walter Stephens reviews the *translatio* of giants into Northern Europe, which was based on the pseudohistory entitled *Commentaries on the Works of Various Authors who spoke of Antiquity* constructed by the Dominican friar Johannes Annius Viterbensis (Giovanni Nanni, 1432–1502). See Stephens, *Giants in Those Days*, 40, 100.

singularitez de Troye (1511–1513), and Rabelais's famous gigantology *Gargantua et Pantagruel* (1509–1532), it appears that giants oscillate between being evil humanoid monsters and being the founding agent of authority in the genealogy of kings and powerful cities.[32] Hovering between the demonic and the benevolently human, Rübezahl's nature, too, remained ambiguous. He reportedly lived most of the time in his mountain wilderness outside the spaces inhabited by humans, except for one instance, when he traveled to England to act as an advisor to Oliver Cromwell. Whether this is meant as an indictment of the English Civil War or a description of the giant's political savvy remains, unfortunately, unresolved.

In keeping with the knowledge that wonders such as the giant are real, Praetorius counts on eyewitnesses to corroborate his work. Their integrity assures veracity. Among them are mail messengers (used repeatedly), apothecaries, pastors, physicians, "glaubwürdige Bürger" (trustworthy burghers), and "ein vornehmer Mann" (a gentleman). Once or twice we even hear from his "very own mother." Truth, to no one's surprise, is best supported by the testimony of the worthy and the educated, which means that the identity of the observer is as important as the sign or phenomenon observed.[33] Specific, descriptive, and atmospheric details about dates, time of day, and places add authenticity to the experience. As it turns out, nationality also plays a role in assessing the reliability of the eyewitness: Praetorius finds Germans, recognizable by their clothes and beards, more trustworthy than the clean-shaven "Welschen" (Italians and French).[34]

Time and again Praetorius returns to Rübezahl's name and nature. At one point, he calls him "Riesenzahl" (counter of giants) and a descendant of the giants whose hubris led them to rebel against God.[35] According to Opitz's *Poetische Wälder*, two of these giants remain buried deep in the caves of the Riesengebirge, witnesses to the fact that giants had indeed been living in Germany, specifically in

---

32. Stephens, *Giants in Those Days*, 128, 160, 183. Descendants of the only giants that survived the Flood, Noah's sons are said to have migrated first to Southern, then to Northern Europe, providing a useful mythology for the nationalist agendas of French and English dynastic historians. Except for one, Noah's son Cham, all evil antediluvian giants had perished in the Flood that punished their assault on the heavens at Babel and their debauchery and unnatural lust. The good giants descended from Noah, far from being the brute, not fully human beasts of the Old Testament and ancient history, were said to have been intelligent and long-lived, endowed with occult knowledge and mystic religiosity.

33. Shapin, *A Social History of Truth*, xvi.

34. Praetorius, *Daemonologia*, 2:163–64.

35. "Unfehlbar sein sol: zum andern / daß durch diesen Geist die Himmelsstürmenden Riesen noch heutigen Tages gleichsam preasentiret würden," Praetorius, *Daemonologia*, 1:79. The name "Rübezahl" has been variously associated with counting, as for instance with counting "beets" (Rüben).

Silesia, since time began, or since Noah's sons crossed the Alps.[36] Opitz tells of coming across the history of these giants when traveling through the *Riesengebirge*. He found this history enhanced by quotes from Hesiod, Apollodorus, and Ovid inscribed on a black stone slab. According to this monument of stone, Riesengebirge is simply the German translation of the Latin *mons Gigantaeus*, not because the mountains are higher than any of the others surrounding them but because the giants remain buried under them.[37] There they guard the gold, silver, and precious stones over which their progeny, the "Gespenst Rübezahl," the "Berggeist" (mountain spirit) reigns. These treasures attract fortune seekers of all kinds, whom Rübezahl rewards as often as he punishes them for their greed. Once again Praetorius struggles with the giant's questionable origin, which makes him kin to spirits generally called "mammonisch" and "Plutisch."[38] Accentuating the postive, Praetorius notes that Rübezahl does not tolerate "Hexenmeister und alte Zoreassen" (witches and old followers of Zoroaster). When black magicians practice their conjurations in the isolation of his mountains, Rübezahl reputedly tears the pages from their books and throws them down the mountain, together with all their magical paraphernalia.[39] The traces of one such encounter, says Praetorius, can still be observed in the mountains. Topography once again confirms demonology.[40]

To gain control over his disparate collection of facts and fancies, Praetorius uses a method he described as "half poetic, half philological": the acrostic. This rhetorical figure is based on the concept that the initials of the words, if written together, make up a word that reflects all or many of the potential meanings of these words. This technique is very congenial to Praetorius's writing style; he favors it because his materials always threaten to disintegrate into textual chaos if they are not structurally controlled. In this way, Praetorius arranges areas of knowledge such as topography, geography, history, mythology, and demonology in a coherent whole, even if this whole is held together only tenuously by the often incongruent arrangement of words whose initials form the concept to be explained.

---

36. *Mart. Opitii Silvarum libri III* (1631).

37. "[D]arunter etwan die Riesen begraben liegen; oder auffs wenigste von solchem Berge / den Himmel stürmen vorgenommen haben," Praetorius, *Daemonologia*, 1:81–82.

38. Ibid., 1:120.

39. "[D]a reißet der Rübezahl dem Kerl das Blat vor der Nase außm Buche / und wirfft ihn mit sammt den Bettel in eltiche hundert Stücke zum Berge herunter," ibid., 1:127.

40. Ibid., 1:127.

Praetorius begins his exploration of Rübezahl's nature and the giant's habitat quite literally with the acrostics GEOGRAPHIA and RUEBEZAHL, using the letters as mnemonic devices to make understanding and memorizing the assembled facts easier. Rejecting the arguments of those who find this method cumbersome, he declares, "If any wise guy finds this upsetting, let them be upset: I am convinced that I am better served by confining things to certain terms and limits than if I were to throw them out willy-nilly and in total confusion."[41] Acrostics are his response to the most important challenge to his writing, which is how to control the often disparate materials for the benefit of a public which might be able to read but is not, on the whole, truly educated.

Aside from acrostics, Praetorius frequently also employs numbers as another type of explanatory design.[42] Adding up the letters in the Latin title of the *DaeMonoLogia rVbInzahLHLesII*, he arrives at the year 1662, the first year "in which Rübezahl came to life" (in welchen Rübezahl lebhafftig gemachet worden), that is, the first printing date of his *Geographia*. He warns his readers that not everyone is able to use this method correctly. He chides "ein ander Klügeler" (some wise guy) who had tallied the letters of Riebendzal (80.9.5.2.5.40.4.500.1.20) in such a way that he arrived at the sum of 666. This is not an inconsequential play with numbers; to the informed, such numerical association brings Rübezahl dangerously close to the Antichrist, whose arrival was, once again, predicted for the year 1666.[43] While always a bit nervous about his hero's status as a giant and/or *Gespenst*, Praetorius vigorously rejects such a very negative affiliation. His Rübezahl may be a *daemon*, but certainly not one of the satanic kind. He is, more fittingly, related to those whom Paracelsus, *der Hochgelehrte Philosophus*, describes as *Elementargeister* (elemental spirits), who dwell in the four elements, fire, air, water, and earth, as "in their own home" (in ihrem eignen Chaos / vnd Wohnung).[44] Following Paracelsus, Praetorius believes that ghosts ("Spücknüsse oder Gespenste") have

---

41. "[U]nd will sich nun einer an die Neoterita [sic] ärgern der mag es immer hin thun: ich vermeine dennoch / daß ich besser gethan habe / weil ich die Sachen also in gewisse *Termino* und Schrancken eingeschlossen habe / as wenn ich sie wie ein *Chaos indigestum, confuse*, oder über Hals über Kopff heraus geworffen hätte," Praetorius, *Daemonologia*, vol. 1, "Die Vorspücknisse."

42. See Gerhard Dünnhaupt, "Chronogramme und Kryptonyme: Geheime Schlüssel zur Datierung und Autorschaft der Werke des Polyhistors Johannes Praetorius," *Philobiblon* 21 (1977): 130–33. See also Lorraine Daston's comments about lists in "The Language of Strange Facts in Early Modern Science," in *Inscribing Science: Scientific Facts and the Materiality of Communication* (Stanford: Stanford University Press, 1998), 36–37.

43. "Merke hiebey / daß in ander Klügeler den Rübezahl zum Antichristlichen Thiere / das in inen Nahmen 666 hat / Cabalistisch habe machen wollen," Praetorius, *Daemonologia*, vol. 1, "Die Vorspücknisse."

44. Praetorius, *Daemonologia*, 1:166.

their rightful place in the cosmos and in nature, a fact that not only common sense but the science of the day confirmed. People who travel or read books know that "in certain places certain ghosts sometimes live and can be seen and heard for longer or shorter periods, generally appearing in the same shape."[45]

This statement is followed by another explanatory acrostic based on the initial letters of the word GESPANSTER (ghosts):

> Gilbertus, ein Zauberer in Ostrogothia:
> Ethneischer Vulcanus in Sicilien.
> Spücknüsse in Böhmen.
> Pilatus bey seiner See in den Schweitzerlande.
> Aeltes Parmenisches Weib.
> Nympfe zu Glatz.
> Schlangen-Jungfer zu Basel.
> Treue Eckart in Thüringen.
> Esel mit drey Beinen zu Leipzig.
> Rübezahl bey den Schlesiern.[46]

Noting the seeming incoherence of this list, Praetorius acknowledges that he understands the category *Gespanst(er)* (ghost(s)) quite broadly, reflecting the magical-scientific worldview and popularizing style that together inform all of his writings. Once again, but this time more obliquely, he refers to Paracelsus's concept of experience (*Erfahrung*) in his attempt to dispel misconceptions and fears about Rübezahl: "What has been confirmed through long experience and what the eyes have seen, the heart can accept."[47] Alluding to Johannes Faustus's pact with the Devil, and once again wishing to distance his giant from association with Satan, Praetorius asks his readers, anticipating a negative answer, "Do you really believe that he will enter into a pact with you so that you might have him as your

---

45. "[A]n unterschiedlichen Oertern gewisse Gespenster sich theils immerdar / theils zu bestimmten Zeiten / hören und sehen laßen / und zwar gemeiniglich in einerley Gestalt und Form," Ibid., 1:161.

46. "Gilbertus, a sorcerer in Östergötland [Sweden]; the volcano, Mount Etna, in Sicily; hauntings in Bohemia; Lake Pilatus in Switzerland; old woman from Parma; nymph from Glatz; snake-girl from Basel; faithful Eckhart from Thuringia; donkey with three legs; Rübezahl of Silesia." Ibid., 1:162.

47. "[W]as aber durch lange Erfahrung bestetiget ist / und die Augen selbst sehen / das kan das Hertze ja glauben." Ibid., 1:54. Paracelsus puts it this way: "Er läßt uns...suchen, die Erde durchwandern und vielerlei erfahren. Und wenn wir es alles erfahren haben, sollen wir, was gut ist, behalten (He [God] lets us search and experience much throughout the earth. And after we have experienced it all, we must keep what is worthwhile)." *De causis morborum invisibilium*, in Theophrastus Paracelsus, *Werke*, ed. Will-Erich Peuckert (1944; reprint, Darmstadt: Wissenschaftliche Buchgesellschaft, 1990), 2:270, cited in Williams, *Defining Dominion*, 47.

servant?"[48] Rübezahl, though sought by many because of his purported reign over much subterranean wealth, is not for hire. He has nothing in common with Faust's Mephistopheles, nor does he lead souls to Satan, even if contemporaries have called him a *Teufflisch Gespenst* (satanic ghost), recalling the negative associations raised by Widman almost a century earlier.[49]

RÜBEZAL, yet another acrostic, enumerates the Giant's pastimes most economically:

> Reisende verführen.
> Vexiren.
> Blitzen und Donnern.
> Ertz besitzen.
> Zaubern.
> Artzneyen.
> Lustig jagen und hetzen.[50]

He is also called a noble Hussite, a mountain deity, a monk, a sorcerer, a nobleman, and a long-eared devourer of donkeys, this last pointing to Praetorius's whimsy—a kind of self-mockery—which often accompanies his more serious scientific explications.

Keeping the giant firmly in the center of the reader's attention, Praetorius constructs a panorama of the vast region that made up the Riesengebirge during a historical period, shortly after the Thirty Years' War, that brought much suffering and death to almost all of central Europe. In this process, the Riesengebirge becomes as much an actor in this tale as does the giant to whom it is home. The mountains take on a maternal persona, sheltering the dead giants as well as the people living in the valleys. The valleys are seen as the mountains' belly, feeding on the mountains' treasures, such as the much-coveted medicinal herbs.[51] A treasure trove of material plenty, the mountains are charged to the care of the giant, who, in times

---

48. "Meinestu / daß er einen Bund mit dir machen werde / daß du ihn immer zum Knechte habest?" Praetorius, *Daemonologia*, 1:142.

49. "[E]in Riese, ein vornehmer Hussit, ein Berggott, ein Mönch, ein Zauberer, Edelman, oder langohriger Eselsfresser." Ibid., 1:78.

50. "Leading travelers astray; vexing [people]; producing thunderstorms; owning ores; practicing magic; making potions; lusty hunting and chasing." Ibid., 1:142.

51. The gendering of nature as topic of recent environmental criticism explores the view of nature as a woman whose wildness needs taming, who is owned and controlled by her male explorers and conquerors. More generally, ecocriticism reviews the relationship between people and nature at certain moments in history. See Richard Kerridge, *Writing the Environment: Ecocriticism and Literature* (London: Zed Books, 1998); "Forum on Literatures of the Environment," *PMLA* 114 (1999): 1089–1104.

of scarcity and want, distributes his gifts to many a poor person who ventures into the wilderness in search of some small share of Rübezahl's legendary wealth.[52] In a brief instance of social criticism, Praetorius chides "princes and lords" (Fürsten und große Herren) because their passion for the hunt destroys the meager harvest of the poor peasant. After their death, they will join the perpetual hunt led by the Devil at night and even in the day, as punishment for their abuse of the people.[53]

As if in a reciprocal relationship, these mountains nurture yet another people, innovative and industrious, who produce linen cloth and glasswares that are coveted far and wide, and receive a share of the giant's golden wealth.[54] They bear the hardships of many solitary walks across the mountains into Bohemia and Poland, peddling their wares, and return home with the means necessary for survival.

Together, the topography of the Riesengebirge and the nature of the giant Rübezahl provide the reader with a guided tour across this isolated area separating Silesia from Bohemia, an area whose remoteness accounts for outsiders' many misconceptions and fears about the people who live in these mountains. Reading nature and understanding culture, Praetorius populates the wild topography of the Riesengebirge with the rich and poor, male and female, young and old, who traverse its landscape and whose encounters with the giant occasionally yield rewards, sometimes punishment, in all cases wonder.[55] Praetorius recounts a myriad of tales about how the giant dispenses leaves, earth, or excrement, which, if carried far enough, turn to gold ducats, gold leaf, or gold dust. Impatient and mistrustful, the people so rewarded more often than not dispose of the gifts long before they turn into the coveted gold or silver. Scraping together the little bits of treasure remaining in empty bags, the (un)happy, now wiser, recipients of the giant's generosity are almost always people of the lower social ranks: journeymen, students, glass merchants, herb gatherers, weavers, peasants. The mountains they traverse are described as dangerous and unforgiving; the highest peak, the

---

52. Lawrence Buell, *The Environmental Imagination: Thoreau, Nature Writing, and the Formation of American Culture* (Cambridge: Belknap Press, 1995), 60.

53. "Hierher gehören auch die Teuffels-Jagten: Da die Teuffel in Gestalt und Person derer / die entwan grausame und unbarmhertzige Jäger gewesen sind / zu Nacht und auch bey hellen Tage sich sehen laßen / hetzen und jagen...bey ihrem Leben mit großer Beschwerung armer Leute. (This also is part of the Devil's hunt: because the Devil appears in the person and shape of those who were [during their lifetimes] cruel and uncharitable hunters / who [now] can be seen by night and by day hunting and chasing...while alive a great burden to the poor)." Praetorius, *Daemonologia*, 1:140–41.

54. "Haben doch die Schlesier gleichsam in ihren Backen Gold. Denn auffs wenigste hat ja bey ihnen ein Bauers-Junge eine güldenen Zahn im Maule" (The Silesians carry some of their gold in their cheeks; there is not a peasant boy among them that does not carry a golden tooth in his mouth). Ibid., 1:66.

55. Buell, *The Environmental Imagination*, 260.

Schneekoppe, more often than not is covered with ice and snow. Nightfall had better find the wanderer in the valley, away from the giant's lair, for he likes people to leave his domain by dusk.

The sole exception to this injunction are a people who dwell in the hidden valleys deep in the interior. Small in stature (as small as children elsewhere), they live peacefully in modest huts, protected from the world by their isolation and their humble and gentle way of life. Though traveling is difficult for them, they nimbly climb over rocks and trails and place stakes into the ground to help them find their way in the deep winter snows. Mindful of Rübezahl's temper, they try not to offend him and, in turn, count on being left in peace. One of Praetorius's reliable witnesses, a letter carrier from Liebenthal, tells of their beautiful cattle and their meadows rich in herbs and lovely flowers.[56] Having to make do without the benefit of regular pastoral visitation and sermons because of their isolation (they have to walk five miles to the next Lutheran church), they are nonetheless content with their lives and remain untouched by the ravages of what Praetorius calls the German War.[57] In the isolation of this mountainous utopia, in the protective arms of their mountain mother, this people coexists with the giant in a relationship of mutual respect and benefit. The harsh natural environment turns out to be their protection against the strife and corruption governing life outside of their mountain home. The maternal, caring, and providing landscape takes on the paternal role of guardian of virtue and (lay) religiosity.

True to his shape-shifting nature, Rübezahl often appears to passersby, walking and talking, mocking, amusing, or frightening them. Like some latter-day Zeus or early modern witch, he expresses his rancor or his preternatural impetuosity by raising violent storms and inclement weather. Presenting, in this way, the inverse of the nurturing mother, the rough mountains and dark valleys, the wildly unpredictable summer and winter weather invite association with another form of gendered topography: with demonology, the actions of Satan and his minions, the witches. This destructive and threatening connection between landscape and demonology stands opposed to the maternal, generous, protective relationship between the mountains and its people. Praetorius, well acquainted with this satanic side, explores it in one of the most widely read and cited witch tracts of the seventeenth century, the *Blockes-Berges Verrichtung / Oder Ausfürlicher Geographischer Bericht... von der Hexenfahrt / und Zauber-Sabbathe...*, published in

---

56. Further, he relates the amazing story of a husband who traveled a considerable distance in order to fetch two pitchers of beer for his wife, who had given birth to their child. Praetorius, *Daemonologia*, 2:153–55.

57. That is, the Thirty Years' War (1618–48).

1668, just around the time that Rübezahl was on his mind. The Blocksberg Rübezahl appears in a different guise. Here he does metamorphose into one of the manifestations of Satan, *der Teufel selbst*. In the Rübezahl volumes under discussion here, Praetorius tries very hard to distance his giant from weather-making witches and from the satanic in general. Theirs is diabolic subterfuge; Rübezahl's weather making is perhaps the expression of anger or impatience, but he does it on his own, without satanic assistance.[58] This weather-making talent shows, once again, that the early modern ambivalence apparent in the uneasy association of the satanic with the wondrous is never very far from the giant's narrative portrait.[59] In the end, Praetorius seeks refuge in a gendered distinction: Rübezahl as a "montanus daemon" (mountain demon) is a much less murderous and threatening creature than a "zauberisches Weib" (witch).[60] The witch's disturbance of Nature is clearly judged an infinitely worse offense than the giant's more innocuous pastime.[61]

On occasion, science helps Praetorius build a more solid case for his giant's nonsatanic nature. While the Devil is indeed real and does have certain divinely ordained powers, natural explanations, based on an understanding of the nature of the elements as they expand and contract ("the nature of the elements is such that they contract and expand"),[62] are at least as plausible to human understanding as any conjectures about Satan's powers.[63] He offers as an example the *Irrwische* (will-o'-the-wisps), little flames that dance across moors and swamps, scaring the lonely wanderer. According to Praetorius, these are natural phenomena: flammable air escapes from wetlands, cemeteries, under gallows, and places where big battles have been fought and many men have died, and as this air ignites, it spawns many small flames that flicker across the dark and desolate

---

58. "So ist es doch unzweiffelbahr / daß sein gewöhnligste Verrichtungen seyn / nach Belieben zu blitzen / donnern und hagelen: wie denu solches über unzehlbahr Erfahrung / von ihm bekrafftiget" (Thus it is not to be doubted, and experience has shown it many times, that he likes nothing more than to throw lightning bolts, to thunder, and to produce hail). Praetorius, *Daemonologia*, 1:119.

59. Daston, "The Nature of Nature," 155.

60. Praetorius, *Daemonologia*, 1:117.

61. Praetorius discusses in some detail who the Devil is and what he is capable of doing. His source is frequently, though not exclusively, the *Theatrum Diabolorum*, a popular and much cited collection of Devil books published in 1569 by the Frankfurt printer Feyerabend. The tract referred to here is *Der Teuffel selbst* by Jodocus Hocker.

62. "Die Natur der Elementischen Sachen / welche sich ausdehnen und zusammenziehen." Praetorius, *Daemonologia*, 1:206.

63. Ibid., 1:204.

landscape and frighten people.[64] It is not the will-o'-the-wisp or some devilish creature that harms people; rather, the danger is in the swamps and wetlands themselves, places whose treacherous topography threatens people, especially the fainthearted and melancholy, who should avoid them at all cost.[65]

In keeping with the authorizing power of canonical texts, and including a liberal amount of self-citation, Praetorius insists that for all their singularity, the wonder and marvels associated with a giant living in the mountains of Silesia are not new or even all that unusual. Throughout the centuries, strange phenomena have confirmed that the meaning of history is hidden in signs and portents, the *prodigia* and *Raritäten* of which Praetorius published a great many.[66] Their historical and geographical ubiquity provide Praetorius with one of his most persuasive arguments for their veracity, namely that the truth of such appearances, or in the case of Rübezahl, such encounters, is based in part on their volume and reported frequency.[67] He pulls phenomena such as giants, bloody rains, plagues, and ghosts from their historical or even mythological niche and positions them within reach of human experience. Thereby he fashions for his readers a panoramic view of the unusual that is disquieting and comforting at the same time: disquieting because Rübezahl and creatures like him, or events like the weather he stirs up, hover just around the corner, presumably up to no good and ready to pounce on unsuspecting humankind; and comforting because the giant can and will do good, strange phenomena do carry good as well as bad messages, and occasionally and quite serendipitously, as is the nature of the preternatural, the giant will reward the weary, protect the weak, and punish the guilty.

64. Ibid., 1:265.
65. Ibid., 1:170.
66. See Robin Barnes, *Prophecy and Gnosis: Apocalypticism in the Wake of the Lutheran Reform* (Stanford: Stanford University Press, 1988); and Eugen Weber, *Apocalypses: Prophecies, Cults, and Millennial Beliefs through the Ages* (Cambridge: Harvard University Press, 1999).
67. Gerhild Scholz Williams, *Confronting the Early Modern Other: Johannes Praetorius (1630–1680) on Wonders and Violence*, Wolfenbüttler Studien (Wiesbaden: Harrassowitz, forthcoming).

DEMONS, NATURAL MAGIC, AND THE VIRTUALLY REAL
Visual Paradox in Early Modern Europe

Stuart Clark

During the two hundred years of the Protestant Reformation and the Scientific Revolution, many European intellectuals seem to have been preoccupied with the question of whether human vision gave reliable access to the real world—of whether vision was veridical. The intellectual history of the period is marked by a pervasive sense of vision's precariousness and fallibility—a feeling that it was not only the noblest of the five senses and the key to all forms of wisdom and enlightenment but also the most vulnerable, the most unreliable, and the most problematic; and a realization, too, that this was a major obstacle that had to be confronted and perhaps overcome before anything else could be achieved. It is striking, for example, that Descartes, Hobbes, and Newton each felt the need, early in their intellectual careers, to become experts in the field of optics. It was as if the agenda for all really serious thought was set by the problems of vision.[1]

The subject also impinged on many different fields of interest in this period—optics itself, obviously, but also philosophy, natural philosophy, religion, moral and social theory, political thought, medicine and psychology, legal theory, and literature. One can read the essays of Montaigne, Pierre Le Loyer's *Quatre livres des spectres*, Martín Del Río's "disquisitions" on magic, Shakespeare's *Macbeth*, the optical works of Kepler, d'Aguilon, and Nicéron, Robert Burton's *Anatomy of Melancholy*, and the *First Meditation* of Descartes—all published within a sixty-year period—and come away feeling that these were contributions to the same

---

1. Richard Tuck, "Optics and Skeptics: The Philosophical Foundations of Hobbes's Political Thought," in *Conscience and Casuistry in Early Modern Thought*, ed. Edmund Leites (Cambridge: Cambridge University Press, 1988), 235–63.

conversation. Nor was the problem merely a matter of abstract and intellectual debate. There were many practical implications in the way it was discussed, notably in the fields of religion, medicine, and the law. This degree of concern does not seem to have existed in medieval culture. Despite the many sophisticated discussions of visual errors and illusions in medieval science and philosophy, they were entirely accounted for in the optical theory of the day. "Their existence," it has been said, "did not cast doubt on the validity of sense experience or the possibility of acquiring knowledge by visual means."[2] Nor does concern seem to have been so marked in Europe after the Scientific Revolution. One of the things the "new" scientists and philosophers did, at least in the now-conventional account, was to put vision back on what was felt to be a more secure basis and help establish what Martin Jay has referred to as "the scopic regime of modernity." After the seventeenth century, the deepest anxieties about vision went away again.

Until modern times, that is. In the late twentieth century the traditional hegemony of vision was deconstructed and many different ways of theorizing about seeing again competed for attention. Western modernity from the eighteenth century onwards is now seen as committed to a particular model of vision designed to secure for it a commanding place in science, in the field of political power, and in the construction of both communal solidarity and personal identity in bourgeois societies. After a century of what Jay has called the "denigration" of this model, particularly in French philosophy, and the emergence of postmodernism throughout Western culture, "ocularcentrism" is now under attack. An iconoclastic history of ways of seeing was one essential component of Michel Foucault's work, and mirroring, imaging, and anamorphosis are all part of the reconceptualizing of vision that has been fundamental to the psychoanalytical theories of Jacques Lacan. Richard Rorty's assault on modern philosophy was built, likewise, on undermining its dependence on the metaphor of the mind as the "mirror" of nature. As a consequence, scholars have become convinced of the constructed nature of vision and the extent to which visual perception and visual meaning are fused.[3] Above all, perhaps, the world is in the middle of a revolution

---

2. David C. Lindberg and Nicholas H. Steneck, "The Sense of Vision and the Origins of Modern Science," in *Science, Medicine and Society in the Renaissance: Essays to Honor Walter Pagel*, ed. Allen G. Debus (New York: Science History Publications, 1972), 1:35; in the course of rebutting Vasco Ronchi's argument that philosophers before Galileo thought of sight as fundamentally untrustworthy. Reprinted in David. C. Lindberg, *Studies in the History of Medieval Optics* (London: Variorum, 1983)

3. From an enormous literature: Martin Jay, *Downcast Eyes: The Denigration of Vision in Twentieth-Century French Thought* (Berkeley: University of California Press, 1993); David Michael Levin, ed., *Modernity and the Hegemony of Vision* (Berkeley: University of California Press, 1993); idem., *Sites of Vision: The Discursive Construction of Sight in the History of Philosophy* (Cambridge, Mass.: MIT

in computer graphics techniques that, in the view of the art historian Jonathan Crary, is bringing about a transformation of the nature of vision more profound than any that has occurred since the coming of perspective. In his book *Techniques of the Observer* he suggests that the established meanings of the terms *observer* and *representation* are being nullified because the visual spaces of the computer are radically different from those of film, photography, and television:

> Computer-aided design, synthetic holography, flight simulators, computer animation, robotic image recognition, ray tracing, texture mapping, motion control, virtual environment helmets, magnetic resonance imaging, and multispectral sensors are only a few of the techniques that are re-locating vision to a plane severed from a human observer.... Most of the historically important functions of the human eye are being supplanted by practices in which visual images no longer have any reference to the position of an observer in a "real," optically perceived world.[4]

One of the items in this list—the development of computer-generated virtual reality environments—relates in particular to the sixteenth- and seventeenth-century debates about veridicality. Simply as a concept, "virtual reality" has implications for one's sense of what makes visual reality different from artifice, fantasy, or illusion, and experiencing it too will presumably affect the codes and categories that govern ways of seeing and so reorganize one's relationship to what is observed. This seems a good moment, therefore, to refocus on another visual culture from the past that was reflective about the relationship between the real and the virtual to a remarkable degree.

■ ■ ■

There are two broad features of virtual reality today that seem to have parallels in late Renaissance Europe. Most obviously, virtual reality is the product of a highly advanced and innovative communications technology. The user wears devices that substitute, for sensory experiences arising from "normal" interaction with

---

Press, 1997); Jacqueline Rose, "The Imaginary," chap. 7 in *Sexuality and the Field of Vision* (London: Verso, 1986); Richard Rorty, *Philosophy and the Mirror of Nature* (Princeton: Princeton University Press, 1979); Chris Jenks, "The Centrality of the Eye in Western Culture," in *Visual Culture*, ed. Richard Rorty (London: Routledge, 1995), 1–25; and Nicholas Davey, "The Hermeneutics of Seeing," in *Interpreting Visual Culture*, ed. Ian Heywood and Barry Sandywell (London: Routledge, 1999), 3–29.

4. Jonathan Crary, *Techniques of the Observer: On Vision and Modernity in the Nineteenth Century* (Cambridge, Mass.: MIT Press, 1990), 1–2.

the world, those produced by a computer, and synthetic experience on this scale has only become possible with the development of extraordinarily complex computer programs. On the other hand, the technology of "virtuality" must also be seen in the much wider context in which the notion of reality has itself become the subject of intense debate. The proposal that reality—including its basis in vision—has cultural rather than natural foundations has become central to research in anthropology and history, particularly in visual anthropology, art history, and cultural history, and in the expanding fields of visual culture and visual hermeneutics. It has also informed a particularly prominent "antirealist" tradition in philosophy, especially the philosophy of language, and in the history of science. The "flight from 'reality' to language" has been most marked of all in critical and cultural theory.[5] Thus, although Western thought and culture have long been subject to a vision-centered model of knowledge, the relationship between vision and reality on which this has been based has now been disrupted and questioned to such an extent that it can no longer be taken for granted.

One has to go back to the early modern centuries to find a comparable situation. At this time, too, the technology of vision was rapidly changing, in ways that contemporaries thought of not only as dynamic and challenging but also as disquieting and problematic. Pictorial space was transformed by the emergence of perspective on exact geometric lines, and artists explored new kinds of visual experience (such as foreshortening and optical correction) in experimental ways, but perspective was widely recognized, indeed defined, as a form of deception, as in the case of Descartes. The French optical theorist Jean-François Nicéron even treated it as a form of magic. The making of truly optical glass and lenses led to striking developments in telescopic and microscopic vision, but, as Catherine Wilson has recently shown, the very artificiality involved could perturb and confuse observers by distorting appearances and inducing a sense of what she calls visual "dislocation."[6] Optics in all its manifestations was crucial for the making of observations of the heavens and so shared in the achievements of astronomy, as well as those of experimental science in general, but what is equally striking is the attention given to the preternatural visual effects that could be achieved in the fields of catoptrics and dioptrics. It was Nicéron again who said that these effects were so prodigious that they were often attributed to witchcraft; to those who did

---

5. The phrase belongs to Gabrielle M. Spiegel, "History, Historicism, and the Social Logic of the Text in the Middle Ages," *Speculum* 65 (1990): 59–86, at 60.

6. Catherine Wilson, *The Invisible World: Early Modern Philosophy and the Invention of the Microscope* (Princeton: Princeton University Press, 1995), 251; cf. chaps. 6 and 7.

not know their optical causes they seemed to be "pure illusions or prestiges of diabolical magic."[7]

In this same period, the very notion of accurate visual perception of external reality—together with the stability of reality itself—were thrown into confusion. Here, three new developments made an enormous impact on late Renaissance intellectual culture, marking its preoccupations off from those of earlier and later periods. First, there was a revival of ancient Greek skepticism, but on a scale unknown in the Greek world. The publication in 1562 of the first Latin edition of the *Outlines of Pyrrhonism* by the ancient skeptic Sextus Empiricus occasioned a skeptical debate throughout Europe of such seriousness that the historian Richard Popkin labeled it a "crisis of Pyrrhonism."[8] With implications throughout the worlds of humanism, theology, and natural philosophy, it was nevertheless grounded in radical doubts about the reliability of the senses, chiefly the evidence of the eyes. Marin Mersenne, for example, acknowledged that of the basic arguments in favor of skepticism, "virtually all of them depended on Optics."[9] Between the 1570s, when Montaigne wrote his *Apology for Raymond Sebond*, and the 1630s, when Descartes composed the *First Meditation*, a whole series of skeptical tropes concerning vision entered intellectual debate.

These tropes did not approach visual problems in the same way as standard late medieval optical theory treated them, as "errors" or "illusions" that were intelligible in relation to correct or normal vision. In standard optics, all manner of sensory weaknesses were allowed for without this compromising the general principle that accuracy was the criterion of sensory success and that this could be measured in terms of fidelity to external reality. What mattered in the Pyrrhonian tropes was no longer the accuracy or inaccuracy of sensory experiences when compared to the external world, but their *difference* when compared to each other. Taken together, the tropes turned *every* aspect of visual experience, "correct" and "incorrect" alike, into something relative, not absolute. They established the contrary principle that for every visual experience deemed to be true it was

---

7. Jean-François Nicéron, *La Perspective curieuse ou magie artificiele des effets merveilleux* (Paris, 1638), 74–75; Nicéron's book was published in Latin in 1646 with the title *Thaumaturgus opticus*, itself part 1 of a projected longer work.

8. Richard Popkin, *The History of Skepticism from Erasmus to Descartes*, rev. ed. (Assen: Van Gorcum, 1964), xii. Illustrative of important recent studies of skepticism's relationship to a wide range of cultural expressions in this period are Stanley Cavell, *Disowning Knowledge in Six Plays of Shakespeare* (Cambridge: Cambridge University Press, 1987); Richard Tuck, *Philosophy and Government 1572–1651* (Cambridge: Cambridge University Press, 1993); and Terence Cave, *Pré-histoires: Textes troublés au seuil de la modernité* (Geneva: Droz, 1999).

9. Tuck, "Optics and Skeptics," 238.

always possible to have a deemed-to-be false visual experience that was indistinguishable from it. They also encouraged the notion that human subjects "make" the objects they perceive, fashioning them out of the qualities that belong intrinsically to perception, not to the objects themselves. "External objects surrender to our mercy," wrote Montaigne; "they dwell in us as we please."[10] Ultimately at issue, as in all attempts to decide between what is virtually real and what *is* real, was a criterion problem—the supposed lack of any independent standard for judging whether sense experience was veridical. Montaigne also argued, for example, that the possible lack of senses additional to the five that human beings possessed meant that they might be experiencing only an approximation of the real world, without having any way of knowing this to be so.

Second, the Protestant Reformation had the effect of severely compromising the reliance normally placed on the sense of sight in the religious sphere. Vision was bound to become a topic of confessional dispute, given the role of images and the exercise of what has been called "sacramental seeing" in contemporary religious worship.[11] In fact, a far-reaching denigration of vision accompanied the religious upheavals of the early modern centuries. But there was a more precise reason for Protestant attempts to demystify the perceptual field of religion—a reason to do with problems of visual reality. Catholic piety, it was realized, was built very successfully on religious experiences with a predominantly visual content—not just images as such, but hosts that bled, statues that wept, souls or saints or angels that appeared in physical form, and so on. Since Protestant theology no longer allowed for such experiences, or for miracles in general, they had to be discredited as visually deceptive—sufficiently good copies of true visual experiences to be virtually indistinguishable from them, and thus convincing as aids to faith, but ultimately not good *enough* copies to save them from exposure as clever counterfeits. In other words, both Catholicism's success and its failure had to be explained in terms of the concept of virtuality.

Protestant polemicists deployed several versions of this concept. Most crudely, they attributed the visual elements of Catholic piety to human deceit. The Catholic clergy were "knaves" and "jugglers" who practiced a vast range of visual deceptions by means of magical techniques, artifices with mirrors, or simply sleight of

---

10. Michel de Montaigne, *The Complete Essays of Montaigne*, trans. Donald M. Frame (Stanford: Stanford University Press, 1958), 422.

11. On this subject, see Bob Scribner, "Popular Piety and Modes of Visual Perception in Late-Medieval and Reformation Germany," *Journal of Religious History* 15 (1989): 448–69; and idem., "Ways of Seeing in the Age of Dürer," in *Dürer and His Culture*, ed. Dagmar Eichberger and Charles Zika (Cambridge: Cambridge University Press, 1998), 93–117.

hand. Much more substantially, Protestant theologians and preachers resorted to the idea of the Antichrist to explain, in particular, the occurrence of false miracles. The notion of the "lying wonder" from 2 Thess. 2:9 was especially apt in this context, once again because it combined an acknowledgment that the Antichrist's deeds were close enough to Christ's to be a convincing reason for centuries of Catholic error with an exposure of them as nevertheless spurious and unreal. Above all, the Devil himself could be brought in—in his virtual form as an "angel of light"—to turn the entire culture of Catholicism into one of false appearances. There was no intention here to destroy the credibility of vision altogether. On the contrary, *Protestant* eyes were needed to detect the implausibility of Catholic miracles in the first place, and they were also required, whenever the authenticity of the visions of radical "enthusiasts" was in question, to discriminate between allowed and disallowed ways of seeing Scripture.[12] Clearly, control of the senses became an important aspect of Protestant social discipline.[13] Nevertheless, these three polemical strategies, undertaken on a huge scale, succeeded in putting visual illusion and trickery at the center of religious debate.

In the case of apparitions—not just of ghosts but of spirits in general—making visual judgments became especially complex and precarious. Having abolished Purgatory, Protestants attributed most apparitions to the impostures of priests, the fabrications of the Devil, or the effects of nature, thus making their visual identification profoundly ambiguous; as Keith Thomas once said, "although men went on seeing ghosts after the Reformation, they were assiduously taught not to take them at their face value."[14] But Catholics, too, conceded demonic interference and natural causes, redoubling the uncertainty by undermining even their own capacity to identify ghosts as souls of the dead. On top of this were the yet more thoroughgoing claims that orthodox churchmen of both denominations had to take notice of—the claims of those they saw as "materialists" and "atheists" that apparitions could *always* be explained away in terms of the hallucinations caused by mental or physical illnesses, or the tricks played by nature on the senses and by

---

12. On miracles see, for example, the arguments in Jean La Placette, *Traité de l'autorité des sens contre la transubstantiation* (Amsterdam, 1700). On the relevance of optics to attacks on "enthusiasm" see Adrian Johns, "The Physiology of Reading and the Anatomy of Enthusiasm," in *"Religio Medici": Medicine and Religion in Seventeenth-Century England*, ed. Ole Peter Grell and Andrew Cunningham (Aldershot: Scolar Press, 1996), 136–70.

13. A point recently reiterated by Charles Zika, "Appropriating Folklore in Sixteenth-Century Witchcraft Literature: The *Nebelkappe* of Paulus Frisius," in *Problems in the Historical Anthropology of Early Modern Europe*, ed. R. Po-chia Hsia and R.W. Scribner (Wiesbaden: Harrassowitz, 1997), 175–218, esp. 178–79.

14. Keith Thomas, *Religion and the Decline of Magic* (London: Weidenfeld & Nicolson, 1971), 590.

the imagination on the mind, or the artifice of magicians and clergy. Once the apparitions debate was fully under way—by the opening of the seventeenth century—nobody fully engaged in it could ignore these various challenges to the veracity of seeing ghosts and spirits or fail to interpret them in terms of the more general problems to do with visual reality that plagued and intrigued the age. Indeed, their writings became the occasion for some of the most sustained and sophisticated of the early modern discussions of truth and illusion in the visual world. In book 1 of his *Quatre livres des spectres ou apparitions et visions d'esprits, anges et demons* of 1586, for example, the Catholic lawyer Pierre Le Loyer offered one of the most complete of all early modern accounts of visual Pyrrhonism, embracing the illusions caused by disease and other natural conditions, those attributable to witches and demons, and those created by jugglers, constructors of marvellous machines, and specialists in catoptrics.[15]

Finally, there was a third development that marked out the late Renaissance period as especially sensitive to the issue of visual reality. This took place in the fields of medicine and psychology. Both the epistemological tropes of the skeptics and the confessional arguments of the warring religions depended in part on the old idea that nature was responsible for a whole range of visual phenomena that undermined the certainties of sight, without any other agency being involved. But there were now many new and independent discussions of these phenomena, particularly medical analyses of pathological impairments to eyesight, natural historical descriptions of objects or substances capable of inducing visual hallucinations, and questions to do with the way vision could be affected by the operation of the human mind. Two themes were explored with almost obsessive enthusiasm. One was the general question of the power of the human imagination to influence visual perception, the other the more precise issue of how the unstable mental states associated with such things as "melancholy," *ephialtes* (nightmare), and lycanthropy could create wholesale delusion. An endless series of theoretical analyses and dramatic individual illustrations circulated through Europe. In this way the philosophical skepticism inherited from ancient thought was broadened and deepened, but a new series of naturalistic benchmarks for illusion was established as well. Debate in these areas meant that contemporary intellectuals were,

---

15. Pierre Le Loyer, *Livres des spectres ou apparitions et visions d'esprits, anges et demons se monstrans sensiblement aux hommes* (Angers, 1586), vol. 4; rev. and expanded in 1605, and trans. (book 1 only) by Zachary Jones as *A treatise of specters or straunge sights, visions and apparitions appearing sensibly unto men* (London, 1605). Le Loyer went on, of course, to *refute* the notion that the eyes could never be trusted to see real ghosts. For standard "atheism" on the subject see Giulio Cesare Vanini, *De admirandis naturae reginae deaeque mortalium arcanis libri quatuor* (Paris, 1616), 368–79.

in effect, looking at an issue that has come to characterize modern discussions of visuality—the extent to which sight is a constructed medium and the eye not the innocent reporter of the world but, in some sense, its creator and, always, its interpreter. Among the classics of this literature were Jerome Nymann's *De imaginatione oratio*, Thomas Fienus's *De viribus imaginationis tractatus*, and Robert Burton's *The Anatomy of Melancholy*, and among other physicians who contributed to it were Riolan, Fincel, Du Laurens, Sennert, Schenckel, and Bartholin.

■ ■ ■

These three developments embraced some of the most fundamental aspects of early modern culture, notably its concepts of knowledge, spirituality, and the human mind. Taken together, they suggest that contemporary intellectuals lost confidence in human vision in a way that does not seem to have been true before or since—until now. Norman Bryson has said (following Lacan) that for human visual experience to be "orchestrated," individual men and women must submit their retinal experiences "to the socially agreed description(s) of an intelligible world." In this way vision is socialized and deviations from the visual norm can be attributed by consensus to things like "hallucination" or "visual disturbance."[16] What seems to have occurred in early modern Europe is that descriptions of the intelligible world—the world of reality—came to be marked by social disagreement, and the problems of vision came to be subject to ideological controversy. What was a hallucination or a piece of theater to some might be a revelation to others, while revelation in its turn could always be compromised if the Devil was at hand.

What is especially noticeable is that many of the occasions for this loss of confidence in vision had to do with not being able to tell the difference between visual experiences that were supposed to be true and those that were supposed to be false—the classic skeptical dilemma. It is this feature of early modern visual culture that makes the concept of virtual reality useful as a historical tool, not just a computer technologist's dream. The parallels are not exact, of course. Virtual reality today aims at a simulated three-dimensional world in which the operator participates using input devices for all the senses, not just vision: "The virtual environment is one in which cybernetic feedback and control systems mimic interaction with real objects, such that the environment appears to be real and

---

16. Norman Bryson, "The Gaze in the Expanded Field," in *Vision and Visuality*, ed. Hal Foster (Seattle: Bay Press, 1988), 91; citing Jacques Lacan, *The Four Fundamental Concepts of Psycho-analysis*, ed. Jacques-Alain Miller, trans. Alan Sheridan (New York: W.W. Norton, 1978), sections 6–9.

can be used as if it were real."[17] Cases of interaction with, or immersion in, the virtual in this full sense can certainly be found in late Renaissance culture—in the world of witchcraft, for example, where, as we will see later, the Devil was granted powers to simulate far greater than those available to any modern cybernetician. Most of the time, however, we are dealing with radical visual indecision about whether the perceived world, or an event in it, was real or an almost exact replica constructed by mental imaging, human artifice, or demonic delusion.

It is true that in some respects our concept of the virtually real has already been trivialized—even before we have acquired the technology to go with it. The word "virtual" is now applied in banal ways not just to computer technology of all kinds but to any simulation of reality, such as in television or film, and even to any not fully achieved effect.[18] Interpreted more strictly, however, virtual reality does seem to contain an illuminating paradox. On the one hand, to mean anything at all, the virtually real must be distinguishable from something it is not—presumably a reality which *is* real in some determinable way (however this is to be determined). On the other hand, the whole point of the technology associated with virtual reality is to make the distinction between the virtual and the real *in*determinable—indeed, radically so. Virtual reality both requires and simultaneously destroys the distinction it contains. This, precisely, is its point; it is built, undecidably so to speak, on the presence and absence of a criterion, of a difference. This is the paradoxical situation that also seems to have puzzled and intrigued the thinkers and writers who reflected on visual reality during the early modern period.

■ ■ ■

That virtuality in the field of vision was seen as a form of paradox in the later Renaissance—and, indeed, that it belonged to a tradition of paradox and paradoxical techniques—was one of the findings of Rosalie Colie's classic study, *Paradoxia Epidemica*, published thirty-five years ago. Colie argued that, whatever their type and context, paradoxical statements necessarily possess common features, notably the self-contradiction and perfectly balanced equivocation encapsulated in the example of the "Cretan liar." They are also always self-referential, both in the sense that the contradictory meanings they contain must be reflections of each other and

---

17. Kevin Robins, *Into the Image: Culture and Politics in the Field of Vision* (London: Routledge, 1996), 44.

18. For complaints about this kind of banality, see Michael Heim, *Virtual Realism* (Oxford: Oxford University Press, 1998), 3–6.

in the sense that each paradox reflects only on itself, leaving no room for external resolution. It comments on its own method and technique as a statement, being qualified by its own technical skill as a paradox to do so but disqualified by its own circularity from ever completely succeeding. In this lies the very paradoxicality, even antirationality, of the paradox, for to be both the subject and object of a statement, to invoke a technique in order to question or examine it, and, above all, to use one's knowledge for self-knowledge, are all inherently tautological, infinitely regressive; they are to be or to do things that cannot be reconciled. All paradoxes, said Colie, "are self-enclosed statements with no external reference point from which to take a bearing upon the paradox itself." Nevertheless, by their very attempt at making statements at all, they are dialectically challenging. They rely on the relativity of opinions and are critical "of absolute and fixed conventional judgments"; they deny commitment and limitation. Although marked by mockery and *serio ludere*, and by the ability to dazzle and cause wonderment, they stimulate further questioning and inquiry and represent, in the case of the Renaissance, an attempt to confront the intellectual phenomenon of many "ideas and systems in competition with one another."[19]

In a book full of remarkable insights, Rosalie Colie traces her "epidemic of paradoxy"—metaphysical, epistemological, moral—in wide-ranging areas of Renaissance thought and writing, including rhetorical defences of "unworthy" subjects, such as folly, utopias and inverted worlds, places where ignorance was learned and opposites coincided, mystical and "negative" theology, mathematical *insolubilia*, devotional and love poetry, and treatments of "nothingness" and negativity (including suicide). This takes her mainly to the writings of Rabelais, Donne, Montaigne, Spenser, and Burton. But she is also acutely aware of how paradox could be expressed both in visual metaphors and in visual practices, especially those relating to mirroring, trompe l'oeil, and illusionism in general. The very equivocation of the paradox made it mirrorlike; the Cretan liar example, she writes, was "literally, speculative, its meanings infinitely mirrored, infinitely reflected, in each other."[20] The mirror, indeed, was the paradox's visual emblem, its images being both themselves "thinkings" ("reflections," "speculations") and an invitation to thought on the part of those observing them. The paradox's mental gymnastics were, likewise, a kind of "prestidigitation" of ideas.[21] Colie sees the art of still life, in particular, as an inherently paradoxical genre, both in content

---

19. Rosalie L. Colie, *Paradoxia Epidemica: The Renaissance Tradition of Paradox* (Princeton: Princeton University Press, 1966), 3–40, at 38, 10, 33.
20. Ibid., 6.
21. Ibid., 22.

and technique. Like the rhetorician's praise of unpraisable things, it was consummate skill contradictorily invested in "lowly," "vain," or "empty" subjects. Like all paradoxes, it combined artistry with duplicity—in this case the triumphant deceit of a naturalism so complete that it tricked men and women into believing in illusions. All artistic illusionism, especially in resorting to the aid of the mirror and the *camera obscura*, risked focusing too much on its own artifice and on the weakness of the human eye. By aiming totally to deceive, it pointed to the relativities of perception. Colie thinks that still life did this in paradoxical abundance: "The still life seeks to transcend its medium in a curious way: by drawing attention to its craft, it flaunts its illusionism, its technical trickery." This made it, again in the self-referential manner of all paradoxes, "an overt commentary on the art of painting," as well as on all ocular experience.[22] Self-portraiture, too, had some of the same qualities: "the more faithful the likeness, the greater the falsity of the picture, the greater its isolation from any reference point outside of the creating, re-creating self."[23]

Something that is said to be "virtually real" clearly partakes of these features of paradoxicality, then as now. Illusion and paradox are closely allied, involving "particularly the paradoxes of relativity, in the sense that one thinks that something is 'real' when in fact it is not."[24] Conceptually speaking, "virtual reality" is only another name for the ocular effects created in the famous story told by Pliny (and known throughout the world of the Renaissance) of the Greek painter Zeuxis, who supposedly tricked birds into pecking at his two-dimensional "grapes" but then was tricked in turn by his competitor Parrhasius into drawing aside his two-dimensional "curtain."[25] In other words, "virtual reality" aims at what Colie calls the tautology of the perfect fit—the better the match between the virtual and the real, the more superfluous and self-contradictory the illusion becomes. Like "still life," it invokes a distinction that it then seeks to deny; in it, "something is declared to be what it manifestly is not."[26] If it is virtual, then it is not real; if it is not virtual, then the whole point of the enterprise is lost. Neither virtual nor real, virtual reality begins an endless circularity (*oscillation* is her term) between real things and

---

22. Ibid., 276, and see 273–99 in general.

23. Ibid., 360.

24. Ibid., 311; and cf. 312: "Deceits of the senses that are two things at once, two-or-more-in-one, are the parallel in natural philosophy of the verbal paradox of contradiction, since they raise and illustrate the same puzzles about the nature of perceived reality."

25. The story comes in bk. XXXV, 65–66 of Pliny's *Natural History* and forms the starting point for Norman Bryson's account of the "natural attitude" in the history of conceptions of painting: see his *Vision and Painting: The Logic of the Gaze* (New Haven: Yale University Press, 1983).

26. Colie, *Paradoxia Epidemica*, 516.

virtual things, from which, hypothetically at least, there is no escape. In doing so, however, it is what she again would have called "self-critical"; it comments on what it is that it is exploiting. It epistemologically questions the processes of human perception—in particular, the conventions, relativities, and other limitations of visuality itself.

■ ■ ■

According to Rosalie Colie, paradoxy was "a major mode of expression" that characterized Renaissance creativity but declined to the point of being mere recreation thereafter.[27] What her study confirms, therefore, is that visual paradox, properly so called, was located in a particular intellectual culture that flourished at a particular historical moment, even if a rather extended one. What it also suggests, moreover, is that visual illusion and deception—or just radical visual uncertainty—did not simply cause anxiety or vulnerability. It was also, without question, an opportunity for exploring the epistemological, psychological, and social attributes of vision and visual representation in original and challenging ways. One of the, again, paradoxical features of the virtual is that it usually becomes more interesting and intriguing than the reality it claims to replace. These disruptive and yet creative aspects of visual paradox in the sixteenth and seventeenth centuries can each be illustrated from the two intellectual fields, one absorbed into *Paradoxia Epidemica* and the other not, where it was most obviously a preoccupation and where, in consequence, it was most carefully discussed—natural magic and demonic magic, the two halves of the preternatural world. An additional aim here will be to set visual paradox—specifically the notion of the *prestige*—alongside the much more widely discussed case of anamorphosis as a second example of the derationalization of vision in a culture which, at least as the standard account again has it, was committed more generally to its rationalization.[28]

There was, of course, a long pre-Renaissance tradition of including *magia praestigiatoria* as one of the branches of magic, where it was invariably discussed as the skill of producing virtual visual effects. A *praestigium* was, in Isidore of Seville's definition, a "false representation by a subtle deception of the senses, especially the eyes, from the word *perstringo*, since it binds them in such a way

---

27. Ibid., 508 ff.

28. This approach to anamorphosis has been the usual one in the years between Jurgis Baltrusaitis, *Anamorphoses; ou, Magie artificielle des effets merveilleux* (Paris: Olivier Perrin, 1969); and Jay, *Downcast Eyes*, 48–49. See also, William M. Ivins, Jr., *On the Rationalization of Sight: With an Examination of Three Renaissance Texts on Perspective* (New York: Da Capo, 1973).

that a thing appears to be what it is not."[29] Like all magical effects, "prestiges" were ambiguous—partly duplicitous and therefore dangerous and immoral, partly exciting technical wonders capable of revealing nature's secrets and God's mysterious ways. But the magical tradition in early modern natural philosophy—and the role of wonder itself—have been reassessed in recent years, and their central importance to scientific change is now well recognized. The fact that natural magicians, in particular, devoted considerable attention to deceiving the eyes therefore testifies to the unusual contemporary significance of this aspect of optics. With typical prescience, Colie herself recognizes the relevance of the paradoxical mode to these visual aspects of natural magic, where intellectuals played with optics in serious ways. She notes, for example, that the natural magicians of "Salomon's House" in Bacon's *New Atlantis* were reported to spend their time creating delusions. In "houses of deceits of the senses" they experimented with juggling, false apparitions, and other "impostures" and "illusions." In "perspective houses," they attempted

> all delusions and deceits of the sight, in figures, magnitudes, motions, colours: all demonstrations of shadows. We find also divers means, yet unknown to you, of producing of light originally from divers bodies. We procure means of seeing objects afar off; as in the heaven and remote places; and represent things near as afar off, and things afar off as near; making feigned distances.[30]

Colie sees this textual description as a reflection of the actual activities and publications of natural magicians and of optics experts interested in the preternatural aspects of their subject. For her, "paradoxical science" was committed specifically to the wonders of vision. This was evidenced in the attention given to visual games and illusions in the work of the Jesuits, Mario Bettini, Gaspar Schott, and Athanasius Kircher, in the "outburst of anamorphosis" in the late Renaissance, and in the appearance of works with titles like that of Giulio Troili's book on optics: *Paradossi per Pratticare la Prospettiva senza saperla* (1683). By these means, "the ambiguous, the paradoxical, the jocose-serious played an essential

---

29. Gaspar Caldera Heredia, *Tribunal magicum, quo omnia quae ad magiam spectant, Accurate tractantur et explanantur, seu tribunalis medici [pars secunda]* (Leiden, 1658), 15, citing Isidore, *Etymologiae*, bk. 8, p. 9. For the standard seventeenth-century definition of *magia praestigiatrix* see Johann Heinrich Alsted, *Encyclopaedia* (Herborn, 1630), 2269.

30. Colie, *Paradoxia Epidemica*, 302; and Francis Bacon, *Works*, ed. J. Spedding, R. L. Ellis, and D. D. Heath (London: Longman, Green, Longman, & Roberts, 1857–74), 3:161–62, cf. 164. See also Bacon's interest in the scientific value of "juggling and conjuring tricks"; ibid., 2:172, 496–97, 655–56; 4:270.

part in their [the Jesuits'] considerations of God, of nature, and of themselves, and gave tone to the wonder and admiration they paid to God's universe."[31]

Rosalie Colie is undoubtedly right. A preoccupation with the marvelous aspects of optical research—especially in the field of catoptrics and its illusionary effects and techniques, and also with regard to the images produced by the camera obscura and the "magic lantern"—is richly evident in the period between Giovanni Battista della Porta's *Magia naturalis* (1558) and Gaspar Schott's *Magia optica* (1671). The subject also impinged on the work of more conventional optics experts such as Christoph Scheiner, François d'Aguilon, Jean-François Nicéron, Emanuel Maignan, Daniel Schwenter, and Johannes Zahn, who themselves treated the prodigious and phantasmagorical aspects of visual perception, such as anamorphosis. The natural philosophical character of this work was embedded in—and thus partly constituted by—religious concerns, especially those of Counter-Reformation piety. Bettini, for example, used illusionist techniques to symbolize the Resurrection and convey its meaning, while anamorphosis was adopted both as a reminder that all specularity was contingent on the medium of reflection and (as Martin Kemp has pointed out) as a way of capturing the hidden spiritual order in a visually chaotic creation.[32]

A special place in this early modern natural magical tradition must be given to the prestiges of Schott's teacher, the Jesuit polymath Athanasius Kircher. In his *Ars magna lucis et umbrae* (1646), Kircher described the construction of a virtual reality machine. A sequence of images of animal heads plus an image of the sun rotating on a hidden drum and aligned with a plain mirror hanging at an angle above would create the illusion of a series of metamorphoses of anyone approaching the drum and looking up at the mirror. The heads would be depicted sitting on human necks and have the same dimensions as human heads. The drum would be turned secretly, and a pulley and rope would make sure the mirror was aligned with both the drum and the observer, who would see first the sun and then the series of animal images alternating with his or her own face. Kircher thought that the realism could be enhanced if the animal heads were carved and given glass eyes and moving jaws and if fur was stuck onto them. A whole series of further specular "metamorphoses" is described in the passage. There was no monstrous form, Kircher thought, into which one could not be changed by using

---

31. Colie, *Paradoxia Epidemica*, 300–28, esp. 307, 312, 315–16.

32. Martin Jay, "Scopic Regimes of Modernity," in *Vision and Visuality*, ed. Foster, 17; and Martin Kemp, *The Science of Art: Optical Themes in Western Art from Brunelleschi to Seurat* (London: Yale University Press, 1992), 167–220. See also Herbert H. Knecht, "Le fonctionnement de la science baroque: Le rationnel et le merveilleux," *Baroque: Revue internationale* 12 (1987): 53–70.

mirrors in various ways; Proteus himself would not have been able to compete. It seems that the machine was actually constructed, since it became an exhibit in one of Europe's main collections of wonders and curiosities, his own *Museo Kircheriano* in Rome.[33]

■ ■ ■

Demonology, yet another defining preoccupation of early modern intellectuals, was also rich in prestiges. A commitment to apocalypticism that lasted well into the seventeenth century, the prosecution of thousands of witches in civil and religious law courts all over Europe, and the onset of major international wars between rival religions gave the Devil a presence in the minds and lives of contemporaries that was not true of their medieval predecessors or true ever again once history and warfare were secularized and witchcraft was no longer a crime. But what kind of Devil? Nothing less than the inventor of virtual reality. In Christian demonology, the Devil, in order to be the supreme and worthy adversary of God, had to come closest to Him and wield almost identical powers, while necessarily falling just short of complete equality. He was thus virtually a deity in the sense of being almost one but actually a creature and so confined within the bounds of nature and its realities. His role in traditional Christianity was to attempt to hide this deficiency and his own evil intentions by *appearing* as God's equal—having the same ends, performing the same things, expecting the same treatment, enjoying the same following. And this appearance of divinity had to be almost successful too. The Devil was the indispensable agent of evil, false miracles, and idolatry, who nevertheless manifested himself as the "angel of light" of 2 Cor. 11:14. Confined to the natural world, the world of sin, he could only pretend to go beyond it—but with powers to deceive that were all but total. In this further sense, the Devil had to be—again there is a necessity here—a master of the virtual and its paradoxes. To give him the total power to deceive was to end in religious and moral absurdity, with no criterion for distinguishing what was really good and true from demonic copies. But to give him something just short of this was to allow for demonically contrived situations where this same criterion might be all but impossible to find. This was one of the great defining themes of early modern theology—and, it seems, of Cartesian philosophy.

33. Athanasius Kircher, *Ars magna lucis et umbrae* (Rome, 1646), 901–6; Filippo Buonanni, *Musaeum Kircherianum sive musaeum a P. Athanasio Kirchero in Collegio Romano Societatis Jesu iam pridem incoeptum nuper restitutum, auctum, descriptum, et iconibus illustratum excellentissimo domino Francisco Mariae ruspolo antiquae urbis agyllinae principi oblatum a P. Philippo Bonanni Societatis Jesu* (Rome, 1709), 311.

What happened to these same considerations when applied to the world of human perception? As a creature, the Devil could only perform natural actions—preternatural at the most, but never supernatural, never miraculous. But for him to camouflage this ultimate weakness meant allowing him powers to deceive the human senses, especially the sense of sight. In conventional Christian theology, extending back through the medieval period and throughout the early modern, the Devil was granted the quite precise ability to interfere in the visual process and create false visual experiences. He could manipulate the world of perceived objects, he could tamper with the medium—the air—through which visual species traveled, and he could alter the workings of both the internal perceptual faculties—the imagination and the "fantasy"—and the external organs of vision—the eyes.[34] He could make men and women think, wrote the German clergyman Theodor Thumm in 1621, "that that which is not, is, and imagine that which is, to be something else," and that the likenesses (*species*) of things were the things themselves.[35] "He can so imitate and counterfeit," said Meric Casaubon in 1672, "that we shall find it a very hard task, to distinguish between the reality of that which he cannot, and the resemblance, which he doth offer unto our eyes."[36]

Accounts of what actually happened could be detailed. In their *Dialogicall discourses of spirits and divels* (1601), the two English divines John Deacon and John Walker (via "Physiologus") explained the transformations attempted by Pharaoh's magicians as successful only "in an outward appeerance." Taking advantage of the deep impression left behind in the sensitive faculties of the onlookers by Aaron's real miracles, Satan superimposed his own images on those produced by the physiology of vision:

> For, much blood descending before into the sensitive facultie, there descends withall, many imagined formes, whereby there is forthwith procured a very lively resemblance of some such things as are not existing at all. By this meanes therefore (there being beforehand procured a commotion of humours, as well in the interiour, as exteriour senses of all the beholders) the Divel might both inwardly and outwardly also,

---

34. Despite this clear capability, the Devil and witchcraft do not seem to enter much into ophthalmological discussions of eye impairments; but see, in a very widely cited text, the remarks on sorcery in Georg Bartisch, *Ophthalmodouleia: Das ist, Augendienst* (Dresden,1583), 231 r–36 v.

35. Theodor Thumm, *Tractatus theologicus, de sagarum impietate, nocendi imbecillitate et poenae gravitate* (Tübingen, 1621), 28.

36. Meric Casaubon, *A treatise proving spirits, witches and supernatural operations by pregnant instances* (London, 1672), 157.

applie certaine apparant formes to the very organons of all the senses; even as effectually, as if they had risen only from outward sensible objects: and (by such a legerdemain) might cause the sorcerers' rods to seeme in appeerance, as though they had beene true serpents in deede.[37]

A century later, John Beaumont was more prosaic; a demonic "prestige" involved presenting to the senses "a thing that is not, as if it were; as that an house, for instance, may seem to be there where there is none."[38]

Comparisons between the Devil's prestiges and those of early modern optics experts—even those of painters—could also be surprisingly exact. Schott, for example, said that demons used the "art of perspective" to deceive the senses, making "square things look round, or scattered things look grouped together, or ugly things look graceful." "If art and nature," he said, "can create such wonders as we see in the fields of *Magia Optica, Catoptrica, Dioptrica,* and *Parastatica,* what cannot the devil produce of the same sort through magicians?"[39] One of the most significant of these portrayals appeared in the course of a detailed description of demonic sensory delusion by the Spanish Dominican theologian Raffaele Della Torre in his *Tractatus de potestate ecclesiae coercendi daemones* (1629). Della Torre explained in the usual way that the Devil could manipulate the object, medium, and organ of vision in any visual encounter, making it, strictly speaking, a "prestige" (*praestigium*). Augustine and Thomas Aquinas were authorities for this fact, but so too was the story in Philostratus's *Life of Apollonius* (bk. IV, 25) of Menippus, who is almost tricked into marriage and death by a *lamia* who appeared to him "in a beautiful likeness" and offered him riches and a wedding feast, which Apollonius promptly exposed as "empty copies" of the real things. The issue here, said Della Torre, was one of representation. By forms of local motion the Devil makes men and women take the images they see "to be the things they represent" (*ut putent esse eas res, quas repraesentant*). It was, indeed,

---

37. John Deacon and John Walker, *Dialogicall discourses of spirits and divels* (London, 1601), 143.

38. John Beaumont, *An historical, physiological and theological treatise of spirits, apparitions, witchcrafts, and other magical practices* (London, 1705), 343.

39. Gaspar Schott, *Magia universalis naturae et artis…opus quadripartitum. Pars I. Continet optica…* (Bamberg, 1677), 1:36–37. François Perrault, a seventeenth-century Huguenot pastor in the Pays de Vaud, even specified the technique: "If humans, by artifice and the use of certain candles, vapours, and smokes, can make it seem that a whole room is full of snakes, even though it is only the eyes that are deceived…then all the more reason to suppose that demons…can deceive and illude our senses by a thousand false and deceiving appearances. This they do either themselves or by magicians and witches." See François Perrault, *Demonologie,* 2d ed. (Geneva, 1656), 121. One of Kircher's and Schott's catoptrical illusions was precisely this—to make a room seem to be full of snakes.

one of *artistic* representation, since the Devil was only doing what sculptors and painters do when, by local motion too, they produce shapes or apply colors and sometimes make the things they depict "seem real and natural" (*ut interdum verae, et naturales videantur*). Della Torre's comparison at this point is telling—precisely the supremely naturalistic representation achieved by Zeuxis and Parrhasius in Pliny's famous story. What, he added, could the Devil *not* represent to sight, being a more skillful artist (*artifex*) than Zeuxis or any other painter or sculptor? Should it be a question of rearranging existing objects in a visual field, rather than creating new ones from air and vapor, he also had the skills to surpass all the jugglers (*circulatores*). And, finally, in the field of optics itself, he could outdo the anamorphisms of the most dedicated perspectivists:

> If those engaged in optics (*perspectivae*) can make the outlines of a long irregular painting look distorted if they are observed directly, but, when they are viewed through an aperture or with the painting rotated, restore the likeness of some artistic image, why cannot the devil, most skilled in every art, do this and stranger things?[40]

Artistry and duplicity in combination—such, indeed, was the Devil's contribution to the Renaissance paradox.[41]

The English philosopher Henry More said that witches, too, got at least their biblical name (Hebrew: *megnonen*;[42] English: *praestigiator*) "from imposing on the sight and making the bystander believe he sees Forms or Transformations of things he sees not."[43] And witchcraft was in fact the area where the demonology

---

40. Raffaele Della Torre, *Tractatus de potestate coercendi daemones circa obsessos et maleficiatos*, in *Diversi tractatus de potestate ecclesiastica coercendi daemones circa energumenos et maleficiatos*, ed. Constantin Munich (Cologne: Constantin Munich, 1629), 214: "Si enim perspectivae dediti homines, efficiunt, ut lineae in tabula longa confuso ordine ductae, si e directo aspiciantur lineae rudes videntur et sunt, si tamen aspicis per foramen, aut tabula rotata speciem reddunt alicuius artificiosae imaginis, cur non hec, et mirabiliora efficiet daemon in omni arte peritissimus." See also 212–22, generally (from the second part of the *Tractatus*, entitled *De potestate daemonum de magorum ad effectus mirabiles et prodigiosos*). Essentially the same arguments and examples appear in Francisco Torreblanca, *Daemonologia sive de magia naturali, daemoniaca, licita, et illicita, deque aperta et occulta, interventione et invocatione daemonis* (Mainz, 1623), 236–40.

41. Claudia Swan, "The Devil as Painter: Demonology and *Fantasia* in the later Sixteenth Century" (paper delivered at Sixteenth Century Studies Conference, St. Louis, 28–31 October 1999), explores the parallel subject of the Devil's impact on the internal senses, notably the "fantasia."

42. Susan Einbinder has kindly explained to me that the correct Hebrew rendering is meᶜonen.

43. "Dr. H.M. his Letter, with the Postscript to Mr. J.G.," in Joseph Glanvill, *Saducismus triumphatus: Or, full and plain evidence concerning witches and apparitions*, 3d ed. (1689; repr. ed. C.O. Parsons, Gainesville, Fla.: Scholars' Facsimiles & Reprints, 1966), 31.

of the senses produced the most dramatic and perplexing results.[44] During the many witchcraft trials of the fifteenth to eighteenth centuries, certain things repeatedly appeared in the confessions of witches that everyone knew, for good theological and natural philosophical reasons, were not true phenomena and that therefore had to be discounted; witches flew off to "sabbats" in spirit only, leaving their physical bodies at home in bed, they gave birth to children fathered by devils, and they changed themselves or other people into animals. It was in the course of the arguments that ensued that the Devil's ability to create virtual worlds—or at least virtual events—where unreal phenomena were scarcely, if at all, distinguishable from their real equivalents was most fully elaborated. Witchcraft writers debated, for example, the extent to which the Devil could create simulacra of innocent people at sabbats, so that guilty witches confessed to seeing them there. His power to impersonate, and the allied question of whether "spectral evidence" could be allowed in the law courts, were key issues in this area of demonology.[45] A very considerable number of such discussions circulated in early modern Europe, adding further to the doubts about the reliability of the eyes that already marked the visual culture of that period.

Eventually these discussions affected the stability of witchcraft theory and of demonology.[46] For two or more centuries writers of demonology compared the Devil metaphorically and literally to a juggler with the consummate skill to deceive the eyes of spectators. Yet this played into the hands of those disbelievers in witchcraft like the Englishmen Reginald Scot, Thomas Ady, and John Webster, who tried to show that witches and devils had *nothing but* the skills of jugglers—certainly nothing more sinister. An attempt to show how great were the Devil's powers ended up demeaning them by comparing them with lesser ones. It was also More who saw the threat in turning the Devil into a trickster in Webster's definition of witchcraft as mere sleight of hand: "As if a Merry Juggler that plays tricks of Legerdemain at a Fair or Market, were such an abomination to either the

---

44. The history of witchcraft and the history of ways of seeing have hardly begun to be considered together, but see Bernd Roeck, "Wahrnehmungsgeschichtliche Aspekte des Hexenwahns—Ein Versuch," *Historisches Jahrbuch* 112 (1992): 72–103.

45. See, for example, Increase Mather, "To the Reader," in *A further account of the tryals of the New-England witches.... To which is added, cases of conscience concerning witchcrafts and evil spirits personating men* (London, 1693), esp. 1–19; cf. Dennis E. Owen, "Spectral Evidence: The Witchcraft Cosmology of Salem Village in 1692," in *Essays in the Sociology of Perception*, ed. Mary Douglas (London: Routledge, 1982), 275–301.

46. These issues are explored more fully in Stuart Clark, *Thinking with Demons: The Idea of Witchcraft in Early Modern Witchcraft* (Oxford: Clarendon Press, 1997), 172–77, 568–70, from which some of the present examples are taken.

God of Israel, or to his Lawgiver Moses; or as if an Hocus-Pocus were so wise a wight as to be consulted as an Oracle."[47]

Much more seriously, giving the Devil so much power to deceive the senses meant that the absurd situation came about—not just in epistemological but in legal terms—in which the difference between reality and virtuality could no longer be guaranteed. At the General Synod of the Hessian clergy in 1582, for example, the Marburg preacher Helferich Herden was arguing, against the views of some of his colleagues, that the confession of a local witch might be discounted because the Devil had the power to represent innocent people at sabbats in the form of their images.[48] A quarter of a century later, during the Inquisition's investigations into witchcraft among the peoples of the Pyrenees, Alonso de Salazar came to the same conclusion in the cases of many hundreds of witches after it had been claimed by witnesses that the Devil could carry off witches to their meetings while they were talking to their neighbors and put counterfeit persons in their places without anyone noticing:

> [I]f we accept the truth of the semblance and metamorphosis, which the witnesses claim that the Devil has effected, the trustworthiness of the witness' statements has been vitiated in advance. That is to say, first the Devil wants to mislead us into thinking that the body of the witch, who is apparently present before the witness, is a counterfeit of the real person who has gone in the meantime to attend the sabbat. Secondly, that witches can pass in front of and approach the witnesses, being invisible when they thus pass through the air before them. In both cases the witness is deprived of the ability to discern the truth, if he relies—as he ought—solely on what he can perceive by his senses.[49]

A quarter century later again, we also find Friedrich Spee building the issue of the Devil's visual simulation of innocent people at sabbats into his wider case for

---

47. "Dr. H.M. his Letter," 31, cf. 41. See also the similar argument against reducing the magic of Pharaoh's magicians (Exod. 7–10) to deception of the eyes in George Sinclair, "To the Reader," in *Satan's invisible world discovered* (1685; repr. Gainesville, Fla.: Scholars' Facsimiles and Reprints, 1969), xxviii, see also xx; and Joseph Glanvill's warning against seeing Moses and Aaron as simply having "more Cunning and Dexterity in the Art of Juggling": *Saducismus triumphatus*, 294.

48. Zika, "Appropriating Folklore," 201.

49. *Second Report of Alonso de Salazar to the Inquisitor General* (Logroño, 24 March 1612), para. 52, see also paras. 9, 46 (English translation kindly provided by Gustav Henningsen). The humanist Pedro de Valencia, who expressed similar doubts to the Inquisition, was a commentator on Cicero's account of philosophical skepticism in the *Academica*; see *Academica sive de iudicio erga verum ex ipsis primis fontibus, opera Petri Valentiae Zafrensis in extreme Baetica* (Antwerp, 1596).

extreme caution in witchcraft trials.[50] As the seventeenth century went by, and certainly after Descartes had started to publish, it was more and more easy to make this kind of skeptical or cautionary case. But the argument was the same. In the 1650s, for example, Ady was saying that orthodox witchcraft belief prevented men and women from believing "their own eyes with confidence"; it brought nothing, he said, "but deceit and cheat upon us, both within and without."[51] In 1677 John Webster argued that to believe in witchcraft led to conditions of perception in which a man would not know "his Father or Mother, his Brethren or Sisters, [or] his Kinsmen or Neighbours."[52] By the early eighteenth century, Francis Hutchinson could add that the Devil's powers over visibility and invisibility reduced the concept of the judicial alibi to "a mere Jest."[53]

■ ■ ■

In many ways, it is indeed Descartes who pulls together the themes discussed here. This essay has suggested that the *praestigium* (prestige), whether human or demonic, was the visual equivalent of the logical and rhetorical paradox. It seems to relate, therefore, both to the "epidemic" of paradox that influenced so many areas of European thought and writing between the mid-sixteenth and mid-seventeenth centuries and to the yet wider interest in veridicality that marked the philosophical, religious, and medical/psychological debates of the period. Descartes most certainly bridged philosophical skepticism and demonology by supposing, in the "demon hypothesis" of the *First Meditation*, that an "evil genius" might be blamed for turning "the heavens, the earth, colors, figures, sound, and all the other external things" into illusions and dreams. The implications of an all-embracing illusion by deity or demon had long been debated, in a skepticism going back to the arguments in Cicero's *Academica*—and they are reminiscent, of course, of modern philosophy's interest in the "brains in a vat" problem.[54] Those among Descartes's near contemporaries who were doubtful of witchcraft's reality as a crime—as well as intellectuals interested in "demonic epistemology" in general—had also established at least the principle of total

---

50. [Friedrich Spee], *Cautio criminalis, seu de processibus contra sagas* (Rinteln, 1631), 331–50.

51. [Thomas Ady], *The doctrine of devils, proved to be the grand apostacy of these later times* (London, 1676), 84, 89, 91–92.

52. John Webster, *The displaying of supposed witchcraft* (London, 1677), 175–76.

53. Francis Hutchinson, *An historical essay concerning witchcraft*, 2d ed. (London, 1720), sig. A4r–v.

54. See esp. Leo Groarke, "Descartes' First Meditation: Something Old, Something New, Something Borrowed," *Journal of the History of Philosophy* 22 (1984): 281–301; and Hilary Putnam, *Reason, Truth and History* (Cambridge: Cambridge University Press, 1981), 1–21.

mental and sensory illusion.[55] Even if Descartes himself thought the idea "metaphysical" and "hyperbolic," he succeeded in giving it a greater philosophical force and a more influential outcome than it had ever had before. Simultaneously, he made *optical* illusion and its remedies a specific focus of his epistemology. For Descartes, Dalia Judovitz has recently written, "illusion in all its forms, be it reflection, *trompe-l'oeil* or artifice, threatens by its deceptive character to impede the search for truth." His *Dioptrics*, published in 1637, may thus be seen as an answer to the twin problems of veridicality and paradoxicality in early modern vision.[56]

Acknowledgments: I would like to thank Claudia Swan of the Department of Art History at Northwestern University for discussions concerning some of the issues raised in this essay and for valuable bibliographical suggestions.

55. Geoffrey Scarre, "Demons, Demonologists and Descartes," *Heythrop Journal* 31 (1990): 3–22.

56. Dalia Judovitz, "Vision, Representation, and Technology in Descartes," in *Modernity and the Hegemony of Vision*, ed. Levin, 63. See also Neil M. Ribe, "Cartesian Optics and the Mastery of Nature," *Isis* 88 (1997): 42–61; and, for a reassessment of the roles of perspective and anamorphosis in Cartesian rationalism, see Lyle Massey, "Anamorphosis through Descartes or Perspective Gone Awry," *Renaissance Quarterly* 50 (1997): 1148–89.

Selected Bibliography of Paracelsiana and Early Modern Science

Collected Works of Paracelsus

Paracelsus. *Bücher und Schriften*. Ed. [Johannes] Huser. 10 vols. Basel, 1589–1590. Also available in a facsimile edition: *Bücher und Schrifften*. Ed. Johannes Huser; foreword by Kurt Goldammer. Hildesheim: Georg Olms, 1971–75.

———. *Chirurgische Bücher und Schrifften*. Ed. Johannes Huser. Strasbourg: Zetzner, 1605. Also available in a facsimile edition: *Bücher und Schrifften*. Ed. Johannes Huser; foreword by Kurt Goldammer. Hildesheim: Georg Olms, 1971–75.

———. *Paracelsus: Sozialethische und sozialpolitische Schriften: Aus dem theologisch-religionsphilosophischen Werk ausgewählt, eingeleitet und mit erklärenden Anmerkungen*. Ed. Kurt Goldammer. Tübingen: J.C.B. Mohr, 1952.

———. *Theophrastus Paracelsus Werke*. 5 vols. Ed. Will-Erich Peuckert. Basel: Schwabe Verlag 1965–68. Reprint, Darmstadt: Wissenschaftliche Buchgesellschaft, 1991.

———. *Theophrast von Hohenheim gen. Paracelsus, Sämtliche Werke. I. Abteilung: Medizinische, naturwissenschaftliche und philosophische Schriften*, 14 vols. Ed. Karl Sudhoff. Munich: Oldenbourg, 1922–33.

———. *Theophrast von Hohenheim gen. Paracelsus, Sämtliche Werke. II. Abteilung: Theologische und religionsphilosophische Schriften*, vol. 1. Ed. Wilhelm Matthießen. Munich: Oldenbourg, 1923.

———. *Theophrast von Hohenheim gen. Paracelsus, Sämtliche Werke. II. Abteilung: Theologische und religionsphilosophische Schriften*, vol. 2–7. Ed. Kurt Goldammer. Wiesbaden: Franz Steiner, 1955–.

Primary Sources in Translation

Paracelsus [Theophrastus Bombastus von Hohenheim], *The Archidoxes of Magic*. London, 1656.

———. *Four Treatises of Theophrastus von Hohenheim called Paracelsus*. Ed. Henry E. Sigerist. Baltimore: Johns Hopkins University Press, 1941. Paperback ed., 1996.

———. *The Hermetic and Alchemical Writings of Aureolus Phillipus Theophrastus Bombast, of Hohenheim, called Paracelsus, the Great....* London: J. Elliott, 1894.

———. *Selected Writings*. Ed. Jolande Jacobi. Princeton: Princeton University Press, 1951.

Bibliographical Aids for Paracelsus Studies

Paulus, Julian. *Paracelsus-Bibliographie 1961–1996*. Heidelberg: Palatina, 1997.

Sudhoff, Karl. *Versuch einer Kritik der Echtheit der Paracelsischen Schriften. I. Theil: Bibliographia Paracelsica: Besprechung der unter Hohenheims Namen 1527–1893 erschienenen Druckschriften*. Berlin, 1894; reprint, Graz: Akademische Druck- und Verlagsanstalt, 1958.

———. *Versuch einer Kritik der Echtheit der Paracelsischen Schriften. 2. Theil: Paracelsus-Handschriften*. Berlin, 1899.

Telle, Joachim. "Bibliographie zum frühneuzeitlichen Paracelsismus unter besonderer Berücksichtigung des deutschen Kulturgebiets." In *Analecta Paracelsica*, ed. Joachim Telle, 557–564. Stuttgart: Franz Steiner, 1994.

Weimann, Karl-Heinz. *Paracelsus-Bibliographie 1932–1960: mit einem Verzeichnis neu entdeckter Paracelsus-Handschriften (1900–1960)*. Wiesbaden: Franz Steiner, 1963.

Secondary Literature

Allen, Don Cameron. *The Star-Crossed Renaissance: The Quarrel about Astrology and Its Influence in England*. Durham: Duke University Press, 1941.

Bakhtin, Mikhail. *Rabelais and His World*. Trans. Helene Iswolsky. Cambridge: MIT Press, 1965.

Barnes, Robin Bruce. *Prophecy and Gnosis: Apocalypticism in the Wake of the Lutheran Reformation*. Stanford: Stanford University Press, 1988.

Bastholm, Eyvind, and Hans Skov, trans. *Petrus Severinus og hans Idea medicinae philosophicae: En dansk paracelsist*. Odense: Odense Universitetsforlag, 1979.

Benzenhöfer, Udo. *Paracelsus*. Reinbek bei Hamburg: Rowohlt, 1997.

———. "Zum Brief des Johannes Oporinus über Paracelsus: Die Bislang älteste bekannte Briefüberlieferung in einer 'Oratio' von Gervasius Marstaller." *Sudhoffs Archiv* 73 (1989): 55–63.

———, ed. *Paracelsus*. Darmstadt: Wissenschaftliche Buchgesellschaft, 1993.

Biegger, Katharina. *"De invocatione Beatae Mariae Virginis": Paracelsus und die Marienverehrung*. Kosmosophie, 6. Stuttgart: Franz Steiner, 1990.

Blaser, Robert-Henri. *Paracelsus in Basel*.... Muttenz /Basel: St. Arbogast Verlag, 1979.

Cassirer, Ernst. *The Individual and the Cosmos in Renaissance Philosophy*. Trans. Mario Domandi. New York: Barnes & Noble, 1963.

Chapman, Allan. "Astrological Medicine." In *Health, Medicine and Mortality in the Sixteenth Century*, ed. Charles Webster, 275–300. Cambridge: Cambridge University Press, 1979.

Clark, Stuart. *Thinking with Demons: The Idea of Witchcraft in Early Modern Europe*. Oxford: Clarendon Press, 1997.

Clulee, Nicholas H. *John Dee's Natural Philosophy: Between Science and Religion*. London: Routledge, 1988.

Cohen, I. Bernard. *Revolution in Science*. Cambridge: Belknap Press of Harvard University Press, 1985.

Conrad, Lawrence I., et al., eds. *The Western Medical Tradition 800 B.C. to A.D. 1800*. Cambridge: Cambridge University Press, 1995.

SELECTED BIBLIOGRAPHY

Copenhaver, Brian P., and Charles B. Schmitt. *Renaissance Philosophy.* Oxford: Oxford University Press, 1992.
Debus, Allen G. *Chemistry, Alchemy and the New Philosophy, 1550–1700: Studies in the History of Science and Medicine.* London: Variorum Reprints, 1987.
———. *The English Paracelsians.* New York: Franklin Watts, 1966.
———. *The French Paracelsians.* Cambridge: Cambridge University Press, 1991.
———. *Man and Nature in the Renaissance.* Cambridge: Cambridge University Press, 1978.
———, ed. *The Chemical Philosophy: Paracelsian Science and Medicine in the Sixteenth and Seventeenth Centuries.* 2 vols. New York: Science History Publications, 1977.
———, ed. *Science, Medicine and Society in the Renaissance.* 2 vols. New York: Science History Publications, 1972.
Debus, Allen G., and Ingrid Merkel, eds. *Hermeticism and the Renaissance: Intellectual History and the Occult in Early Modern Europe.* Washington: Folger Shakespeare Library, 1988.
Debus, Allen G., and Michael Walton, eds. *Reading the Book of Nature: The Other Side of the Scientific Revolution.* Sixteenth Century Essays & Studies, 41. Kirksville, Mo.: Sixteenth Century Journal Publishers, 1998.
Dijksterhuis, E. J. *The Mechanization of the World Picture: Pythagoras to Newton.* Trans. C. Dikshoorn. Oxford: Oxford University Press, 1961.
Dilg, Peter, and Hartmut Rudolph, eds. *Neue Beiträge zur Paracelsus-Forschung.* Stuttgart: Akademie der Diözese Rottenburg-Stuttgart, 1995.
———, eds. *Resultate und Desiderate der Paracelsus-Forschung: Sudhoffs Archiv Beihefte,* no. 31. Stuttgart: Franz Steiner, 1993.
Dilg-Frank, Rosemarie, ed. *Kreatur und Kosmos: Internationale Beiträge zur Paracelsusforschung: Kurt Goldammer zum 65. Geburtstag.* Stuttgart: Fischer, 1981.
Dopsch, Heinz, and Peter F. Kramml, eds. *Paracelsus und Salzburg: Vorträge bei den internationalen Kongressen in Salzburg und Badgasteinan anläßlich des Paracelsus-Jahres 1993.* Salzburg: Gesellschaft für Salzburger Landeskunde, 1994.
Dopsch, Heinz, Kurt Goldammer, and Peter F. Kramml, eds. *Paracelsus (1493–1541): "Keines andern Knecht . . .".* Salzburg: Verlag A. Pustet, 1993.
Domandl, Sepp. *Paracelsus und Paracelsustradition in Salzburg (1524-1976). . . .* Salzburger Beiträge zur Paracelsusforschung 17. Vienna: Verband der wissenschaftlichen Gesellschaften Österreichs, 1977.
———, ed. *Gestalten und Ideen um Paracelsus.* Salzburger Beiträge zur Paracelsusforschung 11. Vienna: Verlag Notring, 1972.
———, ed. *Paracelsus in der Tradition.* Salzburger Beiträge zur Paracelsusforschung 21. Vienna: Verband der wissenschaftlichen Gesellschaften Österreichs, 1980.
———, ed. *Paracelsus—Werk und Wirkung: Festgabe für Kurt Goldammer zum 60. Geburtstag.* Salzburger Beiträge zur Paracelsusforschung 13. Vienna: Verband der wissenschaftlichen Gesellschaften Österreichs, 1975.
Eamon, William. *Science and the Secrets of Nature: Books of Secrets in Medieval and Early Modern Culture.* Princeton: Princeton University Press, 1994.
Easlea, Brian. *Witch-Hunting, Magic and the New Philosophy: An Introduction to the Debates of the Scientific Revolution, 1450–1750.* Atlantic Heights, N.J.: Humanities, 1980.

Selected Bibliography

Eis, Gerhard. *Vor und nach Paracelsus: Untersuchungen über Hohenheims Traditionsverbundenheit und Nachrichten über seine Anhänger.* Medizin in Geschichte und Kultur, 8. Stuttgart: Fischer, 1965.

Evans, R.J.W. *Rudolf II and His World.* 2d ed. Oxford: Oxford University Press, 1984.

Febvre, Lucien. *The Problem of Unbelief in the Sixteenth Century: The Religion of Rabelais.* Trans. Beatrice Gottlieb. Cambridge: Harvard University Press, 1982.

Fellmeth, Ulrich, and Andreas Kotheder, eds. *Paracelsus, Theophrast von Hohenheim: Naturforscher, Arzt, Theologe.* Stuttgart: Wissenschaftliche Verlagsgesellschaft,1993.

French, Peter J. *John Dee: The World of an Elizabethan Magus.* London: Routledge & Kegan Paul, 1972.

Garin, Eugenio. *Astrology in the Renaissance: The Zodiac of Life.* Trans. Carolyn Jackson and June Allen, rev. trans. Clare Robertson and Eugenio Garin. London: Routledge & Kegan Paul, 1984.

———. *Science and Civic Life in the Italian Renaissance.* Trans. Peter Munz. Garden City, N.Y.: Anchor Books, 1969.

Gause, Ute. *Paracelsus (1493–1541): Genese und Entfaltung seiner frühen Theologie.* Tübingen: J.C.B. Mohr, 1993.

Gillispie, Charles Coulston, ed. *Dictionary of Scientific Biography.* New York: Scribner, [1970–80].

Gilly, Carlos. *Cimelia Rhodostaurotica: Die Rosenkreuzer im Spiegel der zwischen 1610 und 1660 entstandenen Handschriften und Drucke.* 2d ed. Amsterdam: Pelikaan, 1995.

———. *Paracelsus in der Bibliotheca Philosophica Hermetica Amsterdam.* Amsterdam: Pelikaan, 1993.

———. "Zwischen Erfahrung und Spekulation: Theodor Zwinger und die religiöse und kulturelle Krise seiner Zeit." *Basler Zeitschrift für Geschichte und Altertumskunde* 77 (1977): 57–137, and 79 (1979): 125–233.

Goldammer, Kurt. *Der göttliche Magier und die Magierin Natur: Religion, Naturmagie und die Anfänge der Naturwissenschaft vom Spätmittelalter bis zur Renaissance; mit Beiträgen zum Magie-Verständnis des Paracelsus.* Kosmosophie, 5. Stuttgart: Franz Steiner, 1991.

———. *Paracelsus: Natur und Offenbarung.* [Hannover-Kirchrode]: T. Oppermann, 1953.

———. *Paracelsus in der deutschen Romantik: eine Untersuchung zur Geschichte der Paracelsus-Rezeption und zu geistesgeschichtlichen Hintergründen der Romantik.* Salzburger Beiträge zur Paracelsusforschung, 20. Vienna: Verband der wissenschaftlichen Gesellschaften Österreichs, 1980.

———. *Paracelsus in neuen Horizonten: Gesammelte Aufsätze.* Salzburger Beiträge zur Paracelsusforschung, 24. Vienna: Verband der wissenschaftlichen Gesellschaften Österreichs, 1986.

Grell, Ole Peter, ed. *Paracelsus: The Man and His Reputation, His Ideas and Their Transformations.* Studies in the History of Christian Thought, 85. Leiden: Brill, 1998.

Grell, Ole Peter, and Andrew Cunningham, eds. *Medicine and the Reformation.* London: Routledge, 1993.

Haas, Alois Maria. "Hohenheims dynamisches Sehen." In *Theophrastus Bombastus von Hohenheim genannt Paracelsus: Standpunkt und Würde: II. Dresdner Symposium,* 12–21. Dresden: Deutsche Bombastus-Gesellschaft, 1999.

———. "Unsichtbares sichtbar machen: Feindschaft und Liebe zum Bild in der Geschichte der Mystik." In *Konstruktionen Sichtbarkeiten. Interventionen*, 8, ed. Jörg Huber and Martin Heller, 265–86. Vienna: Springer, 1999.

Hammond, Mitchell. "The Religious Roots of Paracelsus's Medical Theory." *Archiv für Reformationsgeschichte* 89 (1998): 7–21.

Hannaway, Owen. *The Chemists and the Word: The Didactic Origins of Chemistry*. Baltimore: Johns Hopkins University Press, 1975.

Hemleben, Johannes. *Paracelsus: Revolutionär, Arzt und Christ*. 2d ed. Frauenfeld and Stuttgart: Huber, 1974.

Jütte, Robert. *Ärzte, Heiler und Patienten: Medizinischer Alltag in der frühen Neuzeit*. Munich: Artemis & Winckler, 1991.

———, ed. *Paracelsus heute—im Lichte der Natur*. Heidelberg: Haug, 1994.

Kämmerer, Ernst Wilhelm. *Das Leib-Seele-Geist-Problem bei Paracelsus und einigen Autoren des 17. Jahrhunderts*. Kosmosophie, 3. Wiesbaden: Franz Steiner, 1971.

Keller, Hildegard Elisabeth. "Absonderungen: Mystische Texte als literarische Inszenierung von Geheimnis." In *Deutsche Mystik im abendländischen Zusammenhang: Neu erschlossene Texte, neue methodische Ansätze, neue theoretische Konzepte, Kolloquium, Kloster Fischingen 1998*, ed. Walter Haug and Wolfram Schneider-Lastin. Tübingen: Niemeyer, 2000.

———. *My Secret Is Mine: Studies on Religion and Eros in the German Middle Ages*. Louvain: Peeters, 2000.

———. "Zwo welt in einer Haut: Paracelsische 'Augenlehr' am Beispiel der Frau." In *Nova Acta Paracelsica.... Jahrbuch der Schweizerischen Paracelsus-Gesellschaft*, 14. Bern: Peter Lang, 2000.

Koyré, Alexandre. *From the Closed World to the Infinite Universe*. Baltimore: Johns Hopkins Press, 1957.

———. "Paracelsus." In *The Reformation in Medieval Perspective*, ed. Steven E. Ozment, 185–218. Chicago: Quadrangle Books, 1971.

Kühlmann, Wilhelm, and Wolf-Dieter Müller-Jahncke, eds. *Iliaster: Literatur und Naturkunde in der frühen Neuzeit: Festgabe für Joachim Telle zum 60. Geburtstag*. Heidelberg: Manutius, 1999.

Kühlmann, Wilhelm, and Joachim Telle. "Humanimus und Medizin an der Universität Heidelberg im 16. Jahrhundert." In *Semper Apertus: Sechshundert Jahre Ruprecht-Karls-Universität Heidelberg 1386–1986*, vol. 1, ed. Wilhelm Doerr et al., 255–89. Berlin: Springer, 1985.

———, eds. *Corpus Paracelsisticum: Dokumente frühneuzeitlicher Naturphilosophie in Deutschland*. Bd. 1: *Der Frühparacelsismus*. Erster Teil. Tübingen: Niemeyer, 2001.

Kusukawa, Sachiko. *The Transformation of Natural Philosophy: The Case of Philip Melanchthon*. Cambridge: Cambridge University Press, 1995.

Lindberg, David C. *The Beginnings of Western Science....* Chicago: University of Chicago Press, 1992.

Lindberg, David C., and Ronald L. Numbers, eds. *God and Nature: Historical Essays on the Encounter between Christianity and Science*. Berkeley: University of California Press, 1986.

Lindberg, David C., and Robert S. Westman, eds. *Reappraisals of the Scientific Revolution*. Cambridge: Cambridge University Press, 1990.

Selected Bibliography

Marxer, Norbert. *Praxis statt Theorie! Leben und Werk des Nürnberger Arztes, Alchemikers und Fachschriftstellers Johann Hiskia Cardilucius (1630–1697)*. Heidelberg: Palatina, 2000.

Merkel, Ingrid, and Allen G. Debus, eds. *Hermeticism and the Renaissance: Intellectual History and the Occult in Early Modern Europe*. Washington: Folger Shakespeare Library, 1988.

Midelfort, H.C. Erik. "The Anthropological Roots of Paracelsus' Psychiatry." In *Kreatur und Kosmos*, ed. Rosemarie Dilg-Frank, 67–77. Stuttgart: Fischer, 1981.

———. *A History of Madness in Sixteenth-Century Germany*. Stanford: Stanford University Press, 1999.

Millen, Ron. "The Manifestation of Occult Qualities in the Scientific Revolution." In *Religion, Science, and Worldview...*, ed. Margaret J. Osler and Paul Lawrence Farber, 185–216. Cambridge: Cambridge University Press, 1985.

Miller-Guinsburg, Arlene. "Paracelsian Magic and Theology: A Case Study of the Matthew Commentaries." In *Kreatur und Kosmos*, ed. Rosemarie Dilg-Frank, 125–39. Stuttgart: Fischer, 1981.

Milt, Bernhard. "Conrad Gessner und Paracelsus." *Schweizerische medizinische Wochenschrift* 59 (1929): 486–509.

———. "Paracelsus und Zürich." *Vierteljahrsschrift der Naturforschenden Gesellschaft in Zürich* 86 (1941): 321–354.

Moran, Bruce T. *The Alchemical World of the German Court. Occult Philosophy and Chemical Medicine in the Circle of Moritz of Hessen (1572-1632)*. Stuttgart: Franz Steiner, 1991.

———. *Chemical Pharmacy Enters the University: Johannes Hartmann and the Didactic Care of Chymiatria in the Early Seventeenth Century*. Madison, Wisc.: American Institute of the History of Pharmacy, 1991.

———. "Paracelsus, Religion, and Dissent: The Case of Philipp Homagius and Georg Zimmermann." *Ambix* 43 (1996): 65–79.

———, ed. *Patronage and Institutions: Science, Technology, and Medicine at the European Court, 1500–1750*. Rochester, N.Y.: Boydell Press, 1991.

Müller-Jahncke, Wolf-Dieter. *Astrologisch-Magische Theorie und Praxis in der Heilkunde der frühen Neuzeit*. Sudhoffs Archiv, Beihefte, 25. Stuttgart: Franz Steiner, 1985.

Nauert, Charles Garfield. *Agrippa and the Crisis of Renaissance Thought*. Urbana: University of Illinois Press, 1965.

Newman, William R., and Anthony Grafton, eds. *Secrets of Nature: Astrology and Alchemy in Early Modern Europe*. Cambridge: MIT Press, 2001.

Nutton, Vivian, ed. *Medicine at the Courts of Europe, 1500–1837*. London: Routledge, 1990.

———. "The Seeds of Disease: An Explanation of Contagion and Infection from the Greeks to the Renaissance." *Medical History* 27 (1983): 1–34.

Ohly, Friedrich. *Zur Signaturenlehre der frühen Neuzeit: Bemerkungen zur mittelalterlichen Vorgeschichte und zur Eigenart einer epochalen Denkform in Wissenschaft, Literatur und Kunst*. Stuttgart: Hirzel, 1999.

Pagel, Walter. *Joan Baptista van Helmont: Reformer of Science and Medicine*. Cambridge: Cambridge University Press, 1982.

———. *Paracelsus: An Introduction to Philosophical Medicine in the Era of the Renaissance*. 2d, rev. ed. Basel: Karger, 1982.

———. *Das medizinische Weltbild des Paracelsus: seine Zusammenhänge mit Neuplatonismus und Gnosis.* Wiesbaden: Franz Steiner, 1962.
———. *Religion and Neoplatonism in Renaissance Medicine.* London: Variorum Reprints, 1985.
———. *The Smiling Spleen: Paracelsianism in Storm and Stress.* Basel: Karger, 1984.
Pagel, Walter, and Marianne Winder. "The Higher Elements and Prime Matter in Renaissance Naturalism and in Paracelsus." *Ambix* 21 (1974): 94–127.
Pisa, Karl. *Paracelsus in Österreich: Eine Spurensuche.* St. Pölten: Niederösterreichisches Pressehaus, 1991.
Pörksen, Uwe. "War Paracelsus ein schlechter Schriftsteller?" *Nova Acta Paracelsica* 9 (1995): 25–46.
Principe, Lawrence M. *The Aspiring Adept: Robert Boyle and His Alchemical Quest: Including Boyle's "Lost" Dialogue on the Transmutation of Metals.* Princeton: Princeton University Press, 1998.
Proksch, Johann Karl. *Paracelsus als medizinischer Schriftsteller: Eine Studie.* Vienna: J. Safár, 1911.
Rattansi, Piyo M. "Paracelsus and the Puritan Revolution." *Ambix* 11 (1963): 24–32.
———. "The Helmontian-Galenist Controversy in Restoration England," *Ambix* 12 (1964): 1–23.
Rees, Graham. "Francis Bacon's Semi-Paracelsian Cosmology and the Great Instauration." *Ambix* 22 (1975): 81–101, 161–73.
Rudolph, Hartmut. "Einige Gesichtspunkte zum Thema: Paracelsus und Luther," *Archiv für Reformationsgeschichte* 72 (1981), 34–53.
———. "Theophrast von Hohenheim (Paracelsus)." In *Radikale Reformatoren: Biographische Skizzen von Thomas Müntzer bis Paracelsus,* ed. Hans-Jürgen Goertz, 231–42. Munich: C.H. Beck, 1978.
Schipperges, Heinrich. *Paracelsus: Der Mensch im Licht der Natur.* Stuttgart: Klett, 1974.
Schleiner, Winfried. *Medical Ethics in the Renaissance.* Washington, D.C.: Georgetown University Press, 1995.
Schmidt, C. *Michael Schütz genannt Toxites.* Strasbourg, 1888.
Schmitt, Charles B., et al., eds. *The Cambridge History of Renaissance Philosophy.* Cambridge: Cambridge University Press, 1988.
Schott, Heinz, and Ilana Zinguer, eds. *Paracelsus und seine internationale Rezeption in der frühen Neuzeit: Beiträge zur Geschichte des Paracelsismus.* Brill's Studies in Intellectual History, 86. Leiden: Brill, 1998.
Shackelford, Jole. "Documenting the Factual and the Artifactual: Ole Worm and Public Knowledge." *Endeavour* 23 (1999): 65–71.
———. "Early Reception of Paracelsian Theory: Severinus and Erastus." *Sixteenth Century Journal* 26 (1995): 123–35.
———. "Paracelsianism in Denmark and Norway in the Sixteenth and Seventeenth Centuries." Ph.D. diss., University of Wisconsin, 1989.
———. "Rosicrucianism, Lutheran Orthodoxy, and the Rejection of Paracelsianism in Early Seventeenth-Century Denmark." *Bulletin of the History of Medicine* 70 (1996): 181–204.

———. "Unification and the Chemistry of the Reformation." In *Infinite Boundaries: Order, Disorder, and Reorder in Early Modern German Culture*. Sixteenth Century Essays & Studies, 40, ed. Max Reinhart, 291–312. Kirksville, Mo.: Sixteenth Century Journal Publishers, 1998.

Shumaker, Wayne. *Natural Magic and Modern Science: Four Treatises, 1590–1657*. Binghamton, N.Y.: Center for Medieval and Early Renaissance Studies, 1989.

———. *The Occult Sciences in the Renaissance: A Study in Intellectual Patterns*. Berkeley: University of California Press, 1972.

Siraisi, Nancy G. *Medieval and Early Renaissance Medicine: An Introduction to Knowledge and Practice*. Chicago: University of Chicago Press, 1990.

Smith, Pamela H. *The Business of Alchemy: Science and Culture in the Holy Roman Empire*. Princeton: Princeton University Press, 1994.

Sudhoff, Karl. *Paracelsus: Ein deutsches Lebensbild aus den Tagen der Renaissance*. Meyers kleine Handbücher, 1. Leipzig: Bibliographisches Institut a.g., 1936.

Telle, Joachim. "Bartholomäus Carrichter: Zu Leben und Werk eines deutschen Fachschriftstellers des 16. Jahrhunderts." In *Paracelsus und seine internationale Rezeption in der frühen Neuzeit*. Beiträge zur Geschichte des Paracelsismus, ed. Heinz Schott and Ilana Zinguer, 58–95. Leiden: Brill, 1998. Originally published in *Daphnis* 26 (1997): 715–51.

———. "Kurfürst Ottheinrich, Hans Kilian und Paracelsus: Zum pfälzischen Paracelsismus im 16. Jahrhundert." In *Von Paracelsus zu Goethe und Wilhelm von Humboldt*, Salzburger Beiträge zur Paracelsusforschung, 22, 130–46. Vienna: Verband der wissenschaftlichen Gesellschaften Österreichs, 1981.

———. *Paracelsismus und Alchemie im deutschen Kulturgebiet der Frühen Neuzeit*. Heidelberg: Palatina Verlag, 2002.

———. "Wolfgang Thalhauser: Zu Leben und Werk eines Augsburger Stadtarztes und seinen Beziehungen zu Paracelsus und Schwenckfeld." *Medizinhistorisches Journal* 7 (1972): 1–30.

———, ed. *Analecta Paracelsica: Studien zum Nachleben Theophrast von Hohenheims im deutschen Kultergebiet der frühen Neuzeit*. Heidelberger Studien zur Naturkunde der frühen Neuzeit, Bd. 4. Stuttgart: Franz Steiner, 1994.

———, ed. *Parerga Paracelsica: Paracelsus in Vergangenheit und Gegenwart*. Heidelberger Studien zur Naturkunde der frühen Neuzeit, Bd. 3. Stuttgart: Franz Steiner, 1991.

Temkin, Owsei. "The Elusiveness of Paracelsus." *Bulletin of the History of Medicine* 26 (1952): 201–217.

———. *The Falling Sickness: A History of Epilepsy from the Greeks to the Beginnings of Modern Neurology*. Baltimore, 1945. 2d, rev. ed. Baltimore: Johns Hopkins Press, 1971.

———. *Galenism: Rise and Decline of a Medical Philosophy*. Ithaca: Cornell University Press, 1973.

Tester, S. Jim. *A History of Western Astrology*. Woodbridge, Suffolk: Boydell Press, 1987.

Thomas, Keith. *Religion and the Decline of Magic*. New York: Scribner, 1971.

Thorndike, Lynn. *A History of Magic and Experimental Science*, 8 vols. New York: Macmillan, 1923–58.

Trevor-Roper, Hugh Redwald. "The Court Physician and Paracelsianism." In *Medicine at the Courts of Europe, 1500–1837*, ed. Vivian Nutton, 79–95. London: Routledge, 1990.

———. "The Paracelsian Movement." In *Renaissance Essays*, 149–199. Chicago: University of Chicago Press, 1985.

Vickers, Brian. *Occult and Scientific Mentalities in the Renaissance.* Cambridge: Cambridge University Press, 1984.

Walker, Daniel Pickering. *Spiritual and Demonic Magic from Ficino to Campanella.* London: Warburg Institute, University of London, 1958.

Wear, Andrew, Roger Kenneth French, and Iain M. Lonie, eds. *The Medical Renaissance of the Sixteenth Century.* Cambridge: Cambridge University Press, 1985.

Walton, Michael T. "Boyle and Newton." *Ambix* 27 (1980): 11–18.

Walton, Michael T., and Phyllis J. Walton, "The Geometrical Kabbalahs of John Dee and Johannes Kepler." In *Experiencing Nature...,* ed. Paul H. Theerman and Karen Hunger Parshall, 43–60. Boston: Kluwer Academic, 1997.

Webster, Charles. "Conrad Gesner and the Infidelity of Paracelsus." In *New Perspectives on Renaissance Thought...,* ed. John Henry and Sarah Hutton, 13–23. London: Duckworth, 1990.

———. *From Paracelsus to Newton: Magic and the Making of Modern Science.* Cambridge: Cambridge University Press, 1982.

———. *The Great Instauration: Science, Medicine and Reform, 1626–1660.* New York: Holmes & Meier Publishers, 1976.

———. "Paracelsus: Medicine as Popular Protest." In *Medicine and the Reformation,* ed. Ole Peter Grell and Andrew Cunningham, 57–77. London: Routledge, 1993.

———, ed. *Health, Medicine, and Mortality in the Sixteenth Century.* Cambridge: Cambridge University Press, 1979.

Weeks, Andrew. *Paracelsus: Speculative Theory and the Crisis of the Early Reformation.* Albany: State University of New York Press, 1997.

Williams, Gerhild Scholz. *Defining Dominion: The Discourses of Magic and Witchcraft in Early Modern France and Germany.* Ann Arbor: University of Michigan Press, 1995.

Yates, Frances Amelia. *Giordano Bruno and the Hermetic Tradition.* Chicago: University of Chicago Press, 1964.

Zambelli, Paola, ed. *"Astrologi hallucinati": Stars and the End of the World in Luther's Time.* Berlin: De Gruyter, 1986.

Zedler, J.H., and C.G. Ludovici. *Grosses vollständiges Universal-Lexicon aller Wissenschaften und Künste....* Halle, 1732–50.

Zimmermann, Volker, ed. *Paracelsus: Das Werk—die Rezeption.* Stuttgart: Franz Steiner, 1995.

Index

Note: Page numbers in italics refer to illustrations.

Abendmahlsschriften (Paracelsus), 117–22
Abu Mashar, 158
Academica (Cicero), 243n.49, 244
acrostics, 215
Adam, 57–58, 93, 107, 121
Ady, Thomas, 242, 244
Agricola, Georg, 212
Agrippa, Heinrich Cornelius, xv, 58n.70
Alcocke, Nicholas, 74
Aldrovandi, Ulisse, 48
Alessandrini, Giulio (Julius Alexandrinus von Neustein), 7, 7n.13
"Almanach pour l'an 1541" (Rabelais), 164, 167, 178–79, 179n.69
almanacs, 163–64, 179n.69
Alms House (Augsburg), 24
Alsop, Thomas, 74
amber, 61n.78, 140
am Wald, Georg, 25
anamorphosis, 235, *236*, 237–38
anatomical portrayals, *83*, 83–85
The Anatomical Table of John Banister, 83, 83–84
El Anatomista (Andahazi), 111–12, 112n.87
anatomy, xxi. *See also* Barber-Surgeons' Company of London
Anatomy (Columbo), 84
Anatomy of Melancholy (Burton), 223–24, 231
Andahazi, Federico: *El Anatomista,* 111–12, 112n.87
Andernach, Gunther von, xii
angels, 155
Antichrist, 216, 229
antimony, 19
apocalypticism, 238
Apology for Raymond Sebond (Montaigne), 227
apparitions, 229
Aquinas, Thomas, 240
Arceus, Francisco: *A Most Excellent and Compendious Method of Curing Woundes in the Head,* 79
Aristotelianism
 of Caspar Bartholin, 43, 43nn.21–22, 46, 49–52, 66
 and Paracelsianism, xviii, 37, 42–43, 43nn.21–22, 45–46, 66
Aristotle
 Categories, 46
 on form, matter, and privation, 60
 on imagination, 135–36
 Metaphysica, 46
 Meteors, 46
 on movement/life, 50
 natural philosophy of, 58, 73
 On Generation and Corruption, 46
 On the Heavens, 46
 On the World, 46
 The Organon, 46
 on original knowledge, 58
 Physics, 56–57
 Prior and Posterior Analytics, 46
 Rhetorica, 46
 Topics, 46
Arnald of Villanova, 6
Ars magna lucis et umbrae (Kircher), 237–38
Ars magnesia (Kircher), 141

Index

Asclepius, 167
Aslakssøn, Kort, 38, 40n.15
astrology, xx
   and medicine, 163–64, 164n.4
   predictions via, 164–67, 165n.6
   See also *Disputations against Judicial Astrology; Histories of Gargantua and Pantagruel; 900 Theses*
Astronomia magna (Paracelsus), xix, 117–18, 129–30, 129–30n.46
atomism, 200, 201, 202
Augsburg, 20–21
Augsburg Medical College, xvii, 21–33
   and the apothecary ordinance, 24–25
   barber-surgeons of, 31–32
   and Engelman, 29–30
   and female healers, 31–32
   and Fischer, 30
   and Sebastian Froben, 25–28
   and Herdt, 28–29, 32
   and medical authority, 22–23
   members' responsibilities, 24
   on Paracelsian healers and unofficial medicine, 20–21, 24–33
   and Paracelsus's visit to Augsburg, 21
   and Thalhauser, 21–22
   on women's diseases, 31–32
Augustine, Saint, 124, 128, 189, 240
autopsia, xviii, 72, 72n.2, 79, 92
Avicenna: *Canon,* xii
Ayliffe, Sir John, 74

Bacon, Francis, 53, 53n.54
   New Atlantis, 236
Baker, George
   The Nature and Propertie of Quicksilver, 79
   The Newe Jewell of Health, 79
Bakhtin, Mikhail, 163, 172n.35, 177
Banister, John, 46–47
   on anatomy, 92
   career/influence of, 78, 84
   The Historie of Man, 78–79
   A Needefull, New, and Necessarie Treatise of Chyrurgerie, 78
   on physic vs. surgery, 86–87
   portrait of, *83,* 83–85

baptism, xix, 117–34
   Catholics on, 125, 125n.30, 127, 127n.39
   ceremony of, 118, 123–27, 123n.24, 124–25nn.26–31, 126n.35, 133
   of children, 127–28, 127n.39, 128n.41
   and creation by Christ, 119–22, 120n.11, 132
   effects of, 118, 122, 122nn.21–22, 128–30, 128–30nn.44–46, 128nn.41–42
   exceptions to, 127–28, 127n.39, 133
   by fire, 132–33, 132n.53
   and immortal philosophy, 119–22, 119nn.6–7, 120–21nn.11–13, 121n.16, 122nn.20–22, 133–34
   inclusiveness of, 126–27, 126n.35, 130, 133
   and the *limbus,* 119, 119n.6, 121–22, 122n.20
   and natural philosophy, 129–33
   necessity of, 123–24, 124n.25, 127
   and physicians' sight, 98
   as a sign, 128–29, 128–29nn.44–45
   spiritualist view of, 133, 133n.55
   and virgin birth, 129–30, 129–30n.46
Barber-Surgeons' Company of London, 71–92
   anatomical/surgical regulations by, 71–72, 76–77, 79–80, 81, 91–92
   anatomy/dissection demonstrations by, 79–85, 80–81n.20, 80n.17, 81n.22, *83,* 92
   on apprentices/freemen, training of, 79–81, 81n.22, 92
   autopsia skills of, xviii, 72, 72n.2, 79, 92
   barbers' tasks, 76
   Celsus's influence on, 71
   on empiricism, 76–77, 85–86
   founding of, 72–76, 75
   Galen's influence on, 71, 73, 75–76, 84–86
   involvement in life of London, 73, 73n.4
   and London's health crisis, 73–74

Index

Barber-Surgeons' Company, *continued*
  Masters' election/tasks, 84
  mission of, 73–75, 92
  and physicians vs. surgeons, 72–73, 75–76, 79, 81–82, 86–90
  publications by, 76–77
  on quacks, 71, 72, 77–78, 87–92, 92n.53
  reforms by, 82–83, 85, 91–92
  religious imagery/underpinnings of, 75–76, 85, 92
  reputation/image of, xviii, 71
  size/hierarchy of, 76
  skin diseases treated by, 77
  on surgeons' manners, 81–82
  surgeons' tasks, 76
  syphilis treated by, 77
  on theft/tanning of human skin, 82
  vernacular used by, 85, 87–88
Bartholin, Caspar
  Aristotelianism of, 43, 43nn.21–22, 46, 49–52, 66
  on chemical drugs, 52–53, 52n.52
  De aquis, 51–52
  De mixtione, 51–52
  De philosophiae in medicinae usu et necessitate, 49–50, 52–53
  De studio medico, 46–49
  Enchiridion metaphysicum, 47–48
  Exercitatio disputationis secundae ordinariae, 49, 51
  on fire, 52
  Galenism of, 51–53
  Grell on, 40n.15, 42–44, 46–47, 49, 52, 52n.52
  on iatrochemistry, 46–47, 53
  influences on, 65
  as a medical humanist, 47, 51–52
  on natural philosophy, 41
  and the New Constitutions, 44, 45, 47–48
  Paracelsian influence on, xvii–xviii, 40n.15, 42–43, 46–50, 52–53, 66–67
  Physica generalis praecepta, 50
  Problematum philosophicorum, 50–51
  on Severinus, 46–49, 57
  at University of Copenhagen, 54
  on water, 52, 52n.49
Bartholin, Rasmus, 62, 62n.85
Bartholin, Thomas, 51
  Cista medica hafniensis, 57
Barton, Tamsyn, 164n.4
Basel University, xi–xii
Battus, Levinus, 17
Bauhin, Caspar, 65
Bauhin, Mathiolus, 48
Beaumont, John, 240
Becco, Anne, 193
Beguin, Jean, 48
believers, types of, 120–21n.12
Bellanti, Luca, 159, 160
Benzenhöfer, Udo, xiv
Bernbacher, Mrs. Jacob, 29
Bernheim, Hippolyte, 146
Berrong, Richard M., 165n.7
Bettini, Mario, 236, 237
biblical exegesis, 188–89, 188n.3, 197–98
Biegger, Katharina, xiv
Biel, Gabriel, 127, 127n.39
Biese, Nicolaus, 7, 7n.13
Blockes-Berges Verrichtung (Praetorius), 220–21
Boccaccio, Giovanni: *Genealogy of the Gods,* 213–14
Bodenstein, Adam von, xii, 23
Bodin, Jean, 211
body
  geographic, 178, 178n.68
  interiority of, 181n.86, 182
  mechanical, 178n.68
  as microcosm of the universe, xxi
  and soul, 146
  as a temple that mirrors the cosmos, 184n.95
  as unexplored world, 178
  See also *Histories of Gargantua and Pantagruel*
Boetius de Boodt, Anselm, 48
Bohemian Brethren, 6n.12
Boll, Franz, 160
Bonatti, Antonio, 160
Booke of Observations (Clowes), 79
Bostocke, Richard, 16–17, 71
Boyer, Alain-Michel, 179, 179n.69

Index

Boyle, Robert
  background/education of, 194
  on biblical revelation, xxi
  on borrowing, 198–99
  corpuscularism of, xxi, 188, 200–205, 200n.43, 202n.48
  The Excellency of Theology, 195, 201, 203
  Hebraism of, xxi, 187–88, 194–99, 197–98n.35, 200
  on Jewish readmittance to England, 199
  on Jewish sacrifices, 199–200
  Maimonides's influence on, 199–200
  on mechanism, 201–2, 201n.47
  on natural philosophical basis of Genesis, 202–5
  natural philosophy of, 194–95
  on rabbinical works, 200, 200n.41
  The Sceptical Chymist, 202, 203
  on science, 195
  on *semina,* 188, 202–5, 202n.48
  Some Considerations touching the Style of the Holy Scriptures, 195–97
  theology of, 194–95, 204
Brahe, Tycho, 37–38, 47, 48
"brains in a vat" problem, 244
Brauchius, Balthasar, 14
Brestle, Hans, 26
Brochmand, Jesper, 41, 43n.20, 44, 46
Brosse, Guy de la, 64
Bryson, Norman, 231
Bullard, Melissa, 159n.32
Bullinger, Heinrich, 3, 9–10
Bürgle, David, 25–26, 30
Burke, Peter, 32n.48
Burnett, Duncan: *De praeparatione et compositione medicamentorum chymicorum artificiosa Duncani Bornetti Scoti,* 63–64
Burton, Robert: *Anatomy of Melancholy,* 223–24, 231
Butts, William, 74
Buxtorf, Johannes (Elder and Younger), 188n.3

Cajetan, Tommaso de Vio Gaetani, 127, 127n.39

camera obscura, 234, 237
Campanus, Johannes, 133n.55
Canon (Avicenna), xii
Carlino, Andrea, 84–85
Carlstadt, Andreas Bodenstein von, 133n.55
carnivalesque humor, 172n.35
Cartesian philosophy, 238. *See also* Descartes, René
Casaubon, Meric, 239
Cassirer, Ernst, 160
Categories (Aristotle), 46
Catholics
  on baptism, 125, 125n.30, 127, 127n.39
  Spiritualist, 133, 133n.55
  visual elements of Catholic piety, 228–29
Celestial Hierarchies, 167n.16
celestial spirits/intelligences, 155
Celsus, Aulus Cornelius, 71
Certaine Workes of Chirurgerie (Gale), 77
Cesalpino, Andrea, 48
Cham, 214n.32
Chambre, John, 74
chaos, 142
Charisius, Jonas, 47, 55–56
Charisius, Peter, 47, 48, 56n.62
Charles V, Holy Roman Emperor, 166
chemical philosophy, 35–36
chemistry, xxi
Chirurgia Parva (Lanfranc), 86
Christ
  belief in, 120–21, 120–21nn.12–13
  creation by, 119–22, 120n.11, 132
  and the Trinity, 106, 120
  virgin birth of, 129–30, 129–30n.46
Cicero: *Academica,* 243n.49, 244
Le cinquième livre (Rabelais), 171–72, 171n.34, 176, 181
Cista medica hafniensis (T. Bartholin), 57
civility and education, 82
Clark, Stuart, xxii
Clarke, Samuel, 197
clitoris, 111–12
Clowes, William
  on anatomy, 92
  and Baker, 79

Index

Clowes, William, *continued*
  Booke of Observations, 79
  career/influence of, 78
  De Morbo Gallico, 90
  A Prooved Practice for all Young Chirurgians, 78
  on quacks, 90–91
  A Right Frutefull and Approved Treatise for the Artificiall cure of that malady called in Latin struma, and in English, the evill, cured by Kinges and Queenes of England, 78
Clusius, Carolus (Charles de Lécluse), 48
Colie, Rosalie, 232–35, 234n.24, 236–37
College of Medicine (Augsburg), xvii
College of Physicians, 73–74
Colloquium Ferdinandi regis cum D. Theophrasto Paracelsus Svevo, 8–9, 8–9n.19
Colombo, Mateo Renaldo, 111–12
Columbo, Realdo: *Anatomy,* 84
Commentaries on the Works of Various Authors who spoke of Antiquity (Viterbensis), 213n.31
Company of Barbers, 72, 73. *See also* Barber-Surgeons' Company of London
The Compound of Alchymie (Ripley), 190
contradiction, 234n.24
Controversiarum medicarum exercitationes (O. Worm), 58–60
Coornhert, Dirk, 133n.55
Cop, Martin, 19
Copernicus, Nicolas, 161
corpuscularism, xxi, 188, 200–205, 200n.43, 202n.48
Corpus Hermeticum, 168, 173n.43, 174n.45
Coudert, Allison P., 193
Court of Assistants, 91–92
Crary, Jonathan, 224–25
Crato von Krafftheim, Johannes, 3, 6–9, 8n.17, 14
Craven, William, 161
creation
  by Christ, 119–22, 120n.11, 132
  by God vs. Christ, 119–21
  Helmont on, 193–94

Leibniz on, 193–94
as microcosm, 93, 106–8
vs. procreation, 106–9, 132
Cretan liar, 232, 233
criminals, executed, 80–81, 80nn.17, 19–20
Croll, Oswald, 48, 139–40
Cromwell, Oliver, 214
Cudworth, Ralph, 198
cultural anthropology, 146
curiosity, xi–xii
Curraunce, Helen, 90–91

Daemonologia Rubinzalii (Praetorius), xxi, xxii, 208–22, 217nn.46–47
  acrostics used in, 211, 215–18
  on authenticity of Rübezahl's life, 208–9, 211, 214
  on birth of Rübezahl, 216
  on the Devil, 217–19, 219n.53, 221, 221n.61
  experiential realism of, 209
  eyewitness accounts in, 214
  forms of social distribution used in, 212–13
  on ghosts, 216–17
  on name/nature of Rübezahl, 214, 214n.35
  numbers used in, 216
  on pastimes of Rübezahl, 218, 218n.50
  popularity of, 208
  and the preternatural in early modern natural history, 209–10
  publication of, 207–8
  on the Riesengebirge and its people, 218–20, 219n.54, 220n.56
  on Rübezahl as elemental spirit, 216
  on Rübezahl as mountain sprite/demon, 212, 215, 221
  on Rübezahl as shapeshifter, 220
  shared forms of externalization used in, 212
  shared ideas used in, 212
  sources for, 211–12
  on water spirits, 212
  on wealth of Rübezahl, 218, 219

Index

Daemonologia Rubinzalii (Praetorius), continued
   on weather making by Rübezahl, 221, 221n.58
   on will-o'-the-wisps, 221–22
d'Aguilon, François, 223–24, 237
Daigeler, Hans, 26
da Monte, Giovanni Battista, 7
Daniel, Dane Thor, xix
Daston, Lorraine, 208
Deacon, John: *Dialogicall discourses of spirits and divels,* 239–40
De amore (Ficino), 136
De aquis (C. Bartholin), 51–52
Debus, Allen, xiv, 16, 49
De Caduco Matricis (Paracelsus), 94–95
De causis morborum invisibilium (Paracelsus), 138–39
Dee, John, 192
"La défense et illustration de la langue française" (Du Bellay), 181n.80
De genealogia Christi (Paracelsus), 117–18, 120, 127, 131–32
De imaginatione oratio (Nymann), 231
De la dissection des parties du corps humain (Estienne), 111
Della Porta, Giambattista, 48, 136
   Magia naturalis, 237
Della Torre, Raffaele: *Tractatus de potestate ecclesiae coercendi daemones,* 240–41
Del Río, Martín, 223–24
De magnete, magnetisque corporibus, et de magno magnete tellure (Gilbert), 140
De magnetica vulnerum... curatione (Helmont), 142
De Matrice (Paracelsus), 95, 108–9
De methodo medendi as The institucion of chyrurgerie (Galen of Pergamum), 77
De mixtione (C. Bartholin), 51–52
Democrates, 201
demonology
   and landscape, 220–21
   and vision, xxii, 229, 231, 232, 238–45, 240n.39
De Morbo Gallico (Clowes), 90
Denmark, 41–42. *See also* under Paracelsianism

De philosophiae in medicinae usu et necessitate (C. Bartholin), 49–50, 52–53
De praeparatione et compositione medicamentorum chymicorum artificiosa Duncani Bornetti Scoti (Burnett), 63–64
Descartes, René, 146, 178n.68, 201, 226
   Dioptrics, 245
   First Meditation, 223–24, 227, 244–45
despair, 139
Despasianus, 31
Dessen von Kronenburg, Bernhard: *Medicina veteris et rationalis,* 16
De studio medico (C. Bartholin), 46–49
Devil. *See* demonology
De viribus imaginationis tractatus (Fienus), 231
diagnostic gaze, xix, 95–96, 101–6
Dialogicall discourses of spirits and divels (Deacon and J. Walker), 239–40
digestion, 107, 107n.66
Dioptrics (Descartes), 245
diseases
   belief as producing, 139, 144
   categories/definition of, 59
   seeds of, 142–43
Disputations against Judicial Astrology (Pico), xx, 152–56, 158–62
Disputations on the New Medicine of P. Paracelsus (Erastus), xvi–xvii, 3–17
   on alcoholism/impiety of Paracelsus, 9–12
   Bullinger's account in, 9–10
   on childhood of Paracelsus, 6
   Crato's accounts in, 6–9, 8n.17, 14
   influence of, 15–16, 16n.35
   on magic practiced by Paracelsus, 9–12, 15
   on malpractice by Paracelsus, 7, 13–15
   on Paracelsus in Bavaria, 12–14
   on Paracelsus in Hungary, 7
   on Paracelsus in Moravia, 7–8, 7n.14
   on Paracelsus in Vienna, 8
   on Paracelsus in Zurich, 9–10
   Recklau's account in, 12–14

Index

Disputations on the New Medicine..., continued
  reliability of, 4–5, 15–18
  short testimonies in, 14–15, 15n.34
  sources for, 5–6
  Vetter's account in, 10–12, 11n.23, 17
doctors, astrology used by, 163–64
Donne, John, 93
Dorn, Gerhard, xii
drinking metaphor, 180–81, 181nn.76–77
Du Bellay, Joachim: "La défense et illustration de la langue française," 181n.80
Duden, Barbara, 112–13

Eamon, William, 102
education and civility, 82
Elixir Tychonis Brahei, 48
empiricism
  Barber-Surgeons on, 76–77, 85–86
  of Galen, 63, 66
  and Paracelsianism, 53–54, 62–63, 65–66
  of Ole Worm, 65–66
Encelius (Christoph Entzel), 48
Enchiridion metaphysicum (C. Bartholin), 47–48
Engelman, Matthias, 29–30
English Civil War (1642–49), xiii–xiv
epistemology and loving, 100, 101n.37
Epistola Scripta Theophrasto Paracelso (P. Severinus), 58n.70
Erasmus, Desiderius, xi, 82, 133n.55
Erastianism, 4
Erastus, Thomas, xvii, 4, 67, 191n.11. See also *Disputations on the New Medicine of P. Paracelsus*
Estienne, Charles: *De la dissection des parties du corps humain,* 111
eternal body through baptism. *See* baptism
ether, 141
ethnology, 146
Etliche Tractaten (Paracelsus), *116*
"evil genius" argument, 244. See also *First Meditation*
Evonymus (Gesner), 79

The Excellency of Theology (Boyle), 195, 201, 203
Exercitatio disputationis secundae ordinariae (C. Bartholin), 49, 51
experience, xi–xii
experiential realism, 209
Ezekiel, 182n.82
Fagius, Paul, 188n.3
Febvre, Lucien, 163, 163n.2
Fechner, Johannes, 211
felons, executed, 80–81, 80nn.17, 19–20
Ferdinand I, Holy Roman Emperor, 7, 7n.13, 8–9
Fernel, Jean François, 48
  Pathologia, 78
Ferris, Richard, 74
Festugière, A.J., 172–73n.37
Ficino, Marsilio, 58n.70
  on astral influences, 184n.95
  De amore, 136
  On Obtaining Life from the Heavens, 153, 157, 158, 159n.32, 161
  and Pico, on astrology, 153–61
Fienus, Thomas: *De viribus imaginationis tractatus,* 231
Fincke, Thomas, 41, 44n.24, 56n.62
First Meditation (Descartes), 223–24, 227, 244–45
Fischer, Hans, 30
Flood, 214n.32
Fludd, Robert, 58n.70
  Mosaicall Philosophy, 191
force, concept of, 141
Foss, Niels Christensen, 64n.93
Foucault, Michel, 224
Francis I, king of France, 166
Franck, Sebastian, 133n.55
Freud, Sigmund, 143n.37, 146
  "The Uncanny," 103
Friedrich, Prince, 28, 29
Friis, Christian, 41, 55–56, 56n.62, 57
Froben, Johannes, xi, xii, 14
Froben, Sebastian, 25–28, 30
"fundament" as the grave, 177

Gale, Thomas
  on anatomy, 92
  career/influence of, 77–78

Gale, Thomas, *continued*
    Certaine Workes of Chirurgerie, 77
    empiricism of, 86
    on physic vs. surgery, 86–87
    on quacks, 87
Galenism
    of Caspar Bartholin, 51–53
    vs. Paracelsianism, 36, 37, 42–43, 66
    of Ole Worm, 60, 63, 66
Galen of Pergamum, 48
    Barber-Surgeons as influenced by, 71, 73, 75–76, 84, 85, 86
    De methodo medendi as *The institucion of chyrurgerie,* 77
    empiricism of, 63, 66
    on physic vs. surgery, 87
Gantenbein, Urs Leo, xiv
Gargantua. See *Histories of Gargantua and Pantagruel*
Garin, Eugenio, 156–58
gases, 142
Gassendi, Pierre, 66, 67
Gause, Ute, xiv
Geber: *The Summa Perfectionis,* 201
gendered epistemology, xviii–xix. See also microcosma, seeing of
gendering of nature, 218n.51
Genealogy of the Gods (Boccaccio), 213–14
Genesis, xxi
    and microcosma, 93, 95, 106–9
    natural philosophical basis of, 190–94, 202–5
genitalia, 102–3, 110, 111n.78
Geoffrey of Monmouth: *History of the Kings of Britain,* 213–14
geographic body, 178, 178n.68
Gerard, John: *Herball,* 79
Gerson, Jean de, 127, 127n.39
Gesner, Konrad, 3, 5, 10, 48, 79
    Evonymus, 79
ghosts, 229–30
giants, 213–15, 214n.32. See also *Daemonologia Rubinzalii; Histories of Gargantua and Pantagruel;* Rübezahl
Gilbert, William, xx
    De magnete, magnetisque corporibus, et de magno magnete tellure, 140

Gilly, Carlos, xiv, 67, 68
Glanville, Joseph, 201
Glauber, Johann R., 203n.52
God
    belief in, 120–21nn.12–13
    name of, 167–68, 167–68n.18
    signs of, 168–72, 170nn.29–30, 171n.32, 173, 180, 185–86
    as a sphere without circumference, 185, 185n.97
Gog, 213
Gogmagog, 213n.31
Goldammer, Kurt, xiv, xix
    on divine-spiritual corporeality of Christ, 120
    on knowledge and loving, 100, 101n.37
Goldish, Matt, 189–90, 193
Goliath, 213
gravity, 141
Greaves, Thomas, 197
Greek skepticism, 227–28
Grell, Ole Peter, xiv
    on Bartholin, 40n.15, 42–44, 46–47, 49, 52, 52n.52
    on Danish Lutheranism, 41–42
    on iatrochemistry, 46–47
    on the New Constitutions, 44
    on Paracelsianism, generally, 39n.11, 40, 40n.15, 66, 67
    on Resen, 40n.15
    on Worm, 39n.11, 40n.15, 54–55, 55n.60, 57–59, 62–65, 65n.96
Die Grosse Wundarznei (Paracelsus), *x*, xii, 21–22
Gsell, Monika, 111n.78
Guide for the Perplexed (Maimonides), 199–200
Gunnoe, Charles D., Jr., xvi–xvii
gynecology. See *De Caduco Matricis; De Matrice;* microcosma, seeing of

Habsburg-Valois wars (1494–1559), 166
Halle, John
    career/influence, 77–78
    empiricism of, 86
    on English vs. Latin, 86–88

Index

Halle, John, *continued*
   An Historiall Expostulation: Against the bestlye Abusers, both of Chyrurgerie and Physyke in oure time, 77, 88–90
   on physic vs. surgery, 86–87, 89–90
Hammond, Mitchell, xvii
hangings, 80–81n.20
Harman, Edmund, 74
Harmonices mundi/The Harmony of the World (Kepler), 161, 192–93
Hartmann, Johann, 46–49, 59, 61, 64, 64n.93
Hebrew
   and Genesis and natural philosophy, 190–94, 202–5
   goals of studying, 188–90, 188n.3, 197–98
   See also Boyle, Robert, Hebraism of
Heimliche (arcana), 101–5, 101n.40, 113–14
Helmont, Johann Baptist van. *See* van Helmont, Johann Baptist
Helmontianism, 36n.1
Henry VIII, king of England, 72, 74–75, 75, 77
Heptaplus (Pico), 152, 153, 156–59, 160, 161
Herball (Gerard), 79
Herden, Helferich, 243
Herdt, Georg, 28–29, 32
Hermeticism, 57–58, 58n.70
   dualisms of, 174, 174n.45
   on the Fall, 174, 174n.45
   Hermetic texts, 167–68, 167–68nn.16–17, 172
   on materiality, 171–74, 172–74nn.35, 37, 41, 45, 180–81, 181n.76
   on the spheres, 173, 173n.43
Hess, Tobias, 31
Hilliard, Nicholas, *83*
Hippocrates, 48, 73, 87
An Historiall Expostulation: Against the bestlye Abusers, both of Chyrurgerie and Physyke in oure time (Halle), 77, 88–90
The Historie of Man (Banister), 78–79

The Histories of Gargantua and Pantagruel (Rabelais), 163–86, 213–14
   on the body as microcosm of the universe, 178–79, 185–86
   on the body's physicality, 172–78, 180, 185–86
   on the body's physical/spiritual workings, 179–80
   Le cinquième livre, 171–72, 171n.34, 176, 181
   Her Trippa in, 174–78, 180
   on imagination, 179–80
   Panurge's oracle experience, 183–84, 183n.90
   prophecy in, 170–71
   on signs of God, 168–72, 170nn.29–30, 171n.32, 180, 185–86
   on the spheres/zodiac, 184–85
   on stellar signs vs. human body's divinely inspired knowledge, xx–xxi
   the sybil in, 174, 176–77, 176n.56, 177n.61, 180
   the temple of the oracle in, 184, 184n.95
   Thaumaste's debate with Panurge, 168–71, 170nn.29–30
   Le tiers livre, 163, 165, 171, 173–74, 178–80, 184
   wine metaphor in, 180–86
History of the Kings of Britain (Geoffrey of Monmouth), 213–14
Hobbes, Thomas, 223
Hocker, Jodocus: *Der Teuffel selbst*, 221n.61
Hohenheim, Theophrastus Bombastus von. *See* Paracelsus
Holbein the Younger, Hans, 74–76, *75*, 75n.9
homunculi, 107–8, 108n.67
Horst, Johann Daniel, 61n.78, 69
Hundred Years War (1330s–1450s), 72
Hunnius, Nicolaus, 40n.12
Huntington, Robert, 197
Huser, Johann, 23
Hutchinson, Francis, 244
Hyde, Thomas, 197, 199

Index

hysteria, 97, 97n.18, 143n.37

iatrochemistry, 35, 36, 38, 48, 66
Idea medicinae (P. Severinus), 23, 60, 67, 68, 190–91
Les illustrations de Gaule et singularitez de Troye (Lemaire de Belges), 213–14
imagination, 135–47
    Aristotle on, 135–36
    centrality to history of medicine, 135–36
    and demonic possession, 136
    Helmont on, 137, 142–47
    and magic, 136–40
    and magnetism, xix–xx, 136–37, 137n.9, 140–42, 144–47
    and mass psychology, 139, 146, 146n.48
    in men and women, 143–44, 143n.37
    pejorative meaning of, 140n.24
    power of, 135, 142–47
    Rabelais on, 179–80
    and suggestion, 146
    and vision, 230, 239
immortal corporeality/philosophy. *See* baptism
Isidore of Seville, 235–36

Jay, Martin, 224
Jerome, Saint, 188n.3
Jersin, Jens Dinesen, 64–65, 65n.96
Jesuits, 236–37
Jesus. *See* Christ
Jewish mysticism. *See* Kabbalah
Jews
    conversion by, 199
    conversion of, 188, 188n.3
    readmittance to England, 199
    sacrifices by, 199–200
Jones, G. Lloyd, 189
Jonsson, Arngrim, 61
Jorden, Edward, 52n.49
judicial astrology. *See Disputations against Judicial Astrology*
Judovitz, Dalia, 245
Jütte, Robert, xiv

Kabbalah, 152, 167–68n.18, 188, 188n.3, 191–93
Karthauser, Alexander, 13
Kaske, Carol, 160
Keller, Hildegard Elisabeth, xiv, xviii–xix
Kemp, Martin, 237
Kepler, Johannes, xx, 140, 141, 160, 223–24
    Harmonices mundi, 192–93
    The Harmony of the World, 161
    On the New Star, 161
Kircher, Athanasius, xv, xx, 140, 236
    Ars magna lucis et umbrae, 237–38
    Ars magnesia, 141
Kolbenheyer, Erwin Guido, xiii
Kornerup, Bjørn, 43, 43nn.20–22
Krag, Ander, 38
Krämer (Heinrich Institutoris), 211
Kristeller, Paul Oskar, 160

Lacan, Jacques, 111n.78, 224
Lakoff, George, 209
landscape and demonology, 220–21
Lanfranc, 85
    Chirurgia Parva, 86
Langlot, Joel, 62nn.85–86
languages, ancient vs. modern, 181–82, 181n.80
LeBon, Gustave, 146n.48
Leibniz, Gottfried Wilhelm, 193–94
Leipa, Berthold von, 6, 7, 9
Leipa, Johanna von, 7n.15
Leipa, Johann von, 7
Leipa family, 6n.12, 7n.15
Leipzig, 207
Le Loyer, Pierre: *Quatre livres des spectres*, 223–24
Lemaire de Belges, Jean: *Les illustrations de Gaule et singularitez de Troye*, 213–14
Leucippus, 201
Lewis, Wyndham, 163
Libavius, Andreas, 16, 48–49, 59–60
Libellus de baptismate Christiano (Paracelsus), 117–18, 123, 123n.24, 125
Life of Apollonius (Philostratus), 240
limbus, 119, 119n.6, 121–22, 122n.20
Linacre, Thomas, 73

Index

Lobelius, 48
lodestone, 140
London health crisis, 73–74. *See also* Barber-Surgeons' Company of London
Longomontanus, Christian, 47
Lord's Supper, 121, 121n.16
Lucretius, 201
Ludwig of Hessen-Darmstadt, 54n.56
Lull, Ramon, 6
Lundt, Bea, xiv
Luther, Martin, 7, 124, 125n.30
Lutheranism, 39, 41–42, 41n.16, 46
lychanthropy, 230

Macbeth (Shakespeare), 223–24
macrocosm, *x*, xix
Magenbuch, Johann, 14, 17
Magia naturalis (Della Porta), 237
Magia optica (G. Schott), 237
magic
    astral, xx, 154
    black vs. white, xx, 136
    demonic, 235, 240, 240n.39
    and imagination, 136–40
    and magnetism, 136–37
    natural, 235–38
    and vision, 226, 235–38
magic lanterns, 237
Magirus, Johannes, 47–48
magnetism
    animal (mesmerism), 140n.24, 141, 146
    vs. electrics, 140
    and imagination, xix–xx, 136–37, 137n.9, 140–42, 144–47
    and magic, 136–37
    and natural philosophy, 135
    as world soul, 140–42
Magog, 213
Maignan, Emanuel, 237
Maimonides: *Guide for the Perplexed*, 199–200
male hysteria, 143n.37
Manichaean dualism, 174n.45
Margaritha, Anthonius, 188n.3
Marichal, Robert, 165n.8
Mary, 95

mass psychology, 139, 146, 146n.48
Masten, Jeffrey, 177
Masters, G. Mallary, 181n.77
materiality
    Hermeticism on, 171–74, 172–74nn.35, 37, 41, 45, 180–81, 181n.76
    Pico on, 173, 173n.41, 186
Mauerkirche (walled church), xi
Maximilian II, Holy Roman Emperor, 7, 7n.13
mechanical body, 178n.68
Medicina veteris et rationalis (Dessen von Kronenburg), 16
medicine
    gendered (*see* microcosma, seeing of)
    history of, xii–xiii
    humanist, 63
    orthodox vs. unorthodox, 19, 19n.2
    professionalization of, xviii
    See also Augsburg Medical College; Barber-Surgeons' Company of London
Medigo, Elia del, 152
melancholy, 143–44, 230
Melanchthon, Philipp, 7, 38, 39, 211
Melusine of Lusignan, 212n.23
Menassah ben Israel, 194, 194n.21
Mentzer, Balthasar, 54n.56
Mephistopheles, 210
Mersenne, Marin, 227
Merton, Robert K., 204n.53
Mesmer, Franz Anton, xx, 140n.24, 146
mesmerism, 140n.24, 141, 146
Metaphysica (Aristotle), 46
Meteorology (Paracelsus), 190
Meteors (Aristotle), 46
Mettra, Claude, 178
microcosma, seeing of, 93–115
    creation as microcosm, 93, 106–8
    and creation vs. procreation, 106–9
    and the differences between the sexes, 106–9, 111, 111n.78
    epistemological/ethical implications of, 100
    and the eye as an organ, 96–100, 96n.12

Paracelsian Moments   267

microcosma, seeing of, *continued*
   and gynecology/obstetrics, disputes about, 110–12
   and the heart, 97–98
   and Heimliche, 101–5, 101n.40, 113–14
   and history of anatomy, 112–13
   and hysteria, 97, 97n.18
   and inner seeing, 101–2, 101n.40
   macro- vs. microcosm, xix, 95
   meaning of, 93
   and physicians' blindness, 97–98, 104
   and the physician's gaze, xix, 95–96, 101–6
   and *secreta mulierum*, 94, 103–4
   and secrets, 99, 101–2
   and signs, allocation/reading of, 104–5, 115
   as taboo, 110
   and the Trinity, 106
   and the uterus, 98–100
   and visual competence via experience, 113–15
   and the woman's matrix, xix, 93–94, 98–100, 99n.30
Midelfort, H.C. Erik, xix
midwives, 110
Miller, David, 26
miracles, 228–29
mirrors, 233, 234
Mondeville, Henri de, 85
Monforde, James, 74
Montaigne, Michel de, 223–24, 228
   Apology for Raymond Sebond, 227
Moran, Bruce, xiv
Moran, Jean, 199–200n.41
Moravsky Krumlov (Moravia), 7, 7n.14
More, Henry, 241–43
More, Thomas, 82
Moritz, 54n.56
Mosaicall Philosophy (Fludd), 191
Moschus the Phoenician, 201, 202
Moses, 58n.70
A Most Excellent and Compendious Method of Curing Woundes in the Head (Arceus), 79

A Most Excellent and Compendious Method of Curing Woundes in the Head (Read), 88
mountain sprites/spirits, xv, 212
Muenster, Sebastian, 188n.3
Museo Circheriano (Rome), 238

natural philosophy
   Aristotelian, 58, 73
   and baptism, 129–33
   of Boyle, 194–95
   Genesis as based on, 190–94, 202–5
   magnetism as essential to, 135
   and Paracelsianism, 41–43, 131
The Nature and Propertie of Quicksilver (Baker), 79
A Needefull, New, and Necessarie Treatise of Chyrurgerie (Banister), 78
Neoplatonism, xvi, 57–58, 152, 174n.45
Neustein, Julius Alexandrinus von (Giulio Alessandrini), 7, 7n.13
New Atlantis (Bacon), 236
New Constitutions (Denmark, 1621), 44–48
The Newe Jewell of Health (Baker), 79
Newman, William, 107–8, 201
Newton, Isaac, 141, 193, 194, 223
   Opticks, 205
   Principia, 205, 205n.57
New World, discovery of, 209–10
Nicéron, Jean-François, 223–24, 226–27, 237
Nichols, Stephen, 211n.18
Nicol, Robert, 89
Nicolas of Cusa, 168, 171, 185n.97
Nicolas of Lyra, 188n.3
nightmares, 230
900 Theses (Pico), xx, 151–53, 156, 161–62
Noah's sons, 214n.32
Northern Renaissance, xvi
Nosologia (Petraeus), 48
Nymann, Jerome: *De imaginatione oratio*, 231
nymphs, xv

obstetrics, disputes about, 110–12
Oeser, Erhard, 160

Oldenburg, Henry, 194–95, 194n.21, 197–98n.35
Old-vampirism, 137n.8
On Generation and Corruption (Aristotle), 46
On Obtaining Life from the Heavens (Ficino), 153, 157, 158, 159n.32, 161
On the Heavens (Aristotle), 46
On the invisible diseases (Paracelsus), 140
On the New Star (Kepler), 161
On the Three Languages (Wakefield), 189
On the World (Aristotle), 46
Opitz, Martin, 211
    Poetische Wälder, 214–15
Oporinus, Johannes, 3, 5, 10, 17
Opticks (Newton), 205
Opus Paramirum (Paracelsus), 106
"Oration on the Dignity of Man" (Pico), 168, 173n.41, 180
Oresme, Nicole, 155
The Organon (Aristotle), 46
Orpheus, 58n.70
Ortus medicinae (Helmont), 144–45
Outlines of Pyrrhonism (Sextus Empiricus), 227

Pabst, G. W.: *Paracelsus*, xiii
Pagel, Walter, xiv, 133, 137, 142, 144
Pagnini, Sanctu, 188n.3
palingenesis, 62–63, 62nn.85–86
Paludanus, Johannes, 64n.94
Pantagruel. See *Histories of Gargantua and Pantagruel*
Pantagrueline Prognostication pour l'an 1533 (Rabelais), 163, 166–67, 166n.13
Panthaleon, Dr., 13
Paracelsianism, xvii–xviii, 35–69
    and Aristotelianism, xviii, 37, 42–43, 43nn.21–22, 45–46, 66
    as chemical philosophy, 35–36
    definitions of, 35–37, 66
    and empiricism, 53–54, 62–63
    vs. Galenism, 36, 37, 42–43, 66
    Grell on, 39n.11, 40, 40n.15, 66, 67
    vs. Helmontianism, 36n.1
    as iatrochemistry, 35, 36, 38, 48, 66
    as an ideology, 67
    on life in cadavers, 50
    and Lutheran orthodoxy, 39, 41–42, 41n.16, 46
    and natural philosophy, 41–43, 131 (*see also* natural philosophy)
    and the New Constitutions, 44–48
    and palingenesis, 62–63, 62nn.85–86
    Pumfrey on, 35–36
    and Rosecrucianism, 39–40, 39–40nn.11–12, 40n.15
    on salt, sulphur, and mercury *(tria prima)*, 60
    scientific/philosophical vs. religious, 40, 40n.15, 68, 117n.1, 130–31
    in sixteenth- to seventeenth-century Denmark, 37–40, 39–40nn.11–12, 40n.15
    at University of Copenhagen, 38–39, 41, 67, 69
    See also Bartholin, Caspar; Worm, Ole
Paracelsus, *116*
    anti-Semitism of, 8–9n.19
    at Basel University, xi–xii, 23
    biographies of (see *Disputations on the New Medicine of P. Paracelsus*)
    Catholicism of, xi
    criticism of, 3 (see also *Disputations on the New Medicine of P. Paracelsus*)
    death of, xii
    influence of, xii–xvi, 19–20, 23–24
    as the "Luther of medicine," 17, 17n.39
    medical iconoclasm of, xi–xii, xvii
    piety of, xv–xvi
    scientific vs. religious worldview of, xiii
    solitude of, 105
    wanderings of, xii
    See also Paracelsus, works of
Paracelsus (film; Pabst), xiii
Paracelsus, works of
    Abendmahlsschriften, 117–22
    Astronomia magna, xix, 117–18, 129–30, 129–30n.46
    De Caduco Matricis, 94–95
    De causis morborum invisibilium, 138–39

Paracelsus, works of, *continued*
    De genealogia Christi, 117–18, 120, 127, 131–32
    De Matrice, 95, 108–9
    Etliche Tractaten, 116
    Die Grosse Wundarznei, x, xii, 21–22
    Libellus de baptismate Christiano, 117–18, 123, 123n.24, 125
    Meteorology, 190
    On the invisible diseases, 140
    Opus Paramirum, 106
    Philosophia magna, 139
    Prophecien und Weissagungen, 206
    publication of, xiv–xv, 23, 117n.1
    Three Books of Philosophy to the Athenians, 190
    Vom tauf der Christen, 117–18, 129n.45
    Vom tauf der Christen: De baptismate, 117–18, 123–24, 124n.25, 124nn.27–30, 128nn.41–42
Paradossi per Prattticare la Prospettiva senza saperla (Troili), 236
paradoxy, 232–37, 234n.24, 241, 244
Paré, Ambroise, 103
Parrhasius, 234, 241
Pathologia (Fernel), 78
Paul of Burgos, 188n.3
Payngk, Peter, 38, 65
Payne, Lynda, xviii
Pellican, Conrad, 188n.3
Pelling, Margaret, 73–74
Pen, John, 74
Penotus, 48
Perrault, François, 240n.39
Petraeus, Heinrich: *Nosologia,* 48
Pett, Sir Peter, 197, 198, 200
Pfitzer, Johannes, xvi, 210n.15
philolology, 211n.18
philosophers' stone, 190
Philosophia magna (Paracelsus), 139
Philostratus: *Life of Apollonius,* 240
Physica generalis praecepta (C. Bartholin), 50
physicians. *See* Barber-Surgeons' Company of London
physician's gaze, xix, 95–96, 101–6
Physics (Aristotle), 56–57

Pico della Mirandola, Giovanni, xv, 58n.70
    Disputations against Judicial Astrology, xx, 152–56, 158–62
    and Ficino, 153–61
    Hebrew studied by, 188n.3
    Heptaplus, 152, 153, 156–61
    on materiality, 173, 173n.41, 186
    on natural magic, 172–73
    900 Theses, xx, 151–53, 156, 161–62
    "Oration on the Dignity of Man," 168, 173n.41, 180
    and Savonarola, 160–61
    on stars, physical effects of, xx
    wine metaphor used by, 180–81
Placket, Hewe, 92n.53
plague
    belief as cause of, 139, 144
    metaphor of, xxi
    outbreaks of, 73, 75
Plato, 58, 60
Platter, Felix, 65
Plotinus, 154
Pocoke, Edward, 197
Poetische Wälder (Opitz), 214–15
Poliziano, Angelo, 156
Pontano, Giovanni, 159
Popkin, Richard, 227
popular culture, diffusion of, 32n.48
Pörksen, Gunhild, xiv
Postel, Guillaume, 133n.55, 188n.3
Praetorius, Johannes, xv, xvi
    Blockes-Berges Verrichtung, 220–21
    as populizer of marvels, 207, 208, 222
    See also Daemonologia Rubinzalii
Pratensis, Johannes, 37
prestiges, 235, 237, 240, 244
preternatural beings, place/knowledge of, 210, 210n.13
Principia (Newton), 205, 205n.57
Prior and Posterior Analytics (Aristotle), 46
prisca doctrina, 188
prisca sapientia (original knowledge), 57–58, 58n.70
prisci theologi, 58n.70
Problematum philosophicorum (C. Bartholin), 50–51
procreation vs. creation, 106–9, 132

Index

A Profitable Treatise of the Anatomie of Mans Body (Vicary), 77, 85
Prognostication Perpetuelle, 164
A Prooved Practice for all Young Chirurgians (Clowes), 78
Prophecien und Weissagungen (Paracelsus), 206
Protestant Reformation, xi, 228–30, 230n.15
psychoanalysis, 111n.78, 146, 224
Ptolemy: *Tetrabiblos*, 158
Pumfrey, Stephen, xvii, 35–36
Pyrrhonism, 227–28

quacks, 71, 72, 77–78, 87–92, 92n.53
Quatre livres des spectres (Le Loyer), 223–24
Quercetanus, 48, 62, 62n.85

Rabelais, François, xv, 165n.7
    "Almanach pour l'an 1541," 164, 167, 178–79, 179n.69
    on astrology/almanacs, 163–67, 186
    on conjunctions, eclipse, and comets, 165–66
    as a doctor, 163–64, 178, 186
    Hermeticism of, 165n.8
    Neoplatonism of, 165n.8
    Pantagrueline Prognostication pour l'an 1533, 163, 166–67, 166n.13
    satirical almanacs of, 165–66
    See also The Histories of Gargantua and Pantagruel
Rabin, Sheila, xx, 183n.90
Randall, Catharine, 171n.32
Rausner, Georg, 13
Rawcliffe, Carole, 73n.4
Rawsworme, Valentine, 90–91
Read, John, 79, 86–88
    A Most Excellent and Compendious Method of Curing Woundes in the Head, 88
reality and vision, 225–31
Recklau, Johann, 12n.27
Recklau, Markus, 12–14, 12n.27
Rees, Graham, 53n.54
Reformation. *See* Protestant Reformation
Reich, Wilhelm, 146n.48

Reichenbach, Carl, 137n.8
religion and science, 204, 204n.53
religious dissidence, 209–10
Resen, Hans P., 40n.15, 41–43, 43nn.20–22
resurrection, 50, 63
Reuchlin, Johannes, 188n.3
Rhetorica (Aristotle), 46
Rhodius, Ambrosius, 56n.62, 68–69
Riesengebirge, 207, 208, 214–15, 218–20, 219n.54, 220n.56
A Right Frutefull and Approved Treatise for the Artificiall cure of that malady called in Latin struma, and in English, the evill, cured by Kinges and Queenes of England (Clowes), 78
Ripley, George: *The Compound of Alchymie*, 190
Ronchi, Vasco, 224n.2
Roper, Lyndal, 102–3
Rorty, Richard, 224
Rosecrucianism
    and Paracelsianism, 39–40, 39–40nn.11–12, 40n.15
    on *prisci theologi*, 58n.70
    Ole Worm on, 39–40, 39n.11, 40n.15, 54–55
Rosenkrantz, Holger, 40n.15, 41
Rübezahl, xxi–xxii
    as a demon/ghost/monk/shapeshifter, 208–10, 210n.15, 214
    in literature, 210 (see also *Daemonologia Rubinzalii*)
    as a marvel, 208
    in the Riesengebirge, 207, 208
    Widman on, 210
Rudolf II, Holy Roman Emperor, 7, 141
Rudolph, Hartmut, xiv, xix, 130, 131
Ruland, Martin, 61n.78

sacramental seeing, 228
sacraments, 117–18n.2. *See also* baptism, Lord's Supper
Sala, Angelus, 48
Salazar, Alonso de, 243
salt, sulphur, and mercury *(tria prima)*, 60
Saturn, 144
Saulnier, V. L., 164

Index

Savonarola, Girolamo, 160–61
Sawday, Jonathon, 75n.9, 178, 178n.68, 181n.86, 184n.95
The Sceptical Chymist (Boyle), 202, 203
Scheiner, Christoph, 237
Schepelern, H.D., 54n.56, 55, 55n.60
Schickfuss, Jakob, 211
Schott, Gaspar, 236, 240
    Magia optica, 237
Schott, Heinz, xix–xx
Schwenckfeldt, Caspar von, 211
Schwenter, Daniel, 237
science and religion, 204, 204n.53
Scoggins, Dené, xx–xxi
Scot, Reginald, 242
secrets, 99, 101–2
    secreta mulierum (women's secrets), 94, 99, 101–4
self-portraiture, 234
semina
    Boyle on, 188, 202–5, 202n.48
    Petrus Severinus on, 52, 52n.49, 58–59, 68
Sennert, Daniel, 16, 47–48
Severinus, Anna, 47, 56n.62
Severinus, Frederik, 48, 55–57, 56n.62, 56n.65, 68–69
Severinus, Petrus, xvii–xviii, 37
    Aristotelianism of, 57
    Caspar Bartholin on, 46–49, 57
    on disease, 68
    on the elements, 60, 60n.76
    Epistola Scripta Theophrasto Paracelso, 58n.70
    and Erastus, 67, 191n.11
    on experience/practical knowledge, 53
    as an iatrochemist, 59, 64–65, 64n.94
    Idea medicinae, 23, 60, 67, 68, 190–91
    Paracelsianism of, 55–59, 66, 67–68
    on *prisca sapientia,* 57–58, 58n.70
    on salt, sulphur, and mercury *(tria prima),* 60
    on *semina,* 52n.49, 58–59, 68
    on vital philosophy, 59
    Ole Worm as influenced by, 55–60, 55n.60, 60n.76, 68–69

Sextus Empiricus: *Outlines of Pyrrhonism,* 227
Shackelford, Jole, xiv, xvii–xviii
Shakespeare, William: *Macbeth,* 223–24
Shumaker, Wayne, 159
Sigerist, Henry, xiv
signs
    allocation/reading of, 104–5, 115
    of God, 168–72, 170nn.29–30, 171n.32, 173, 180, 185–86
Silesia, 208–9
Simpson, Nicholas, 74
Smith, Thomas, 197
Socrates, 86
Some Considerations touching the Style of the Holy Scriptures (Boyle), 195–97
soul, 141, 146
Spee, Friedrich, 243–44
sperm, 107–9, 108n.67
spleen, 143–44
Sprengel, Kurt, 35
still life, in art, 233–34
Sudhoff, Karl, xii–xiii, 11
suggestion, 146
The Summa Perfectionis (Geber), 201
surgeons, fraternity of, 72–73. *See also* Barber-Surgeons' Company of London

Tabernamontanus, 48
Telle, Joachim, xiv
Temkin, Owsei, xiv
Tester, S.J., 153
Tetrabiblos (Ptolemy), 158
Der Teuffel selbst (Hocker), 221n.61
Thalhauser, Wolfgang, xvii, 14, 17, 21–22
Theatrum Diabolorum, 221n.61
Theophrastia Sancta movement, xiii–xiv
Third Reich, xiii
Thomas, Keith, 229
Thorndike, Lynn, xiv
Three Books of Philosophy to the Athenians (Paracelsus), 190
Thumm, Theodor, 239
Le tiers livre (Rabelais), 163, 165, 171, 173–74, 178–80, 184
Topics (Aristotle), 46
Toxites, Michael, xii, 11, 11n.23

Index

Tractatus de potestate ecclesiae coercendi daemones (Della Torre), 240–41
Tragus (Hieronymus Bock), 48
transmutation, 107
Tremellius, Immanuel, 188n.3
Trinity, 106
Trismegistus, Hermes, 58n.70
Trismosin, Solomon, 190
Troili, Giulio: *Paradossi per Pratticare la Prospettiva senza saperla*, 236
Tudor Reformation, 71. *See also* Barber-Surgeons' Company of London
Tumulus pestis (Helmont), 142–43

"The Uncanny" (Freud), 103
Unheimliche (alienation), 103
United Company of Barber-Surgeons, 72. *See also* Barber-Surgeons' Company of London
University of Copenhagen
    anatomy at, 43–45
    botany at, 43–45
    chemical medicine at, 43–46
    medical reforms at, 43–46
    Paracelsianism at, 38–39, 41, 67, 69
Ussher, James, 194
uterus, 98–100

Valdés, Juan de, 133n.55
Valencia, Pedro de, 243n.49
Valentine, Basil, 48
vampires, 136, 137n.8
van Helmont, Johann Baptist, xx, 36n.1
    on creation, 193–94
    De magnetica vulnerum…curatione, 142
    on *gas* and *blas*, 191–92
    Hebrew knowledge of, 191–92
    on imagination, 137, 142–47
    influence of, 141
    on the Jesuits, 142
    Ortus medicinae, 144–45
    Tumulus pestis, 142–43
Vesalius, Andreas, 48, 101
Vetter, Georg, 10–12, 11n.23, 17
Vicary, Thomas, 74, 77–78
    A Profitable Treatise of the Anatomie of Mans Body, 77, 85

virtual reality, 225–26, 231–32, 234–35, 237–38
vision, xxii, 223–45
    and anamorphosis, 235, 236, 237–38
    anxieties about reliability of, 223–24, 224n.2, 235
    and apparitions, 229
    and artistic illusionism, 233–34
    and computer graphics, 224–25
    as constructed, 224, 228, 230–31
    and demonology, xxii, 229, 231, 232, 238–45, 240n.39
    the eye as an organ, 96–100, 96n.12
    and hallucination, 231
    imagination's influence on, 230, 239
    and magic, 226, 235–38
    mental states's influence on, 230
    ocularcentrism, 224
    and paradoxy, 232–35, 234n.24, 236–37, 241, 244
    and perspective, 226
    and the Protestant Reformation, 228–30, 230n.15
    and Pyrrhonism, 227–28
    and reality, 225–31
    and revelation, 231
    and sacramental seeing, xxii
    and virtual reality, 225–26, 231–32, 234–35, 237–38
    and witchcraft, xxii, 226, 232, 238, 241–45
    See also microcosma, seeing of
vital balsam, 50
vital philosophy, 59
Viterbensis, Johannes Annius:
    Commentaries on the Works of Various Authors who spoke of Antiquity, 213n.31
Vogt, Johann, 13
Vom tauf der Christen (Paracelsus), 117–18, 129n.45
Vom tauf der Christen: De baptismate (Paracelsus), 117–18, 123–24, 124n.25, 124nn.27–30, 128nn.41–42
von Pfeten family, 12–13, 13n.28
Vossius, Dionysius, 188n.3, 199–200n.41
Vossius, Gerardus, 188n.3
Vossius, Isaac, 188n.3

Index

Wagner, Georg, 25–26
Waite, Arthur Edward, xiv
Wakefield, Robert: *On the Three Languages,* 189
Walker, D.P., 57, 153, 154
    Dialogicall discourses of spirits and divels, 239–40
Walton, Michael, xxi
Warhafftige Historien... So D. Iohannes Fausts (Widman), 210, 210n.15
Wars of the Roses (1455–85), 72
water spirits, 212
Wear, Andrew, 72n.2
Webster, Charles, xiv, xix
Webster, John, 242, 244
Weidner, Ursula, 31–32
Weinberg, Florence, 176, 176n.56, 181n.77
Westman, Robert, 161
Weyer, Johann, 211
Widemann, Karl, 30–33
Widman, Georg, xvi
    Warhafftige Historien... So D. Iohannes Fausts, 210, 210n.15
Wilhelm of Bavaria, Prince, 28
Wilhelm of Württemberg, Prince, 28, 29
Williams, George Huntston, 133n.55
Williams, Gerhild, xxi–xxii
Wilson, Catherine, 226
wine metaphor, 180–81, 181nn.76–77
witchcraft and vision, xxii, 226, 232, 238, 241–45
witchhunts, 209–10, 238
Wittelsbach family, 13
Witzel, George, 133n.55
women. *See* microcosma, seeing of
Worm, Ole, 54–66
    Aristotelianism of, 58–60, 66
    Controversiarum medicarum exercitationes, 58–60
    on disease, 59
    on the elements, 60, 60n.76, 62
    empiricism of, 65–66
    Galenism of, 60, 63, 66
    Grell on, 39n.11, 40n.15, 54–55, 55n.60, 57–59, 62–65, 65n.96
    on Hartmann, 64, 64n.93
    as an iatrochemist, 59–61, 63–65, 65n.96
    influence of, 54
    laboratory skills of, 60–61, 61n.78
    Lutheranism of, 54n.56
    on palingenesis, 62–63, 62nn.85–86
    Paracelsian influence on, xvii–xviii, 42, 54–58, 62–67
    on Rosecrucianism, 39–40, 39n.11, 40n.15, 54–55
    on salt, sulphur, and mercury *(tria prima),* 60
    and Frederik Severinus, 56–57, 56n.62, 56n.65
    Petrus Severinus's influence on, 55–60, 55n.60, 60n.76, 68–69
    at University of Copenhagen, 41, 54, 66
    on urine distillation, 60, 60n.77
Worm, William, 61

Yates, Frances, 57, 58n.70, 154, 184n.95

Zahn, Johannes, 237
Zapletal, Vladimir, 7n.14, 8n.17
Zeidner, Lisa, 111–12, 112n.87
Zeiler, Martin, 211
Zerotín, Johann III von, 7, 7n.15
Zerotín family, 7n.15
Zeuxis, 234, 241
Zurich Paracelsus Project, 117n.1
Zwinger, Theodor, xii
Zwingli, Huldrych, 9